Lecture Notes in Mathematics

Edited by A. Dold and B. Eckmann

582

J. M. G. Fell

Induced Representations and Banach *-Algebraic Bundles

with an Appendix due to A. Douady and L. Dal Soglio-Hérault

Springer-Verlag
Berlin · Heidelberg · New York 1977

Author
J. M. G. Fell
Department of Mathematics, University of Pennsylvania
Philadelphia, PA 19174/USA

Library of Congress Cataloging in Publication Data

Fell, James Michael Gardner, 1923-
Induced representations and Banach *-algebraic bundles.

(Lecture notes in mathematics ; 582)
Bibliography: p.
Includes index.
1. Locally compact groups. 2. Representations of groups. 3. Banach algebras. 4. C*= algebras.
I. Title. II. Series: Lecture notes in mathematics (Berlin) ; 582.
QA3.L28 no. 582 [QA387] 510'.8s [512'.55] 77-5113

AMS Subject Classifications (1970): 22D10, 22D30, 43A65, 46H15, 46H25, 46K10, 46L05

ISBN 3-540-08147-X Springer-Verlag Berlin · Heidelberg · New York
ISBN 0-387-08147-X Springer-Verlag New York · Heidelberg · Berlin

Printed in Germany
Printing and binding: Beltz Offsetdruck, Hemsbach/Bergstr. 2141/3140-543210

TABLE OF CONTENTS

CHAPTER III. INDUCED REPRESENTATIONS AND IMPRIMITIVITY FOR BANACH *-ALGEBRAIC BUNDLES

The author gratefully acknowledges support by NSF Grant No. MCS 74-07556 during part of the preparation of this volume.

INTRODUCTION

The theory of induced representations of groups was born in 1898, when Frobenius [1] defined the representation T of a finite group G induced by a representation S of a subgroup H of G. While Frobenius' original definition was couched in matrix terminology, the modern formulation (equivalent to that of Frobenius) uses the terminology of linear spaces. We choose to phrase the definition in such a way that T acts by <u>left</u> translation operators: Let Y be the linear space on which S acts. Then the induced representation T acts on the linear space X of all those functions $f : G \longrightarrow Y$ which satisfy

$$f(xh) = S_{h^{-1}}(f(x)) \quad (x \in G; h \in H);$$

and the operators of T are given by left translation:

$$(T_y f)(x) = f(y^{-1}x) \quad (f \in X; x,y \in G).$$

Why are induced representations important? It is because they enable us to give a more penetrating analysis of a group representation than is afforded by a mere direct sum decomposition of it into irreducible parts. To see this it is helpful to discuss systems of imprimitivity. Let G be a finite group and T any finite-dimensional representation of G; and suppose that the space X of T can be written as a linear space direct sum of non-zero subspaces X_m:

$$X = \sum_{m \in M}^{\oplus} X_m , \qquad \ldots(1)$$

where the (finite) index set M is itself a left G transformation space, the action of G on X being "covariant" with the action of G

on M :

$$T_x(X_m) \subset X_{xm} \quad (x \in G; m \in M). \qquad \ldots(2)$$

Such a decomposition (1) is called a _system of imprimitivity for_ T _over_ M .

We say that T is _primitive_ if the only system of imprimitivity for T is that in which M consists of a single point. Obviously one-dimensional representations are primitive; and in general G will have primitive representations of dimension greater than 1. A direct sum decomposition of T is clearly a special kind of system of imprimitivity, in which the action of G on M is trivial. Thus a primitive representation is automatically irreducible. However the converse of this is far from true: A system of imprimitivity for an irreducible representation of G may be highly non-trivial -- though it must be at least _transitive_ in the sense that the index set M is acted upon transitively by G .

Now any induced representation T of G automatically carries with it a transitive system of imprimitivity. Indeed, let T be induced as above from a representation S of the subgroup H of G ; and let M be the left coset space G/H . The space X of T can then be written as a direct sum in the form (1) by setting

$$X_m = \{f \in X : f \text{ vanishes outside } m\}$$

for each coset m in G/H . Since (2) clearly holds with respect to the natural G-space structure of M , we have a (transitive) system of imprimitivity for G over M .

The basic Imprimitivity Theorem for finite groups asserts just the converse of this. Let H be a subgroup of G, and M the left G-space G/H. (Thus M is essentially the most general transitive G-space.) The Imprimitivity Theorem says that every system of imprimitivity for G over M is equivalent to that obtained by inducing from an (essentially unique) representation of the subgroup H. In the context of finite groups the proof of this result is a simple and routine matter.

From the Imprimitivity Theorem one deduces easily that every irreducible representation T of G is obtained by inducing up to G from some primitive representation S of some subgroup of G. This gives us an important means of analyzing the structure of the irreducible representations of G. Special interest attaches to those T for which S is one-dimensional, that is, those irreducible representations of G which are induced from representations of dimension 1; such representations are called <u>monomial</u>. Blichfeldt has shown (see Lang [1], p. 478) that if G is supersolvable, then every irreducible representation of G is monomial. In the opposite direction we have Taketa's theorem (see Curtis and Reiner [1], Theorem 52.5): Every finite group whose irreducible representations are all monomial must be solvable.

As a point of interest we remark that, according to Brown [1], there exist <u>infinite</u> nilpotent discrete groups whose irreducible unitary representations are not all monomial.

The 1940's saw the birth of the general theory of unitary representations of arbitrary locally compact groups. If such a group is neither compact nor abelian, its irreducible unitary representations are in general infinite-dimensional. Several major papers of that period were devoted to classifying the irreducible unitary representations of specific non-compact non-abelian groups. We mention the work of Wigner [1] on the Poincaré group (the symmetry group of special relativity); the work of Gelfand and Naimark on the "ax+b" group (Gelfand and Naimark [1]) and on the classical complex semisimple Lie groups (Gelfand and Naimark [2,3]); and the work of Bargmann [1] on $SL(2,\mathbb{R})$. In each of these investigations, the actual exhibition of the irreducible representations of the group in question required an ad hoc adaptation to the continuous group context of the notion of induced representation as developed by Frobenius for finite groups. A systematic treatment of infinite-dimensional induced representations of arbitrary locally compact groups was thus clearly called for, and was largely provided by Mackey [1,3,4].

We remark incidentally that Mackey's formulation of the infinite-dimensional unitary inducing process was not quite adequate to embrace _all_ the ad hoc inducing processes needed to construct unitary representations in the works of Bargmann, Gelfand and Naimark mentioned above. To catch the so-called supplementary series of irreducible representations of $SL(2,\mathbb{C})$ and other semisimple groups, one must generalize Mackey's definition by allowing the representations of the subgroup from which we induce to be non-unitary (see Mackey [6], §8). Similarly, to catch the so-called discrete series of irreducible unitary representations of $SL(2,\mathbb{R})$ and other semisimple groups, one has to require the functions on which the induced representations act to be "partially holomorphic." The resulting generalized inducing process is called holomorphic induction (see Blattner [2], Dixmier [2]). Interestingly enough, it was proved by Dixmier [2] that, if G is a connected solvable algebraic linear group, every irreducible unitary representation of G is monomial in the sense of holomorphic induction, that is, is holomorphically induced from a one-dimensional representation of some subgroup. This is an obvious analogue of the result of Blichfeldt mentioned earlier.

The fact that, at least for certain semisimple Lie groups G, all the irreducible unitary representations of G are obtained by inducing (in a suitably generalized sense) from one-dimensional representations of a subgroup shows (by comparison with Taketa's Theorem mentioned above) that induced representations are if anything even more important in the context of continuous groups than they are for finite groups.

Returning to Mackey's unitary inducing construction, we find in Mackey [1,5] a generalization of the Imprimitivity Theorem to the context of an arbitrary separable locally compact group G. As with finite groups, this result asserts that to describe a unitary representation T of G as induced from a unitary representation of a closed subgroup H of G is the same as to specify a system of imprimitivity for T based on the transitive G-space G/H. Of course the definition of a system of imprimitivity over a continuous G-space like G/H is much more sophisticated than in the case of finite groups; and, correspondingly, the proof of the Imprimitivity Theorem, which for finite groups is a simple routine argument, becomes for continuous groups a difficult technical feat. Mackey's generalized Imprimitivity Theorem gave rise almost immediately to two very important applications. The first was Mackey's generalization of the Stone-von Neumann theorem on the uniqueness of the operator representation of the Heisenberg commutation relations of quantum mechanics (see Mackey [2]). The second was the so-called normal subgroup analysis (sometimes called the 'Mackey machine') by which, given a locally compact group G and a closed normal subgroup N of G, one analyzes the irreducible representations of G in terms of the irreducible representations of N and of subgroups of G/N. This normal subgroup analysis, as applied to finite groups, was almost certainly familiar to Frobenius. For finite-dimensional representations of arbitrary groups it was worked out by Clifford [1]. For (infinite-dimensional) unitary representations of separable locally compact groups, it was systematically developed by Mackey [4,5]. The methods that had been applied earlier by Wigner [1] to the Poincaré group and by Gelfand and Naimark [1] to the "ax+b" group were simply special cases of Mackey's normal subgroup analysis.

Substantial steps forward were taken by Loomis [1] and Blattner [1,3]. In these articles the authors define induced representations for non-separable groups, removing Mackey's standing hypothesis of separability. More importantly, they give a new proof of the Imprimitivity Theorem for locally compact groups which is completely different from Mackey's (and makes no use of separability). In a subsequent paper (Blattner [4]), Blattner also extended most of Mackey's normal subgroup analysis to the non-separable context.

Starting in the late 1960's it began to be realized that the theory of induced representations, the Imprimitivity Theorem, and Mackey's normal subgroup analysis have natural generalizations to contexts much wider than that of locally compact groups. The first, purely algebraic, steps in this direction were taken by Dade [1], Fell [5], and Ward [1]. According to Fell [5], the appropriate setting for this generalization is the notion of an algebraic bundle. Let us briefly sketch the simplest version of this notion (in its involutory form); for it plays a dominant role in the present work.

Let G be a finite group (with unit e). A <u>*-algebraic bundle</u> $\underset{\sim}{B}$ <u>over</u> G is a *-algebra A together with a collection $\{B_x\}$ $(x \in G)$ of linear subspaces of A indexed by G, with the following properties:

The B_x are linearly independent and

$$\sum_{x \in G} B_x = A \; ; \qquad \ldots(3)$$

$$B_x B_y \subset B_{xy} \quad (x,y \in G) \; ; \qquad \ldots(4)$$

$$(B_x)^* = B_{x^{-1}} \quad (x \in G) . \qquad \ldots(5)$$

Conditions (4) and (5) say that multiplication and involution in A are "covariant" with multiplication and inverse in G. We call B_x the <u>fiber over</u> x. Note from (4) and (5) that B_e is a *-subalgebra of A, but that the remaining B_x are not. More generally, if H is a subgroup of G, (4) and (5) imply that $B_H = \Sigma_{x \in H} B_x$ is a *-subalgebra of A.

Let us suppose for simplicity that A has a unit 1 (necessarily in B_e). We call $\underset{\sim}{B}$ <u>homogeneous</u> if every fiber B_x contains an element u which is unitary in A (i.e., $u^*u = uu^* = 1$). Assume that $\underset{\sim}{B}$ is homogeneous; let H be a subgroup of G; and let S be a *-representation of the *-subalgebra B_H, acting in some Hilbert space Y. One can then easily generalize the classical inducing process for groups to obtain an induced *-representation T of A. Indeed, let U_x stand for the set of unitary elements in B_x; thus $U = \bigcup_{x \in G} U_x$ is a group under the multiplication in A. Let X be the Hilbert space of all functions $f : U \longrightarrow Y$ satisfying

$$f(ut) = S_{t^{-1}} f(u) \quad (u \in U; t \in \cup_{x \in H} U_x), \qquad \ldots(6)$$

the inner product in X being given by

$$\langle f,g \rangle_X = \sum_{\alpha \in G/H} \langle f(u_\alpha), g(u_\alpha) \rangle_Y \qquad \ldots(7)$$

where $u_\alpha \in \cup_{x \in \alpha} U_x$ for each α in G/H. (Notice that the particular choice of u_α is immaterial in view of (6).) Then the equation

$$(T_{au} f)(v) = S_{v^{-1}av}(f(u^{-1}v)) \quad (a \in B_e; u,v \in U; f \in X) \ \ldots(8)$$

determines a *-representation T of A acting on X (cf. Fell [5], Prop. 7). We refer to T as <u>induced from</u> S. The formal similarity of this definition with the classical inducing process on groups is evident. On the basis of this definition one can prove the Imprimitivity Theorem and develop the whole Mackey normal subgroup analysis. (This is done in Fell [5], §§7-10, in the purely algebraic, non-involutive, setting.)

The most obvious example of a homogeneous *-algebraic bundle is the case when A is the group *-algebra of G, and for each x in G B_x is the one-dimensional subspace of all elements of A which vanish except at x. The unitary elements of B_x are then just those f in B_x for which $|f(x)| = 1$. We call $\underset{\sim}{B}$ in this case the <u>group bundle</u> of G. Induction of representations within the group bundle, as defined in the preceding paragraph, is identical with the classical construction of induced representations of a finite group with which we began.

The next step was to extend this purely algebraic work to a topological setting. This is done in the Memoir of Fell [6]. This

Memoir begins with a definition of a Banach $*$-algebraic bundle $\underset{\sim}{B}$ over an arbitrary topological group G. Let us roughly describe this definition (for its exact form see §11). To begin with, recall that in the above definition of a $*$-algebraic bundle over a finite group G we began with a $*$-algebra A, and the fibers B_x were linear subspaces of A. In the case of a continuous group G, however, it will turn out that the fibers do <u>not</u> arise in a natural way as linear subspaces of a Banach $*$-algebra. Therefore we discard the A, and begin with merely a collection $\{B_x\}$ $(x \in G)$ of Banach spaces indexed by the elements of G. We call B_x again the <u>fiber</u> over x, and make the assumption that the B_x are pairwise disjoint as sets. The union $B = \bigcup_{x \in G} B_x$ is called the <u>bundle space</u>, and is assumed to carry a topology having the following reasonable properties with respect to the fibers and their Banach space structures: (i) The map $\pi : B \to G$ defined by $\pi^{-1}(x) = B_x$ (the <u>bundle projection</u>) is continuous and open; (ii) the norm-function $b \to \|b\|$ is continuous on B; (iii) scalar multiplication $b \mapsto \lambda b$ is continuous on B for each complex λ; (iv) addition $(a,b) \mapsto a+b$ is continuous on the subset $\{(a,b): \pi(a) = \pi(b)\}$ of $B \times B$ on which it is defined; and (v) If $\{b_i\}$ is a net of elements of B such that $\pi(b_i) \to x$ in G and $\|b_i\| \to 0$, then $b_i \to 0_x$ (the zero of the Banach space B_x). The structure B, $\{B_x\}$ thus described is called a <u>Banach bundle</u> over G.

Notice that so far the multiplication in G has not been invoked. One can thus consider Banach bundles over an arbitrary topological space M. The general theory of Banach bundles over locally compact Hausdorff spaces was initiated by Godement [1]. Especially important for us is his construction of the $\underset{\sim}{L}_p$ spaces of measurable cross-sections of a Banach bundle with respect to a measure on the base space M, which we sketch in §10. It should be observed that Godement [1] takes the space of continuous cross-sections of the bundle as the object to be axiomatized, rather than (as we do) the topology of the bundle space. The two axiomatizations are equivalent, if the base space is locally compact, however, in view of an important unpublished result of Douady and dal Soglio-Hérault [1]: _Every Banach bundle (in our sense) over a base space which is either paracompact or locally compact has "enough" continuous cross-sections_. With the consent of A. Douady and L. dal Soglio-Hérault, we present their proof of this result in the Appendix of this work.

Suppose again that we have a Banach bundle B over a topological group G. To define a Banach *-algebraic bundle, we must have some more structure, namely a continuous multiplication operation $\cdot$ on $B \times B$ to B, and a continuous involution * on B to B. These must satisfy the same covariance properties (4), (5) that were assumed for *-algebraic bundles over finite groups. They must satisfy other axioms too which we shall not write out explicitly here -- namely the axioms of multiplication and involution in a Banach *-algebra in so far as these make sense in a system where addition is only partially defined. Such a system $\underset{\sim}{B} = (B, \{B_x\}, \cdot, *)$ is called a _Banach *-algebraic bundle over_ G.

The property of homogeneity can again be defined in this very general context (see Fell [6], §6). The last half of Fell [6] is devoted to constructing induced representations of homogeneous Banach *-algebraic bundles (following the same pattern as (6), (7), (8)), and to showing that the Imprimitivity Theorem and the Mackey normal subgroup analysis go through just as naturally for homogeneous Banach *-algebraic bundles over locally compact groups as they do for locally compact groups themselves.

Now let $\underset{\sim}{B}$ be any Banach *-algebraic bundle over a locally compact group G with left Haar measure λ and modular function Δ. It turns out (see Fell [6], §8) that the space $\underset{\sim}{L}_1(\underset{\sim}{B})$ of all λ-summable cross-sections of $\underset{\sim}{B}$ is a Banach *-algebra under the convolution $*$ and involution * defined by the following familiar-looking formulae:

$$(f*g)(x) = \int_G f(y)g(y^{-1}x)d\lambda y \ , \qquad \ldots(9)$$

$$f^*(x) = \Delta(x^{-1})(f(x^{-1}))^* \qquad \ldots(10)$$

($f,g \in \underset{\sim}{L}_1(\underset{\sim}{B})$; $x \in G$). (Notice by (4) that the values of the integrand in (9) all lie in the Banach space B_x. The integral in (9) can in fact be interpreted as a B_x-valued Bochner integral. Likewise, by (5), the $f^*(x)$ of (10) lies in B_x. Therefore (9) and (10) both define cross-sections of $\underset{\sim}{B}$.) If $\underset{\sim}{B}$ happens to be the group bundle of G (defined in analogy with the finite group case), then $\underset{\sim}{L}_1(\underset{\sim}{B})$ is just the ordinary $\underset{\sim}{L}_1$ group algebra of G. In general $\underset{\sim}{L}_1(\underset{\sim}{B})$ can be regarded as the continuous analogue of the A with which we started the definition of *-algebraic bundles over finite groups. We call $\underset{\sim}{L}_1(\underset{\sim}{B})$ the $\underset{\sim}{L}_1$ <u>cross-sectional</u> <u>algebra</u> of $\underset{\sim}{B}$.

A great deal of literature in functional analysis has been concerned with more or less special cases of these $\underset{\sim}{L}_1$ cross-sectional algebras of homogeneous Banach *-algebraic bundles. For example, the σ-group algebras of Mackey [5] are special cases of them. We may also mention Johnson [1], Glimm [1], Effros and Hahn [1], Edwards and Lewis [1], Turumaru [1], Zeller-Meier [1], and Doplicher, Kastler and Robinson [1]. Takesaki [1] developed the Mackey normal subgroup analysis for the cross-sectional algebras of semidirect product bundles (that is, the "covariance algebras" of Doplicher, Kastler and Robinson [1]).

In addition to the works cited in the last paragraph, two concepts have arisen in the literature which are almost co-extensive with homogeneous Banach *-algebraic bundles. These are Leptin's "generalized $\underset{\sim}{L}_1$-algebras" (see Leptin [1,2,3,4,5]) and the "twisted group algebras" of Busby and Smith [1]. Leptin [5] develops the theory of induced representations, the Imprimitivity Theorem, and part of the Mackey normal subgroup analysis for his generalized $\underset{\sim}{L}_1$-algebras (subject to a certain restriction). It has been shown by Busby [1] that both twisted group algebras and generalized $\underset{\sim}{L}_1$-algebras are essentially the same as homogeneous Banach *-algebraic bundles in the separable context.

At this stage it is natural to ask whether the construction of induced representations, the Imprimitivity Theorem, and the Mackey normal subgroup analysis can be developed for *arbitrary* Banach *-algebraic bundles, without the restriction of homogeneity. The answer, it turns out, is that induced representations can be constructed and an Imprimitivity Theorem proved for quite arbitrary Banach *-algebraic bundles over locally compact groups. *The main object of this work, in fact, is to establish this generalization*. For the Mackey normal subgroup analysis, however, some restriction of the Banach *-algebraic bundle appears to be necessary. The natural restriction seems to be the property of saturation (which is weaker than homogeneity; see §11). We shall postpone the extension of the Mackey normal subgroup analysis to a later publication.

The extension of the theory of induced representations beyond the domain of homogeneous Banach *-algebraic bundles is, we suspect, by no means a purely academic exercise. There is reason to suppose that Banach *-algebraic bundles which are saturated but not homogeneous may eventually prove to have substantial applications in physics. See Remark 10, §33.

To introduce the theory of induced representations and the Imprimitivity Theorem for non-homogeneous Banach *-algebraic bundles, it is useful to consider an even wider generalization. We ask the question: Can induced *-representations be defined on more or less arbitrary *-algebras?

In the non-involutory setting this question was first asked and answered by Higman [1]. His construction (or rather one of his constructions) was as follows: Let A be an arbitrary (complex associative) algebra, B a subalgebra of A, and S an arbitrary representation of B (in the purely algebraic sense) acting on a linear space X. By the B-tensor product $A \otimes_B X$ we mean as usual the quotient of $A \otimes X$ by the linear span of the set of all $ab \otimes \xi - a \otimes S_b\xi$ $(a \in A\ ;\ b \in B\ ;\ \xi \in X)$. If $a \otimes_B \xi$ means the image of $a \otimes \xi$ in $A \otimes_B X$, the equation

$$T_c(a \otimes_B \xi) = ca \otimes_B \xi \qquad \ldots(11)$$

defines a representation T of A acting in $A \otimes_B X$, which we refer to as algebraically induced by S. If A is the group algebra of a finite group G and B is the group algebra of a subgroup H of G, this definition specializes to Frobenius' original definition of an induced group representation.

An obstacle arises if we try to adapt Higman's construction to the involutory context. Indeed, given a *-algebra A, a *-subalgebra B of A, and a *-representation S of B acting on X, there seems to be no natural way to introduce an inner product into $A \otimes_B X$ so as to make the T defined by (11) a *-representation. It was apparently Rieffel [1] who first saw the way out of this difficulty. He considers not merely a $*$-algebra A and a $*$-subalgebra B of A, but also a piece of extra structure, namely a linear map $p : A \longrightarrow B$ satisfying (i) $p(a^*) = p(a)^*$ and (ii) $p(ab) = p(a)b$ $(a \in A ; b \in B)$. Such a p is called an A, B <u>conditional expectation</u>. Suppose now that S is a $*$-representation of B acting on a Hilbert space X; and let S be p-<u>positive</u> in the sense that

$$S_{p(a^*a)} \text{ is a positive operator for all } a \text{ in } A. \qquad \ldots(12)$$

One can then introduce into $A \otimes X$ the conjugate-bilinear form $\langle\ ,\ \rangle_0$ given by

$$\langle a_1 \otimes \xi_1, a_2 \otimes \xi_2 \rangle_0 = \langle S_{p(a_2^* a_1)} \xi_1, \xi_2 \rangle_X \qquad \ldots(13)$$

$(a_i \in A ; \xi_i \in X)$; and it follows from (12) that $\langle\ ,\ \rangle_0$ is positive. So $A \otimes X$ can be completed to a Hilbert space Y with respect to $\langle\ ,\ \rangle_0$ (after factoring out the null space of $\langle\ ,\ \rangle_0$); and under rather general conditions the equation

$$T_c(a \tilde{\otimes} \xi) = ca \tilde{\otimes} \xi \quad (c, a \in A ; \xi \in X) \qquad \ldots(14)$$

will define a $*$-representation T of A on Y. (Here $a \tilde{\otimes} \xi$ is the image of $a \otimes \xi$ in Y. Notice the similarity of (14) with (11).) We refer to T as <u>induced from</u> S <u>via</u> p.

If A and B are the group $*$-algebras of a finite group G and of its subgroup H respectively, and if p is the A, B conditional expectation which sends each f in A into its restriction to H, then every unitary representation S of H is p-positive; and (13) and (14) exactly duplicate the classical definition of T as the unitary (and not merely algebraic) representation of G induced by S.

Actually, the construction just described is only a very special case of the inducing process developed in Rieffel [1]. Indeed, it is evidently not essential that B be a $*$-subalgebra of A, but merely that it act to the right on A. Also, it is not p that is important but the conjugate-bilinear mapping $(a_1,a_2) \mapsto p(a_2^*a_1)$ of $A \times A$ into B. Finally, A plays two roles in this construction. In one role it is a $*$-algebra; in the other it is a linear space acted upon to the left by A and to the right by B. It is natural to separate these two roles. One is thus led to the notion of a rigged module, which is the basis of Rieffel's general involutory inducing process. This notion is defined as follows: Let A and B be any two $*$-algebras whatever. A B-<u>rigged</u> A-<u>module</u> is a linear space L which is both a left A-module and a right B-module, together with a "rigging" map $(r,s) \mapsto [r,s]$ of $L \times L$ into B such that

$[r,s]$ is linear in s and conjugate-linear in r,

$$[r,s]^* = [s,r],$$

$$[r,sb] = [r,s]b$$

$$[ar,s] = [r,a^*s]$$

$(r,s \in L;\ a \in A;\ b \in B)$.

In particular, if B is a $*$-subalgebra of A and p is an A, B conditional expectation, then A (considered as a left A-module and a right B-module under multiplication), together with the rigging $[a_1,a_2] = p(a_1^*a_2)$, is a B-rigged A-module.

We remark here that it is sometimes useful to consider B-<u>rigged spaces</u>; these are obtained from B-rigged A-modules by disregarding A (see §4). Objects similar to B-rigged spaces were studied by Paschke [1] (a year before Rieffel [1]), and long before him by Kaplansky [1] in the case that B is abelian.

Now take an arbitrary B-rigged A-module $\underset{\sim}{L} = (L, [\ ,\ ])$; and let S be a *-representation of B (acting in a Hilbert space X) which is <u>$\underset{\sim}{L}$-positive</u> in the sense that

$$S_{[r,r]} \text{ is a positive operator for all } r \text{ in } L\,. \qquad \ldots(15)$$

A process almost identical with that of (13) and (14) enables us to build an induced *-representation T of A : One first defines a conjugate-bilinear form $\langle\ ,\ \rangle_0$ on $L \otimes X$ as follows:

$$\langle r_1 \otimes \xi_1, r_2 \otimes \xi_2 \rangle_0 = \langle S_{[r_2, r_1]}\xi_1, \xi_2 \rangle_X\,. \qquad \ldots(16)$$

This is positive in virtue of (15). Factoring and completing $L \otimes X$ with respect to $\langle\ ,\ \rangle_0$, we obtain a Hilbert space Y . Under broad conditions (for example, if A is a Banach *-algebra; see Prop. 4.2) the equation

$$T_a(r \tilde{\otimes} \xi) = ar \tilde{\otimes} \xi \qquad \ldots(17)$$

($a \in A$; $r \in L$; $\xi \in X$; $r \tilde{\otimes} \xi$ is the image of $r \otimes \xi$ in Y) defines a *-representation T of A on Y . We refer to T as <u>induced from</u> S <u>via</u> $\underset{\sim}{L}$. This is the fundamental Rieffel inducing process, discussed in §4 of this work.

The familiar Gelfand-Naimark-Segal construction of a *-representation of a *-algebra A , starting from a positive functional p on A , is evidently a special case of the Rieffel process (obtained by taking $L = A$, $B = \mathbb{C}$, and $[a_1, a_2] = p(a_1^* a_2)$). The close relation between the Gelfand-Naimark-Segal process and the inducing process on groups seems first to have been noticed by Blattner [3].

When the classical inducing process for finite groups is formulated in terms of the Rieffel construction (as in (13), (14)), the appropriate generalization to an inducing process for arbitrary *-algebraic bundles over finite groups becomes almost obvious. Indeed, let $\underset{\sim}{B} = (A, \{B_x\})$ be a *-algebraic bundle over a finite group (in the sense of (3), (4), (5)); let H be a subgroup of G and B_H the corresponding *-subalgebra $\sum_{x \in H} B_x$ of A. Let p be the obvious analogue of restriction: $p(a) = a$ if $a \in B_x$ for some $x \in H$; $p(a) = 0$ if $a \in B_x$ for some $x \notin H$. Then p is an A, B_H conditional expectation; and relations (12), (13), (14) define an inducing process for passing from *-representations of B_H to *-representations of A.

The main difference between this general *-algebraic bundle situation and the more special group situation is that in the general situation the property (12) of p-positivity need not hold for all *-representations of B_H, and so not all *-representations of B_H need be inducible to A. If $\underset{\sim}{B}$ is homogeneous, however, p-positivity always holds; and the Rieffel inducing process gives the same result as that of (6), (7), (8).

It is very noteworthy that the remarks of the preceding paragraphs can be extended to arbitrary locally compact groups. Thus the Rieffel inducing process includes as a special case a natural inducing process on arbitrary Banach *-algebraic bundles $\underset{\sim}{B}$ over arbitrary locally compact groups (see §§21-25). Again, if $\underset{\sim}{B}$ is homogeneous, this process reduces to the inducing construction developed in the second half of Fell [6]. In particular, if $\underset{\sim}{B}$ is the group bundle, it reduces to the classical Mackey-Blattner construction (see §26).

We note that Moscovici [1] has developed an interesting generalization of Mackey's inducing construction on groups. Moscovici's construction is a special case of the Rieffel inducing process, though not of the bundle construction in §25.

There are other representation-theoretic processes which up till now have not been thought of as related to the inducing process, but which can be formulated as special cases of the Rieffel inducing process. Most notable perhaps among these is the conjugation of representations discussed in §33. Observe that, to describe conjugation as a Rieffel inducing process, conditional expectations will not suffice; we need the general conjugate-bilinear rigging maps allowed in the definition of a rigged module.

It should be mentioned that our formulation of the Rieffel inducing process diverges substantially from that of Rieffel [1]. He deals throughout with C^*-algebras rather than with arbitrary $*$-algebras; and his B-rigged A-modules $\underset{\sim}{L}$ are so defined that all $*$-representations of B are automatically $\underset{\sim}{L}$-positive.

The reader may recall that Higman's original non-involutory a gebraic inducing process has the advantage of a built-in Frobenius Reciprocity relation. This relation is lost in the Rieffel inducing process; but the loss is compensated by the existence of a very elegant and general formulation of the Imprimitivity Theorem for the Rieffel process, which we shall now sketch. This formulation, and its implications for the understanding of the Imprimitivity Theorem for locally compact groups (and more generally for Banach $*$-algebraic bundles over locally compact groups) has been the guiding motivation for the form of the present work.

Let A and B be two $*$-algebras. Suppose that L is a B-rigged A-module with respect to the B-valued rigging $[\ ,\ ]$; and suppose also that, with the same module structures, L is an A-rigged

B-module with respect to an A-valued rigging $[\ ,\ ]'$. (To make sense of the last supposition we must interchange left and right in the postulates of a rigged module.) We also assume the associative law

$$[r,s]'t = r[s,t] \quad (r,s,t \in L).$$

Then L, $[\ ,\ ]$, $[\ ,\ ]'$ is what we call an A, B <u>imprimitivity bimodule</u> (see §7). If now S is a *-representation of B which is $[\ ,\ ]$-positive (i.e., $S_{[r,r]} \geq 0$ for all r in L), we can hope to induce it via L, $[\ ,\ ]$ to a *-representation T of A. Similarly, if T is a $[\ ,\ ]'$-positive *-representation of A, we can hope to induce it via L, $[\ ,\ ]'$ to a *-representation S of B. The Imprimitivity Theorem in its abstract form now makes the following valuable assertion (see Thm. 8.3): Suppose in addition that the linear spans of the ranges of $[\ ,\ ]$ and $[\ ,\ ]'$ are equal to B and A respectively. Then the above constructions are possible and are each other's inverses. They thus set up a very natural one-to-one correspondence between the $[\ ,\ ]$-positive *-representations of B on the one hand and the $[\ ,\ ]'$-positive *-representations of A on the other hand.

The above paragraph should be compared with Definition 6.10 and Theorem 6.23 of Rieffel [1]. Actually, imprimitivity bimodules and the abstract Imprimitivity Theorem sketched above have been familiar to algebraists for some time in the purely algebraic non-involutory context (see Morita [1]). In fact, Morita's work shows that under suitable conditions a converse holds: Given two algebras A and B having "isomorphic representation theories", there is a canonical A, B imprimitivity bimodule (in a non-involutory sense) which implements this isomorphism of their representation theories. In the C*-algebraic and W*-algebraic contexts this converse has been developed by Rieffel [2].

To see how this abstract Imprimitivity Theorem is related to the Mackey Imprimitivity Theorem for groups, let us recall what the latter asserts. Given a locally compact group G and a closed subgroup H of G , it asserts a one-to-one correspondence between equivalence classes of unitary representations of H and equivalence classes of systems of imprimitivity for G over G/H . Now, as Glimm [1] was the first to point out, systems of imprimitivity for G over G/H are essentially just *-representations of a certain Banach *-algebra E called the <u>transformation algebra</u> for G and G/H (see §30). So the Imprimitivity Theorem for groups can be regarded as asserting that two Banach *-algebras -- namely E and the L_1 group algebra B of H -- have isomorphic *-representation theories. According to the abstract Imprimitivity Theorem, this will be proved if we can set up an E , B imprimitivity bimodule of the right sort. The construction of (roughly) such an imprimitivity bimodule, not only for the group context but also for Banach *-algebraic bundles, will occupy us in §31, and the resulting Imprimitivity Theorem for Banach *-algebraic bundles will emerge in §32.

In this connection we should like to acknowledge the vital role played in our investigation of this subject by the Loomis-Blattner proof (mentioned earlier) of the Imprimitivity Theorem for groups. A key step in the growth of our understanding was the realization that the Loomis-Blattner construction of a unitary subgroup representation which will give rise by induction to a given system of imprimitivity was nothing but a special case of the Rieffel inducing process.

We have still not reached the level of greatest generality in our discussion of the abstract Imprimitivity Theorem. In its most general form it is not a statement about *-algebras at all, but about what we shall call operator inner products. Given a linear space L and a Hilbert space X, by an <u>operator inner product on</u> L, <u>acting in</u> X, we mean a conjugate-bilinear map V of $L \times L$ into the space $\mathcal{O}(X)$ of all bounded linear operators on X, such that the conjugate-bilinear form $\langle\ ,\ \rangle_0$ on $L \otimes X$ given by

$$\langle r \otimes \xi, s \otimes \eta \rangle_0 = \langle V_{r,s}\xi, \eta \rangle_X$$

$(r,s \in L;\ \xi,\eta \in X)$ is positive. Such a V of course gives rise, by factoring and completing $L \otimes X$ with respect to $\langle\ ,\ \rangle_0$, to a new Hilbert space Y. It also gives rise in a canonical manner to a new operator inner product $W : \bar{L} \times \bar{L} \longrightarrow \mathcal{O}(Y)$ on the complex-conjugate space $\bar{L}$, acting in Y. We shall say that W is <u>deduced from</u> V. The germ of the abstract Imprimitivity Theorem now lies in the following assertion: Let the operator inner product V be non-degenerate; then, if W is deduced from V and V' is deduced from W, V and V' must be unitarily equivalent. Thus the construction $V \longmapsto W$ becomes a duality (see Theorem 2.2). The abstract Imprimitivity Theorem is an easy consequence of this.

Our §§1 and 2 owe of course a great deal to the ground-breaking article of Stinespring [1].

In a recent thesis, Bennett [1] has generalized the notion of a B-rigged A-module (in the C*-algebra context) by eliminating the right B-module structure and replacing it with a complete positivity condition on the rigging. His construction thus lies, roughly speaking, halfway in generality between rigged modules and our operator inner products.

It seems likely that, just as the inducing construction and the

Imprimitivity Theorem can be formulated quite abstractly, without any reference to groups or even to Banach *-algebraic bundles, so the Mackey normal subgroup analysis can be given a quite abstract formulation. However we have not yet found this formulation. This is the main reason (apart from a reluctance to increase the bulk of the present work) why we are postponing to a later publication the extension of the Mackey normal subgroup analysis to non-homogeneous bundles.

It should be mentioned that a good part of the present work amounts to a detailed exposition of the results announced in Fell [7].

This work is organized in three chapters. Chapter I (§§1-9) deals with *-representations of abstract *-algebras, without any reference to groups or Banach *-algebraic bundles. It centers around the Rieffel inducing process and the abstract Imprimitivity Theorem. Chapter II (§§10-20) introduces the reader to the elementary theory of Banach *-algebraic bundles and their *-representations. It contains much new material that is not found in Fell [6]. It is self-contained as regards definitions and statement of theorems, though for brevity we refer to Fell [6] for proofs wherever possible. Chapters I and II are independent of each other. Chapter III (§§21-33) combines the results of the first two chapters to obtain a general theory of induced representations and imprimitivity for arbitrary Banach *-algebraic bundles over locally compact groups, generalizing the classical work of Mackey (in the form given to it by Loomis and Blattner).

The conjugation of *-representations of saturated Banach *-algebraic bundles, discussed in the last §33, will be of vital importance for the extension of Mackey's normal subgroup analysis to saturated bundles, to be presented in a future publication.

Terminology

We conclude this Introduction with several items of terminology:

Pairs are denoted by (,), inner products by < , >.

The restriction of a function f to a subset H of its domain is $f|H$.

If W is a set, Ch_W is the characteristic function of W.

δ_{xy} is the Kronecker delta ($= 1$ if $x = y$, $= 0$ if $x \neq y$). By δ_x we mean the function $y \mapsto \delta_{xy}$.

If X is a set, $\mathbb{1}_X$ is the identity function on X.

The ordering relation in a general directed set is denoted by $\prec$.

$\mathbb{R}$, $\mathbb{C}$, $\mathbb{Z}$, $\mathbb{Z}_+$ stand for the set of all reals, complexes, integers and positive integers respectively. $\mathbb{C}^n$ is the standard n-dimensional complex vector space. $\mathbb{E}$ is the circle group $\{z \in \mathbb{C} : |z| = 1\}$.

All linear spaces and algebras are over the complex field, in the absence of explicit mention to the contrary.

If W is a subset of a linear space L, $[W]$ is the linear span of W in L.

If L is a linear space, $\bar{L}$ is the linear space complex-conjugate to L; that is, $\bar{L}$ has the same underlying set and addition as L, but has modified scalar multiplication:

$$(rx)_{\bar{L}} = (\bar{r}x)_L .$$

Let M be a locally compact Hausdorff space, and X a Banach space. Then $\mathcal{C}(M;X)$ is the linear space of all continuous functions on M to X; $\mathcal{C}_0(M;X)$ is the subspace of $\mathcal{C}(M;X)$ consisting of those f for which $\lim_{m\to\infty} \|f(m)\| = 0$; $\underset{\sim}{L}(M;X)$ is the subspace of $\underset{\sim}{C}_0(M;X)$ consisting of those f which vanish outside some compact set. We write $\underset{\sim}{C}(M)$, $\underset{\sim}{C}_0(M)$, $\underset{\sim}{L}(M)$ instead of $\underset{\sim}{C}(M;\mathbb{C})$, $\underset{\sim}{C}_0(M;\mathbb{C})$, $\underset{\sim}{L}(M;\mathbb{C})$. When $\underset{\sim}{C}(M)$ and $\underset{\sim}{C}_0(M)$ are treated as *-algebras, it is the pointwise operations of multiplication and complex-conjugation that we have in mind. The supremum norm on $\underset{\sim}{C}_0(M;X)$ is denoted by $\| \|_\infty$.

If G is a group and H a subgroup of G , G/H denotes the coset space of all left cosets xH ($x \in G$). A <u>transversal</u> for G/H is a subset of G containing exactly one element from each coset in G/H .

If N and H are topological groups, a <u>group extension</u> γ <u>of</u> N <u>by</u> H is a topological group G together with a topological isomorphism i of N into G and a continuous open homomorphism j of G onto H such that $\ker(j) = i(N)$. We write

$$\gamma : N \xrightarrow[i]{} G \xrightarrow[j]{} H . \qquad \ldots(18)$$

If i(N) is contained in the center of G , the extension is <u>central</u>. If

$$\gamma' : N \xrightarrow[i']{} G' \xrightarrow[j']{} H$$

is another group extension of N by H , γ and γ' are <u>isomorphic</u> if there is a topological isomorphism φ of G onto G' such that $i' = \varphi \circ i$ and $j = j' \circ \varphi$. If $G = N \times H$, $i(n) = (n, e_H)$, and $j(n,h) = h$, (18) is the <u>trivial</u> <u>extension</u>.

A <u>*-algebra</u> is of course a (complex) associative algebra having an involution operation $* : A \longrightarrow A$ such that (i) * is conjugate-linear, (ii) $a^{**} = a$, and (iii) $(ab)^* = b^* a^*$.

If A is a *-algebra, by a <u>left</u> [<u>right</u>] A-<u>module</u> we mean a left [right] A_0-module, where A_0 is the associative algebra underlying A. In other words the involution is disregarded in speaking of A-modules.

A *-<u>ideal</u> of a *-algebra is a left (or right) ideal which is closed under involution, and which is therefore automatically a two-sided ideal.

The term <u>Banach</u> *-<u>algebra</u> will always mean that the involution is an isometry: $\|a^*\| = \|a\|$.

If Y is a subset of a Hilbert space X, then $Y^\perp$ is the orthogonal complement of Y in X.

If X is a linear space, a <u>seminorm</u> on X is a function $p : X \to \mathbb{R}$ such that $p(x) \geq 0$, $p(rx) = |r|p(x)$, and $p(x+y) \leq p(x) + p(y)$ $(x,y \in X ; r \in \mathbb{C})$. If P is a collection of seminorms on X which separates points of X, then P determines a topology on X in which convergence is given as follows: $x_i \to x$ if and only if $p(x_i - x) \to 0$ for all p in P. This topology is said to be <u>generated by</u> P. With this topology X becomes a (Hausdorff) <u>locally convex space</u> (see for example Kelley and Namioka [1]).

If X and Y are locally convex spaces (in particular Banach or Hilbert spaces), $\mathcal{O}(X,Y)$ is the family of all continuous linear maps of X into Y. We abbreviate $\mathcal{O}(X,X)$ to $\mathcal{O}(X)$.

If X is a Banach space, $\mathcal{O}_c(X)$ is the family of all operators in $\mathcal{O}(X)$ which are <u>compact</u> (i.e., the norm-closure of the image of the unit ball is norm-compact).

If A is a *-algebra and X is a pre-Hilbert space (i.e. a Hilbert space without the completeness axiom), a <u>pre-*-representation</u> of A on X is a homomorphism T of A into the algebra of all (not necessarily continuous) linear operators on X, such that T satisfies

$$\langle T_a \xi, \eta \rangle = \langle \xi, T_{a*} \eta \rangle$$

$(a \in A\,;\ \xi,\eta \in X)$. We call X the <u>space</u> of T and denote it by $X(T)$. If X is a Hilbert space and the T_a are continuous, T is a *-<u>representation</u> of A. If $\mathrm{Ker}(T)$ $(= \{a \in A : T_a = 0\})$ is $\{0\}$, T is <u>faithful</u>.

From any positive linear functional p on the *-algebra A, one constructs a pre-*-representation of A in the well-known Gelfand-Naimark-Segal manner: Put $N = \{x \in A : p(x^*x) = 0\}$; then N is linear, and $X = A/N$ is a pre-Hilbert space under the inner product $\langle a+N, b+N \rangle = p(b^*a)$; and the equation $T_b(a+N) = ba + N$ defines T as a pre-*-representation of A on X. We say that p is <u>admissible</u> if each T_b is continuous on X, so that T can be extended to a *-representation of A on the completion of X.

Let T be a *-representation of the *-algebra A, acting in X. The <u>essential space</u> of T is the closed linear span in X of $\{T_a\xi : a \in A,\ \xi \in X\}$. If the essential space of T is X, T is <u>non-degenerate</u>; otherwise it is <u>degenerate</u>. If $T_a = 0$ for all a, T is a <u>zero representation</u>. A closed linear subspace Y of X is T-<u>stable</u> if $T_a(Y) \subset Y$ for all a in A. In that case the *-representation $a \mapsto T_a|Y$ of A acting in Y is a <u>subrepresentation</u> of T. If there are no closed T-stable subspaces of X except $\{0\}$ and X, and if T is not the one-dimensional zero representation, T is <u>irreducible</u>.

Given two *-representations S and T of A, an S, T <u>intertwining operator</u> is an element F of $\underset{\sim}{O}(X(S),X(T))$ satisfying $F \circ S_a = T_a \circ F$ for all a in A. If this F is an isometry of $X(S)$ onto $X(T)$, S and T are <u>unitarily equivalent</u> (in symbols, $S \cong T$).

The set of all T, T intertwining operators is a von Neumann algebra on $X(T)$, and is denoted by $\underset{\sim}{I}(T)$.

If $\{T^i\}$ $(i \in I)$ is an indexed collection of $*$-representations of A such that $\sup\{\|T^i_a\|: i \in I\} < \infty$ for every a in A, we can form the Hilbert direct sum $*$-representation $T = \sum^{\oplus}_{i\in I} T^i$ of A, acting on the direct sum Hilbert space $X(T) = \sum^{\oplus}_{i\in I} X(T^i)$, by means of the operators

$$(T_a\xi)_i = T^i_a\xi_i \quad (a \in A \,;\, \xi \in X(T)).$$

If G is a topological group, a unitary representation of G, acting on a Hilbert space X, is a homomorphism T of G into the group of all unitary operators on X, such that $x \mapsto T_x\xi$ is continuous on G to X for all ξ in X.

The definitions of stable subspace, irreducibility, etc., made above for $*$-representations of a $*$-algebra, can clearly be repeated for unitary representations of groups.

CHAPTER I

THE ABSTRACT INDUCING PROCESS AND ABSTRACT IMPRIMITIVITY

§1. Operator inner products.

In this section we shall study operator inner products, which generalize ordinary inner products by being operator-valued instead of complex-valued. What we ultimately have in mind are such operator inner products as $(f,g) \mapsto S_{p(g^* * f)}$, where f and g run over the convolution group *-algebra A of a finite group G, S is a unitary representation of a subgroup H of G, and $p(h)$ is the restriction to H of an element h of A.

Let L be a linear space and X a Hilbert space.

Definition. An operator inner product on L, acting in X, is a map $V: L \times L \longrightarrow \underset{\sim}{O}(X)$ such that (i) $V_{s,t}$ is linear in s and conjugate-linear in t, and (ii) we have

$$\sum_{i,j=1}^{n} \langle V_{t_i,t_j}\xi_i,\xi_j\rangle \geq 0 \qquad \ldots(1)$$

for all $n \in \mathbb{Z}_+$, all $t_1,\ldots,t_n$ in L, and all $\xi_1,\ldots,\xi_n$ in X.

We call X the space of V and denote it by $X(V)$. Property (1) is called the complete positivity of V.

It follows from (1) that

$$V_{t,t} \geq 0 \quad (t \in L), \qquad \ldots(2)$$

and hence by a standard argument that

$$V_{t,s} = (V_{s,t})^* \quad (s,t \in L). \qquad \ldots(3)$$

In general (see Stinespring [1]) (1) is a stronger condition that (2). If X is one-dimensional, however, so that $X \cong \underset{\sim}{O}(X) \cong \mathbb{C}$, then (i) and (2) imply that

$$\sum_{i,j=1}^{n} \langle V_{t_i,t_j}\xi_i,\xi_j\rangle = \sum_{i,j} V_{t_i,t_j}\xi_i\overline{\xi_j}$$

$$= V_{\Sigma_i \xi_i t_i, \Sigma_j \xi_j t_j} \geq 0 .$$

Thus (i) and (2) imply (1) if X is one-dimensional.

In view of (3) the range of an operator inner product V is a self-adjoint collection of operators; and we can apply to it a number of standard definitions. Thus V will be called <u>non-degenerate</u> if $\{V_{s,t}\xi : s,t \in L ; \xi \in X(V)\}$ has dense linear span in $X(V)$ - or, equivalently, if $\xi = 0$ whenever $V_{s,t}\xi = 0$ for all s,t. In general, if Y is the closed linear span of $\{V_{s,t}\xi : s,t \in L ; \xi \in X(V)\}$, the result of restricting each $V_{s,t}$ to Y is called the <u>non-degenerate part</u> of V. Let V and V' be two operator inner products on L. An operator F in $\underset{\sim}{O}(X(V),X(V'))$ satisfying $F \cdot V_{s,t} = V'_{s,t} \cdot F$ $(s,t \in L)$ is a V, V' <u>intertwining operator</u>. If there exists an isometric V, V' intertwining operator whose range is $X(V')$, V and V' are <u>unitarily equivalent</u> (in symbols, $V \cong V'$). The von Neumann algebra of all V, V intertwining operators is called the <u>commuting algebra</u> of V, and will be denoted by $\underset{\sim}{I}(V)$. If $\underset{\sim}{I}(V)$ consists of the scalar operators only, then V is irreducible (that is, $\{0\}$ and $X(V)$ are the only closed subspaces stable under all the $V_{s,t}$), and conversely.

<u>Proposition 1.1</u>. (Generalized Schwarz Inequality) <u>Let</u> V <u>be an</u>

<u>operator inner product on</u> L . <u>If</u> $t_1,\dots,t_n,s \in L$ <u>and</u> $\xi_1,\dots,\xi_n \in X(V)$, <u>then</u>

$$\left\| \sum_{i=1}^{n} V_{t_i,s}\xi_i \right\|^2 \le \|V_{s,s}\| \sum_{i,j=1}^{n} \langle V_{t_i,t_j}\xi_i,\xi_j\rangle. \qquad \dots(4)$$

<u>Proof</u>. Let η be any unit vector in $X(V)$, λ a real number, and μ a complex number such that $|\mu| = 1$ and $\bar{\mu}\,\Sigma_{i=1}^{n}\langle V_{t_i,s}\xi_i,\eta\rangle = -|\Sigma_{i=1}^{n}\langle V_{t_i,s}\xi_i,\eta\rangle|$. In (1) let us replace n by $n+1$; $t_1,\dots,t_n$ by $\lambda t_1,\dots,\lambda t_n,\mu s$; and $\xi_1,\dots,\xi_n$ by $\xi_1,\dots,\xi_n,\eta$. Putting $k = \Sigma_{i,j=1}^{n}\langle V_{t_i,t_j}\xi_i,\xi_j\rangle$, we obtain:

$$0 \le \lambda^2 k + |\mu|^2\langle V_{s,s}\eta,\eta\rangle + 2\mathrm{Re}[\lambda\bar{\mu}\sum_{i=1}^{n}\langle V_{t_i,s}\xi_i,\eta\rangle]$$

$$= \lambda^2 k - 2\lambda\left|\sum_{i=1}^{n}\langle V_{t_i,s}\xi_i,\eta\rangle\right| + \langle V_{s,s}\eta,\eta\rangle.$$

Since this holds for all real λ ,

$$\left|\sum_{i=1}^{n}\langle V_{t_i,s}\xi_i,\eta\rangle\right|^2 \le k\langle V_{s,s}\eta,\eta\rangle.$$

Letting η run over all unit vectors in this inequality we obtain (4). □

Taking $n = 1$ in (4), and letting ξ_1 run over all unit vectors, we obtain

$$\|V_{t,s}\|^2 \le \|V_{s,s}\|\,\|V_{t,t}\| \quad (s,t \in L). \qquad \dots(5)$$

<u>The deduced Hilbert space</u>.

Now fix an operator inner product V on L , acting in a Hilbert space X (with inner product $\langle\ ,\ \rangle$). Since the expression $\langle V_{s,t}\xi,\eta\rangle$ is linear in s and ξ and conjugate-linear in t and η , the equation

$$\langle s \otimes \xi, t \otimes \eta \rangle_0 = \langle V_{s,t}\xi, \eta \rangle \qquad \ldots(6)$$

defines a conjugate-bilinear form $\langle\ ,\ \rangle_0$ on $M = L \otimes X$ (algebraic tensor product). By (1) $\langle\ ,\ \rangle_0$ is positive (that is, $\langle \zeta, \zeta \rangle_0 \geq 0$ for $\zeta \in M$). (In fact, the positivity of $\langle\ ,\ \rangle_0$ is equivalent to the complete positivity of V.) Thus, factoring out the null space of $\langle\ ,\ \rangle_0$ and completing, we obtain a new Hilbert space called the Hilbert space deduced from V and denoted by $\underset{\sim}{X}(L,V)$ or simply $\underset{\sim}{X}(V)$. The inner product and norm in $\underset{\sim}{X}(V)$ are called $\langle\ ,\ \rangle_0$ and $\|\ \|_0$; and the image of $s \otimes \xi$ in $\underset{\sim}{X}(V)$ is written $s \tilde{\otimes} \xi$. Thus (6) becomes

$$\langle s \tilde{\otimes} \xi, t \tilde{\otimes} \eta \rangle_0 = \langle V_{s,t}\xi, \eta \rangle \qquad \ldots(7)$$

$(s,t \in L; \xi, \eta \in X(V))$. Notice that

$$\|s \tilde{\otimes} \xi\|_0^2 \leq \|V_{s,s}\| \|\xi\|^2 . \qquad \ldots(8)$$

It follows that for each s in L the map $\xi \mapsto s \tilde{\otimes} \xi$ of $X(V)$ into $\underset{\sim}{X}(V)$ is continuous. In particular, if Y is a dense subset of $X(V)$, then $\{s \tilde{\otimes} \xi : s \in L, \xi \in Y\}$ has dense linear span in $\underset{\sim}{X}(V)$.

Here is an important example of an operator inner product. Let X be a non-zero Hilbert space (with inner product $\langle\ ,\ \rangle$); and for each ξ, η in X let $V_{\xi,\eta}$ be the operator on X of rank 1 (or 0) given by

$$V_{\xi,\eta}\zeta = \langle \zeta, \xi \rangle \eta \quad (\zeta \in X). \qquad \ldots(9)$$

If $\xi_1, \ldots, \xi_n, \eta_1, \ldots, \eta_n \in X$ we have $\Sigma_{i,j} \langle V_{\eta_i,\eta_j}\xi_i, \xi_j \rangle = |\Sigma_i \langle \xi_i, \eta_i \rangle|^2 \geq 0$.

Notice that $V_{\xi,\eta}$ is linear in η and conjugate-linear in ξ. Hence V is an operator inner product not on X but on the complex-conjugate linear space $\bar{X}$ (and acting in X). The reader will verify that $\underset{\sim}{X}(V)$ is one-dimensional; in fact, identifying $\underset{\sim}{X}(V)$ with $\mathbb{C}$, we have

$$\xi \tilde{\otimes} \eta = \langle \eta, \xi \rangle \qquad (\xi \in \bar{X}; \eta \in X).$$

Remark. It might at first sight seem more reasonable to interchange ξ and η in (9) and so avoid introducing the complex-conjugate space $\bar{X}$. The trouble is that, if we do that, V is no longer completely positive. Indeed: Let ρ and σ be orthogonal unit vectors, and put $\xi_1 = \eta_2 = \rho$, $\xi_2 = -\eta_1 = \sigma$. Then by (9) $\sum_{i,j=1}^{2} \langle V_{\eta_j,\eta_i} \xi_i, \xi_j \rangle$ turns out to be -2. So V ceases to be completely positive if we interchange its variables.

We next observe that operators in $\underset{\sim}{I}(V)$ give rise to operators on the deduced Hilbert space $\underset{\sim}{X}(V)$.

Theorem 1.2. Let V and V' be two operator inner products on L; and put $X = X(V)$, $X' = X(V')$, $\underset{\sim}{X} = \underset{\sim}{X}(V)$, $\underset{\sim}{X}' = \underset{\sim}{X}(V')$. If $F : X \to X'$ is a V, V' intertwining operator, the equation

$$\tilde{F}(s \tilde{\otimes} \xi) = s \tilde{\otimes} F(\xi) \qquad (s \in L; \xi \in X) \qquad \ldots(10)$$

determines a unique bounded linear map $\tilde{F} : \underset{\sim}{X} \to \underset{\sim}{X}'$.

In particular, if $F \in \underset{\sim}{I}(V)$, then $\tilde{F} \in \underset{\sim}{O}(\underset{\sim}{X})$.

Proof. Let $I : L \to L$ be the identity map; and let $\langle\ ,\ \rangle$, $\langle\ ,\ \rangle'$, $\langle\ ,\ \rangle_0$ and $\langle\ ,\ \rangle_0'$ be the inner products of X, X', $\underset{\sim}{X}$ and $\underset{\sim}{X}'$. For $r, s \in L$, $\xi \in X$, and $\eta \in X'$, we have

$$\langle r \otimes F\xi, s \otimes \eta\rangle_0' = \langle V'_{r,s}F\xi,\eta\rangle'$$
$$= \langle FV_{r,s}\xi,\eta\rangle' = \langle V_{r,s}\xi, F^*\eta\rangle$$
$$= \langle r \otimes \xi, s \otimes F^*\eta\rangle_0 .$$

It follows that

$$\langle (I \otimes F)\zeta,\zeta'\rangle_0' = \langle \zeta, (I \otimes F^*)\zeta'\rangle_0 \qquad \text{...(11)}$$

for all ζ in $L \otimes X$ and ζ' in $L \otimes X'$.

To prove the theorem it is enough to show that

$$\langle (I \otimes F)\zeta, (I \otimes F)\zeta\rangle_0' \le \|F\|^2 \langle \zeta,\zeta\rangle_0 \qquad \text{...(12)}$$

for all ζ in $L \otimes X$. Now $\|F\|^2 - F^*F$ is a positive operator in $\underset{\sim}{I}(V)$, and so has a positive square root S belonging to $\underset{\sim}{I}(V)$. Applying (11) twice (once with F replaced by S), we have for each ζ in $L \otimes X$

$$0 \le \langle (I \otimes S)\zeta, (I \otimes S)\zeta\rangle_0$$
$$= \langle (I \otimes S^2)\zeta,\zeta\rangle_0 = \langle (I \otimes (\|F\|^2 - F^*F))\zeta,\zeta\rangle_0$$
$$= \|F\|^2\langle \zeta,\zeta\rangle_0 - \langle (I \otimes F^*F)\zeta,\zeta\rangle_0$$
$$= \|F\|^2\langle \zeta,\zeta\rangle_0 - \langle (I \otimes F)\zeta, (I \otimes F)\zeta\rangle_0' .$$

But this is (12). □

Formula (12) shows that

$$\|\tilde{F}\| \le \|F\| . \qquad \text{...(13)}$$

The correspondence $F \mapsto \tilde{F}$ is of a functorial nature: It is linear and clearly preserves composition of operators; and an easy calculation shows that it preserves adjoints, that is, $(\tilde{F})^* = (F^*)^\sim$. If V is non-degenerate, then $F \mapsto \tilde{F}$ is an isometry on $\underset{\sim}{I}(V)$. Indeed, suppose that $F \in \underset{\sim}{I}(V)$ and $\tilde{F} = 0$. Then $0 = \langle \tilde{F}(s \otimes \xi), t \otimes \eta \rangle_0 = \langle V_{s,t} F\xi, \eta \rangle = \langle V_{s,t}\xi, F^*\eta \rangle$ for all s, t in L and ξ, η in $X(V)$. Fixing η and letting s, t, ξ vary we conclude from the non-degeneracy of V that $F^*\eta = 0$. Since η is arbitrary this implies $F^* = 0$, or $F = 0$. Thus $F \mapsto \tilde{F}$ is one-to-one, hence a $*$-isomorphism between C^*-algebras, hence an isometry.

We conclude this section with remarks on direct sums and tensor products of operator inner products.

For each i in an index set I let V^i be an operator inner product on the fixed linear space L; and assume that for each s, t in L the collection $\{V^i_{s,t} : i \in I\}$ is norm-bounded. Thus for each s, t we can form the direct sum operator $V_{s,t} = \sum^{\oplus}_{i \in I} V^i_{s,t}$ acting on the direct sum Hilbert space $X = \sum^{\oplus}_{i \in I} X(V^i)$.

<u>Proposition 1.3</u>. *$V : (s,t) \mapsto V_{s,t}$ is an operator inner product on L; we call it the Hilbert direct sum of the V^i. We have*

$$\underset{\sim}{X}(V) \cong \sum^{\oplus}_{i \in I} \underset{\sim}{X}(V^i) \quad \text{(Hilbert direct sum)}. \qquad \ldots(14)$$

The simple proof of this proposition is left to the reader.

Since any operator inner product V is the direct sum of its non-degenerate part and its zero part, it follows from the last proposition that $\underset{\sim}{X}(V)$ *is unaltered when* V *is replaced by its non-degenerate part*.

As regards tensor products, let V and W be operator inner products on linear spaces L and M respectively; and let $P = L \otimes M$ (algebraic tensor product). Let $X = X(V)$, $Y = X(W)$; and let Z be the Hilbert space tensor product $X \otimes Y$.

<u>Proposition 1.4</u>. <u>The equation</u>

$$U_{s_1 \otimes t_1, s_2 \otimes t_2} = V_{s_1, s_2} \otimes W_{t_1, t_2} \qquad \ldots (15)$$

$(s_i \in L; t_i \in M)$ <u>defines an operator inner product</u> U <u>on</u> P, <u>acting in</u> Z. <u>This</u> U <u>is called the tensor product of</u> V <u>and</u> W. <u>We have</u>

$$\underset{\sim}{X}(U) \cong \underset{\sim}{X}(V) \otimes \underset{\sim}{X}(W) \quad \text{(Hilbert space tensor product)}. \qquad \ldots (16)$$

<u>Proof</u>. Assume first that $X = Y = \mathbb{C}$. If V and W are inner products (i.e., have zero null-spaces), the proposition amounts simply to the justification of the ordinary definition of the Hilbert space tensor product. Once this is observed, it is a simple matter to extend the proposition to the case that the null-spaces of V and W are not both zero.

We now discard the assumption that $X = Y = \mathbb{C}$. Let V^0, W^0, and U^0 be the conjugate-bilinear forms on $L \otimes X$, $M \otimes Y$, and $P \otimes Z$ respectively given by

$$V^0(s \otimes \xi, s' \otimes \xi') = \langle V_{s,s'} \xi, \xi' \rangle,$$

$$W^0(t \otimes \eta, t' \otimes \eta') = \langle V_{t,t'} \eta, \eta' \rangle,$$

$$U^0(q \otimes \zeta, q' \otimes \zeta') = \langle U_{q,q'} \zeta, \zeta' \rangle$$

$(s,s' \in L; t,t' \in M; q,q' \in P; \xi,\xi' \in X; \eta,\eta' \in Y; \zeta,\zeta' \in Z)$. Now $(L \otimes X) \otimes (M \otimes Y)$ can be identified with a subspace Q of $P \otimes Z$, and with this identification one verifies that

$$U^0(a \otimes b, a' \otimes b') = V^0(a,a')W^0(b,b') \qquad \ldots(17)$$

$(a,a' \in L \otimes X; b,b' \in M \otimes Y)$. The complete positivity of V and W imply that V^0 and W^0 are positive, and so by (17) (and the first paragraph of this proof) that U^0 is positive on Q. Hence by continuity (and the denseness of Q in $P \otimes Z$ with respect to its second factor) U^0 is positive on $P \otimes Z$. So U is completely positive and hence an operator inner product. Since $X(V^0) = X(V)$, $\underset{\sim}{X}(W^0) = \underset{\sim}{X}(W)$ and $\underset{\sim}{X}(U^0) = \underset{\sim}{X}(U)$, equation (16) now follows from the first paragraph of the proof applied to V^0, W^0, and U^0. □

§2. Duality for operator inner products.

The main result of this section is Theorem 2.2, which will be the basis of the proof of the abstract Imprimitivity Theorem in §8.

We fix a linear space L and an operator inner product V on L, and write X and Y for $X(V)$ and $\underset{\sim}{X}(V)$ respectively. $\langle\ ,\ \rangle_0$ and $s \tilde{\otimes} \xi$ have the same meanings as in §1.

By Prop. 1.1 each element s of L gives rise to a unique bounded linear map $U_s : Y \longrightarrow X$ such that

$$U_s(t \tilde{\otimes} \xi) = V_{t,s}\xi \quad (t \in L; \xi \in X). \qquad \ldots(1)$$

U_s is conjugate-linear in s (since $V_{t,s}$ is), and by §1(4)

$$\|U_s\|^2 \leq \|V_{s,s}\|. \qquad \ldots(2)$$

Given $s,t \in L$, let us compose U_s with the bounded linear map $\xi \mapsto t \tilde{\otimes} \xi$ of X into Y (see §1(8)); we then obtain a bounded linear map $W_{s,t} : Y \longrightarrow Y$:

$$W_{s,t}(\zeta) = t \tilde{\otimes} U_s(\zeta) \quad (\zeta \in Y), \qquad \ldots(3)$$

that is,

$$W_{s,t}(r \tilde{\otimes} \xi) = t \tilde{\otimes} V_{r,s}\xi \quad (r \in L; \xi \in X). \qquad \ldots(4)$$

Combining (2) and §1(8), we get

$$\|W_{s,t}\|^2 \leq \|V_{s,s}\|\|V_{t,t}\|. \qquad \ldots(5)$$

Notice that $W_{s,t}$ is linear in t and conjugate-linear in s. An easy calculation based on §1(3), (7) shows that

$$(W_{s,t})^* = W_{t,s} \quad (s,t \in L). \qquad \ldots(6)$$

Thus $\{W_{s,t} : s,t \in L\}$ is self-adjoint. I claim that this collection of operators acts non-degenerately on Y. To see this, recall that V may be replaced by its non-degenerate part without altering Y or the $W_{s,t}$; so we may assume from the beginning that V is non-degenerate. Then by (4) the linear span of $\{\text{range}(W_{s,t}): s \in L\}$ consists of the $t \tilde{\otimes} \eta$ where η runs over a dense subset of X. It follows that $W : (s,t) \mapsto W_{s,t}$ is non-degenerate.

Proposition 2.1. (I) The above W is an operator inner product on the complex-conjugate space $\bar{L}$. We shall call it the operator inner product deduced from V.

(II) In view of (I) (replacing L, V by $\bar{L}$, W) we can form the operator inner product $\tilde{V}$ on $\bar{\bar{L}} = L$ deduced from W. If V is non-degenerate we have $\tilde{V} \cong V$.

<u>Proof</u>. If $s,s',t,t' \in L$ and $\xi,\xi' \in X$,

$$\langle W_{s,s'}(t \tilde{\otimes} \xi), t'\tilde{\otimes} \xi'\rangle_0$$

$$= \langle s'\tilde{\otimes} V_{t,s}\xi, t'\tilde{\otimes} \xi'\rangle_0$$

$$= \langle V_{s',t'} V_{t,s}\xi, \xi'\rangle$$

$$= \langle V_{t,s}\xi, V_{t',s'}\xi'\rangle \quad \text{(by §1(3))}. \qquad \ldots(7)$$

Now to prove (I) it is sufficient (by the boundedness of the $W_{s,s'}$) to show that $\sum_{i,j}\langle W_{t_i,t_j}\zeta_i,\zeta_j\rangle \geq 0$ when the ζ_i run over a dense subset of Y, for example over the linear combinations of the $s \tilde{\otimes} \xi$. Thus it is sufficient to prove that

$$\sum_{i,j=1}^{n} \langle W_{t_i,t_j}(s_i\tilde{\otimes} \xi_i), s_j\tilde{\otimes} \xi_j\rangle_0 \geq 0 \qquad \ldots(8)$$

whenever $n = 1,2,\ldots$, the s_i and t_i are in L, and the ξ_i are in X. But by (7) the left side of (8) is

$$\sum_{i,j=1}^{n} \langle V_{s_i,t_i}\xi_i, V_{s_j,t_j}\xi_j\rangle = \|\sum_{i=1}^{n} V_{s_i,t_i}\xi_i\|^2 \geq 0 .$$

So (I) has been proved.

By (I) we may form the Hilbert space $\tilde{X} = \underset{\sim}{X}(W)$ deduced from W. This is obtained by factoring and completing the space $\bar{L} \otimes Y$ with respect to the positive conjugate-bilinear form $\langle\ ,\ \rangle_{00}$ on $\bar{L} \otimes Y$ given by:

$$\langle s \otimes \zeta, t \otimes \zeta'\rangle_{00} = \langle W_{s,t}\zeta,\zeta'\rangle_0 \qquad \ldots(9)$$

($s,t \in \bar{L}; \zeta,\zeta' \in Y$). As usual the image of $s \otimes \zeta$ in $\tilde{X}$ is called

$s \tilde{\otimes} \zeta$. Combining (7) and (9) we get:

$$\langle s \tilde{\otimes}(t \tilde{\otimes} \xi), s'\tilde{\otimes}(t'\tilde{\otimes} \xi')\rangle_{00} = \langle V_{t,s}\xi, V_{t',s'}\xi'\rangle \qquad \ldots(10)$$

$(s,s',t,t' \in L; \xi,\xi' \in X)$. In view of the denseness in $\tilde{X}$ of the linear span of $\{s \tilde{\otimes}(t \tilde{\otimes} \xi): s,t \in L; \xi \in X\}$, (10) implies that the equation

$$F(s \tilde{\otimes}(t \tilde{\otimes} \xi)) = V_{t,s}\xi \quad (s,t \in L; \xi \in X) \qquad \ldots(11)$$

defines a linear isometry F of $\tilde{X}$ into X. Since V is non-degenerate, F is onto X.

Thus, to prove (II), it is sufficient to show that F intertwines $\tilde{V}$ and V. But, if $r,s,p,q \in L$ and $\xi \in X$, the equation (4) applied to $\tilde{V}$ and W gives:

$$\begin{aligned} F\tilde{V}_{p,q}(r \tilde{\otimes}(s \tilde{\otimes} \xi)) &= F(q \tilde{\otimes} W_{r,p}(s \tilde{\otimes} \xi)) \\ &= F(q \tilde{\otimes}(p \tilde{\otimes} V_{s,r}\xi)) \\ &= V_{p,q} V_{s,r} \xi \qquad \text{(by (11))} \\ &= V_{p,q} F(r \tilde{\otimes}(s \tilde{\otimes} \xi)). \end{aligned}$$

Since the $r \tilde{\otimes}(s \tilde{\otimes} \xi)$ span a dense subspace of $\tilde{X}$, this shows that $F \circ \tilde{V}_{p,q} = V_{p,q} \circ F$. So F is $\tilde{V}$, V intertwining; and (II) is proved. □

The preceding proposition gives rise immediately to the following duality theorem. Let L be a fixed linear space.

<u>Theorem 2.2</u>. (Duality for operator inner products) <u>The map sending each operator inner product</u> V <u>on</u> L <u>into the operator inner product</u> W <u>on</u> $\bar{L}$ <u>deduced from</u> V <u>sets up a bijection</u> $\Phi : V \longmapsto W$ <u>between</u>

the set $\underset{\sim}{V}$ *of all unitary equivalence classes of non-degenerate operator inner products on* L *, and the set* $\underset{\sim}{W}$ *of all unitary equivalence classes of non-degenerate operator inner products on* $\bar{L}$ *. The inverse* Φ^{-1} of Φ *is the map which sends an element of* $\underset{\sim}{W}$ *into the element of* $\underset{\sim}{V}$ *deduced from it.*

If $\Phi(V) = W$, we often call W the operator inner product *dual to* V, and vice versa.

Here is an important example. Let L itself be a Hilbert space and V its (complex-valued) inner product. In that case $\underset{\sim}{X}(V) = L$. Comparing (4) with §1(9), we see that the operator inner product dual to V coincides with the V of §1(9). Conversely, the operator inner product on L dual to the V of §1(9) is just the (complex-valued) inner product on L.

The Remark following §1(9) shows clearly why the complex-conjugate space $\bar{L}$ figures in the above theorem.

By (5) $\|W_{s,s}\| \leq \|V_{s,s}\|$; and by Theorem 2.2 the roles of V and W can be interchanged. So we have

$$\|W_{s,s}\| = \|V_{s,s}\| \quad (s \in L) \qquad \ldots(12)$$

whenever V and W are related as in Proposition 2.1. (Notice however that in general $\|W_{s,t}\| \neq \|V_{s,t}\|$ when $s \neq t$.)

The duality of Theorem 2.2 preserves direct sums and tensor products. More precisely, we have:

Proposition 2.3. Let $\{V^i : i \in I\}$ *be an indexed collection of operator inner products on* L *whose direct sum* V *exists. For each* i *let* W^i *be dual to* V^i*. Then the dual of* V *is unitarily equivalent to the direct sum of the* W^i *.*

Proposition 2.4. Let L *,* M *,* P *,* V *,* W *,* U *be as in Proposition 1.4. If* V' *,* W' *,* U' *are dual to* V *,* W *,* U *respectively, then* U' *is unitarily equivalent to the tensor product of* V' *and* W' *.*

The proofs of these two propositions are left to the reader.

Another very important property preserved by the duality of Theorem 2.2 is the isomorphism type of the commuting algebras.

Let V be an operator inner product on L, acting in $X = X(V)$, and W the operator inner product deduced from V, acting in $Y = \underset{\sim}{X}(V)$. If $A \in \underset{\sim}{I}(V)$, an easy calculation shows that the bounded linear operator $\tilde{A}$ on Y constructed in Theorem 1.2 actually lies in $\underset{\sim}{I}(W)$.

Theorem 2.5. Assume that V *is non-degenerate. Then* $A \mapsto \tilde{A}$ *is a *-isomorphism of* $\underset{\sim}{I}(V)$ *onto* $\underset{\sim}{I}(W)$.

Proof. In view of the observation following Theorem 1.2 we have only to show that $A \mapsto \tilde{A}$ maps $\underset{\sim}{I}(V)$ *onto* $\underset{\sim}{I}(W)$.

Let $\tilde{V}$ be the operator inner product on L deduced from W. We shall keep the notation of the proof of Proposition 2.1. If $B \in \underset{\sim}{I}(W)$ we will also denote by $\tilde{B}$ the operator in $\underset{\sim}{I}(\tilde{V})$ constructed from B:

$$\tilde{B}(t \tilde{\otimes} \zeta) = t \tilde{\otimes} B\zeta \quad (t \in \bar{L}; \zeta \in Y).$$

Because of the symmetric roles of V and W, it will be enough to show that $B \mapsto \tilde{B}$ maps $\underset{\sim}{I}(W)$ onto $\underset{\sim}{I}(\tilde{V})$.

Now, according to the proof of Proposition 2.1, $\tilde{V}$ and V are unitarily equivalent under the F defined in (11). Thus the most general element of $\underset{\sim}{I}(\tilde{V})$ is of the form $F^{-1}AF$, where $A \in \underset{\sim}{I}(V)$. Hence, to prove that $B \mapsto \tilde{B}$ maps $\underset{\sim}{I}(W)$ onto $\underset{\sim}{I}(\tilde{V})$, it will suffice to show that

$$\tilde{\tilde{A}} = F^{-1}AF \quad \text{for all } A \text{ in } \underset{\sim}{I}(V). \qquad \ldots(13)$$

To see this, let $A \in \underset{\sim}{I}(V)$, $s,t \in L$, $\xi \in X$. Then

$$\begin{aligned} F\tilde{\tilde{A}}(s \,\tilde{\otimes}(t \,\tilde{\otimes}\, \xi)) &= F(s \,\tilde{\otimes}\, \tilde{A}(t \,\tilde{\otimes}\, \xi)) \\ &= F(s \,\tilde{\otimes}(t \,\tilde{\otimes}\, A\xi)) \\ &= V_{t,s}A\xi \qquad \text{(by (11))} \\ &= AV_{t,s}\xi \\ &= AF(s \,\tilde{\otimes}(t \,\tilde{\otimes}\, \xi)). \end{aligned}$$

Since the $s \,\tilde{\otimes}(t \,\tilde{\otimes}\, \xi)$ span a dense subspace of $\tilde{X}$, this implies that $F \circ \tilde{\tilde{A}} = A \circ F$, which is (13). □

Thus $\underset{\sim}{I}(V)$ and $\underset{\sim}{I}(W)$ are *-isomorphic whenever V and W are dual to each other under the duality of Theorem 2.2. In particular W is irreducible if and only if V is.

We conclude this section with an interesting observation on compact operator inner products.

Definition. An operator inner product V on L is compact if $V_{s,t}$ is a compact operator for all s, t in L.

Proposition 2.6. *Let V be an operator inner product on L; and let W be the operator inner product on $\bar{L}$ deduced from V. If $t \in L$ and $V_{t,t}$ is a compact operator, then $V_{s,t}$ and $W_{s,t}$ are compact for all s in L.*

Proof. We recall that a bounded linear operator F on a Hilbert space

Z is compact if and only if $\|F\zeta_\alpha\| \longrightarrow 0$ whenever $\{\zeta_\alpha\}$ is a norm-bounded net of vectors in Z such that $\zeta_\alpha \longrightarrow 0$ weakly.

We shall first show that $W_{s,t}$ is compact for each s in L. Let $\{\zeta_\alpha\}$ be a norm-bounded net of vectors in $Y = \underset{\sim}{X}(V)$ approaching 0 weakly. By (3)

$$\|W_{s,t}\zeta_\alpha\|^2 = \langle t \tilde{\otimes} U_s\zeta_\alpha, t \tilde{\otimes} U_s\zeta_\alpha\rangle_0$$

$$= \langle V_{t,t}U_s\zeta_\alpha, U_s\zeta_\alpha\rangle. \qquad \ldots(14)$$

Since U_s is continuous, the net $\{U_s\zeta_\alpha\}$ is norm-bounded and converges weakly to 0. So the assumed compactness of $V_{t,t}$ implies that $\|V_{t,t}U_s\zeta_\alpha\| \underset{\alpha}{\longrightarrow} 0$. From this, (14), and the norm-boundedness of $\{U_s\zeta_\alpha\}$, it follows that $\|W_{s,t}\zeta_\alpha\| \underset{\alpha}{\longrightarrow} 0$. This and the first sentence of the proof establish the compactness of $W_{s,t}$.

In particular $W_{t,t}$ is compact. Applying to this the result already proved, with V and W reversed, we conclude that $V_{s,t}$ is compact for all s. □

As a corollary of this we get immediately:

Theorem 2.7. *If V and W correspond under the duality of Theorem 2.2, then W is compact if and only if V is compact.*

§3. Hermitian modules.

In this section we consider operator inner products on an A-module, where A is a *-algebra.

The following simple and important lemma seems not to be well known.

<u>Lemma 3.1</u>. <u>Let</u> T <u>be a pre-*-representation of a Banach *-algebra</u> A , <u>acting on a pre-Hilbert space</u> X . <u>Then</u> T_a <u>is continuous on</u> X <u>for each</u> a ; <u>that is</u>, T <u>extends to a *-representation of</u> A <u>on the completion of</u> X .

<u>Proof</u>. Without loss of generality we may assume that A has a unit 1 and that T_1 = identity. It is enough to show that

$$\|T_a\xi\| \leq \|a\|\|\xi\| \quad (a \in A; \xi \in X). \qquad \text{...(1)}$$

Fix $\xi \in X$; and let p be the positive linear functional on A given by $p(a) = \langle T_a\xi, \xi\rangle$. So (see Rickart [1], p. 219, Corollary 4.5.13) there is a cyclic *-representation S of A acting on a Hilbert space Y , with cyclic vector η in Y , such that $p(a) = \langle S_a\eta, \eta\rangle$ $(a \in A)$. We then have, for each a in A ,

$$\|T_a\xi\|^2 = \langle T_{a^*a}\xi, \xi\rangle = p(a^*a)$$
$$= \langle S_{a^*a}\eta, \eta\rangle = \|S_a\eta\|^2 . \qquad \text{...(2)}$$

Putting $a = 1$ in (2) we get $\|\xi\| = \|\eta\|$. Therefore by (2), for any a in A ,

$$\|T_a\xi\| \leq \|S_a\|\|\eta\| = \|S_a\|\|\xi\| \leq \|a\|\|\xi\| ;$$

and this is (1). □

Now fix a *-algebra A ; and let L be a linear space which is also a left A-module. Furthermore let V be an operator inner product on L .

<u>Definition</u>. We say that the A-module L is <u>Hermitian with respect to</u>

V if

$$V_{as,t} = V_{s,a^*t} \quad (a \in A; s,t \in L). \qquad \ldots(3)$$

Let us assume that L is Hermitian with respect to V. We ask the question: Can the action of A on L be "expanded" to a *-representation of A on $\underset{\sim}{X}(V)$?

To begin with, adopting the notation of §1(7), we denote by $\underset{\sim}{X}_0(V)$ the linear span of $\{s \tilde{\otimes} \xi : s \in L, \xi \in X(V)\}$; this is of course dense in $\underset{\sim}{X}(V)$. Suppose that $a \in A$ and $\sum_{i=1}^{n} t_i \tilde{\otimes} \xi_i = 0$ $(t_i \in L;$ $\xi_i \in X(V))$. Then for all s in L and η in $X(V)$ we have by (3) and §1(7):

$$\begin{aligned}\sum_{i=1}^{n} \langle at_i \tilde{\otimes} \xi_i, s \tilde{\otimes} \eta\rangle_0 &= \sum_{i=1}^{n} \langle V_{at_i,s}\xi_i, \eta\rangle \\ &= \sum_{i=1}^{n} \langle V_{t_i,a^*s}\xi_i, \eta\rangle \\ &= \sum_{i=1}^{n} \langle t_i \tilde{\otimes} \xi_i, a^*s \tilde{\otimes} \eta\rangle_0 = 0 .\end{aligned}$$

By the arbitrariness of s and η this implies that $\sum_{i=1}^{n} at_i \tilde{\otimes} \xi_i = 0$. Thus $\sum_i t_i \tilde{\otimes} \xi_i = 0 \Rightarrow \sum_i at_i \tilde{\otimes} \xi_i = 0$; and it follows that for each a in A the equation

$$T_a(t \tilde{\otimes} \xi) = at \tilde{\otimes} \xi \quad (t \in L; \xi \in X(V)) \qquad \ldots(4)$$

defines a linear endomorphism T_a of $\underset{\sim}{X}_0(V)$. In view of (3) $T : a \mapsto T_a$ is a pre-*-representation of A; it is called the <u>pre-*-representation of</u> A <u>deduced from</u> L, V.

In general the T_a are not continuous on $\underset{\sim}{X}_0(V)$. If however it happens that each T_a is continuous on $X_0(V)$, then L, as an A-module, is said to be V-*bounded*. In that case each T_a extends to a continuous linear operator T'_a on $X(V)$; and $T' : a \mapsto T'_a$ is called the **-representation of* A *deduced from* L, V.

Obviously L is V-bounded if $X(V)$ is finite-dimensional. More importantly, by Lemma 3.1, L *is automatically* V-*bounded if* A *is a Banach *-algebra*.

Assume that L is V-bounded; and let T be the deduced *-representation of A on $X(V)$. If $F \in I(V)$, it is obvious that the operator $\tilde{F}$ on $X(V)$ constructed in Theorem 1.2 belongs to the commuting algebra $I(T)$ of T. Thus, if V is non-degenerate, $F \mapsto \tilde{F}$ is a *-isomorphism of $I(V)$ into $I(T)$. In particular, T cannot be irreducible unless V is irreducible.

To conclude this section we observe one important special case of the context discussed in this section. Let A be a *-algebra and X a Hilbert space. A linear map $P : A \to O(X)$ is called *completely positive* if the map $V : (a,b) \mapsto P(b^*a)$ is an operator inner product, that is, if

$$\sum_{i,j=1}^{n} \langle P(a_j^* a_i)\xi_i, \xi_j \rangle \geq 0 \qquad \ldots(5)$$

whenever n is a positive integer, $a_1,\ldots,a_n \in A$, and $\xi_1,\ldots,\xi_n \in X$. Suppose that P is completely positive. Then obviously A, as a left A-module under left multiplication, is Hermitian with respect to V. So, if A as a left A-module is V-bounded (in particular, if A is a Banach *-algebra), we can form the *-representation T of A deduced from A, V; we call T the *-representation of A *generated*

<u>by</u> P . This generated *-representation is essentially what is studied by Stinespring [1]. If P is complex-valued, it amounts of course to the long-familiar Gelfand-Naimark-Segal construction.

§4. <u>Rigged modules and the abstract inducing process</u>.

We come now to Rieffel's abstract inducing process, sketched in the Introduction.

The essential ingredient of the abstract inducing process is the notion of a rigged module.

<u>Rigged spaces</u>.

We begin with rigged spaces. Fix a *-algebra B .

<u>Definition</u>. A B-<u>rigged</u> <u>space</u> is a linear space L , together with a right B-module structure for L and a map $[\ ,\ ]: L \times L \longrightarrow B$, such that the following properties hold:

(i) $[s,t]$ is linear in t and conjugate-linear in s ;

(ii) $[s,t]^* = [t,s]$ $(s,t \in L)$;

(iii) $[s,tb] = [s,t]b$ $(s,t \in L; b \in B)$.

<u>Remark</u>. The tb of property (iii) refers of course to the right B-module structure of L . Notice that, unlike an operator inner product, [,] is linear in its <u>second</u> variable.

The above mapping [,] is called a B-<u>rigging</u> of L .

Conditions (ii) and (iii) clearly imply that

$$b[s,t] = [sb^*,t] \quad (s,t \in L; b \in B). \qquad \ldots(1)$$

They also imply that the linear span of the range of [,] is a

*-ideal of B .

Rigged spaces are closely related to operator inner products. Let $\underset{\sim}{L} = (L, [\ ,\ ])$ be an arbitrary B-rigged space, and S a *-representation of B ; and let us put

$$V_{s,t} = S_{[t,s]} \qquad (s,t \in L). \qquad \ldots(2)$$

Thus $V : L \times L \longrightarrow \underset{\sim}{O}(X(S))$ is linear in s and conjugate-linear in t .

<u>Proposition 4.1</u>. V <u>is an operator inner product if and only if</u>

$$S_{[t,t]} \geq 0 \quad \underline{\text{for all}} \quad t \quad \underline{\text{in}} \quad L . \qquad \ldots(3)$$

<u>Proof</u>. Condition (3) is obviously necessary. We shall assume (3) and show that V is completely positive. Now S can be written in the form

$$S \cong \sum_{i \in I}^{\oplus} S^i ,$$

where each S^i is either a cyclic representation or a zero representation of B . By Proposition 1.3 it is enough to show that $V^i : (s,t) \longmapsto S^i_{[t,s]}$ is completely positive for each i . This is obvious if S^i is a zero representation. So we may as well assume from the beginning that S is cyclic with cyclic vector ξ .

In proving §1(1) it is enough to have the ξ_i run over the dense subset $\{S_b \xi : b \in B\}$ of $X(S)$. If $t_1, \ldots, t_n \in L$ and $\xi_i = S_{b_i} \xi$ $(b_i \in B; i = 1,2,\ldots,n)$, then

$$\sum_{i,j=1}^{n} \langle V_{t_i,t_j}\xi_i,\xi_j\rangle = \sum_{i,j=1}^{n} \langle S_{[t_j,t_i]}S_{b_i}\xi,S_{b_j}\xi\rangle$$

$$= \sum_{i,j=1}^{n} \langle S_{b_j^*[t_j,t_i]b_i}\xi,\xi\rangle$$

$$= \sum_{i,j=1}^{n} \langle S_{[t_jb_j,t_ib_i]}\xi,\xi\rangle \qquad \text{(by (iii) and (1))}$$

$$= \langle S_{[r,r]}\xi,\xi\rangle \geq 0 \qquad \text{(by (3))},$$

where $r = \sum_{i=1}^{n} t_ib_i$. So §1(1) holds for V . □

<u>Definition</u>. A *-representation S of B such that (3) holds is called <u>positive with respect to</u> $\underset{\sim}{L}$. In this case, by the preceding proposition, (2) defines an operator inner product V on L .

Suppose that the *-representation S of B is positive with respect to $\underset{\sim}{L}$; and construct the operator inner product V as in (2). Let $\underset{\sim}{X}(V)$ be the Hilbert space deduced from V as in §1; and adopt the rest of the notation of §1(7). We then notice the following important identity:

$$rb \tilde{\otimes} \xi = r \tilde{\otimes} S_b\xi \quad (r \in L; b \in B; \xi \in X(S)). \qquad \ldots(4)$$

Indeed: Fixing r , b , and ξ , we have for any s in L and η in X(S)

$$\langle rb \tilde{\otimes} \xi - r \tilde{\otimes} S_b\xi, s \tilde{\otimes} \eta\rangle_0$$

$$= \langle S_{[s,rb]}\xi,\eta\rangle - \langle S_{[s,r]}S_b\xi,\eta\rangle = 0$$

by (iii). Since the $s \tilde{\otimes} \eta$ span a dense subspace of $\underset{\sim}{X}(V)$, this implies

(4).

Rigged modules.

We now fix, in addition to B, a second *-algebra A.

Definition. A B-rigged A-module is a B-rigged space L, $[\ ,\ ]$ which (in addition to its right B-module structure) carries a left A-module structure satisfying the "Hermitian" relation:

$$[as,t] = [s,a^*t] \quad (s,t \in L; a \in A). \qquad \ldots(5)$$

Remark. We do not assume the associative law $(as)b = a(sb)$ $(a \in A;$ $b \in B; s \in L)$, though this will hold in the important applications.

An important special case of a rigged module is provided by a so-called conditional expectation. As before A and B are *-algebras; and we assume that the underlying linear space of A has the structure of a right B-module satisfying the associative law $(a_1a_2)b = a_1(a_2b)$ $(a_1,a_2 \in A; b \in B)$. (This would automatically be the case, for example, if B were a *-subalgebra of A.)

Definition. An A, B conditional expectation is a linear map $p : A \longrightarrow B$ such that (i) $p(a^*) = (p(a))^*$ $(a \in A)$, and (ii) $p(ab) = p(a)b$ $(a \in A; b \in B;$ ab refers of course to the right B-module structure of $A)$.

Now let p be an A, B conditional expectation; regard A also as a left A-module under multiplication; and define

$$[a_1,a_2] = p(a_1^*a_2) \quad (a_1,a_2 \in A). \qquad \ldots(6)$$

One verifies without difficulty that $(A,[\ ,\ ])$ is then a B-rigged

A-module.

Rigged modules obtained in this way will be of considerable importance in the last chapter.

Let us now fix an arbitrary B-rigged A-module $\underset{\sim}{L} = (L, [\ ,\ ])$, and a $*$-representation S of B which is *positive with respect to* $\underset{\sim}{L}$ (by which we mean, of course, that S is positive with respect to the B-rigged space underlying $\underset{\sim}{L}$). We construct the operator inner product V as in (2), denote by $\underset{\sim}{X}(V)$ the Hilbert space deduced from V as in §1, and adopt the rest of the notation of §1(7). By (5) the left A-module structure of L is Hermitian with respect to V.

Definition. If in addition the left A-module structure of L is V-bounded (see §3), we say that the $*$-representation S of B is *inducible to* A *via* $\underset{\sim}{L}$. In that case the $*$-representation of A on $\underset{\sim}{X}(V)$ deduced from L, V (see §3) is said to be (*abstractly*) *induced from* S *via* $\underset{\sim}{L}$, and is denoted by $\mathrm{Ind}^{\underset{\sim}{L}}_{B\uparrow A}(S)$, or simply $\mathrm{Ind}_{B\uparrow A}(S)$ or even $\mathrm{Ind}(S)$ if no ambiguity can arise.

This is Rieffel's fundamental abstract inducing process, whose ramifications will occupy us throughout much of this work.

Proposition 4.2. *If* A *is a Banach* $*$-*algebra, then a* $*$-*representation* S *of* B *will be inducible to* A *via* $\underset{\sim}{L}$ *provided only that it is positive with respect to* $\underset{\sim}{L}$.

Proof. In that case the left A-module L is automatically V-bounded in virtue of §3. ◻

Remark. Suppose that $\underset{\sim}{L}$ was constructed from an A, B conditional expectation p as in (6). Then the phrases '*positive with respect to* p' and '*inducible via* p' will mean 'positive with respect to $\underset{\sim}{L}$'

and 'inducible via $\underset{\sim}{L}$'; and we may write $\mathrm{Ind}_{B\uparrow A}^{p}(S)$ instead of $\mathrm{Ind}_{B\uparrow A}^{\underset{\sim}{L}}(S)$.

<u>Example; finite groups</u>. Our first concern is to verify that this abstract inducing process generalizes the classical construction of induced representations of finite groups.

Let G be a finite group and H a subgroup of G; and let A and B stand for the group *-algebras of G and H respectively (under the standard convolution $*$ and involution). B will be identified with a *-subalgebra of A, so that A becomes a natural right B-module. One easily verifies the important fact that the restriction map $p : f \mapsto f|H$ $(f \in A)$ is an A, B conditional expectation.

Now let S be a finite-dimensional unitary representation of H; and denote by S' the corresponding *-representation of B (in other words, the integrated form of S). We shall see that (a) S' is positive with respect to p, hence inducible to A via p; (b) $\mathrm{Ind}_{B\uparrow A}^{p}(S')$ is the integrated form of the representation T of G classically induced from S.

Put $X = X(S)$; and fix a transversal Γ for G/H. We recall the classical definition of T: $X(T)$ is the Hilbert space of all functions $\varphi : G \to X$ satisfying

$$\varphi(xh) = S_{h^{-1}}(\varphi(x)) \quad (x \in G; h \in H), \qquad \ldots(7)$$

with the inner product

$$\langle \varphi, \psi \rangle_{X(T)} = \sum_{x \in \Gamma} \langle \varphi(x), \psi(x) \rangle_X ; \qquad \ldots(8)$$

and the operators of T are just those of left translation:

$$(T_x\varphi)(y) = \varphi(x^{-1}y) \qquad (\varphi \in X(T); x,y \in G). \qquad \ldots(9)$$

We now define a linear map $F : A \otimes X \longrightarrow X(T)$ as follows:

$$(F(f \otimes \xi))(x) = \sum_{h\in H} f(xh)S_h\xi \qquad \ldots(10)$$

$(f \in A; \xi \in X; x \in G)$. To begin with, one checks that range(F) is indeed contained in $X(T)$. If φ is any element of $X(T)$, a simple calculation shows that $\gamma = \sum_{x\in\Gamma} \delta_x \otimes \varphi(x)$ satisfies $F(\gamma) = \varphi$. So range$(F) = X(T)$.

Let $[\ ,\ ]_B$ stand for the B-rigging on $A \times A$ given by (6); that is,

$$[f,g]_B = p(f^* * g) = (f^* * g)|H . \qquad \ldots(11)$$

Then, if $f,g \in A$ and $\xi,\eta \in X$

$$\langle S'_{[g,f]_B}\xi,\eta\rangle = \sum_{h\in H} (g^* * f)(h)\langle S_h\xi,\eta\rangle$$

$$= \sum_{h\in H}\sum_{x\in G} \overline{g(x)}f(xh)\langle S_h\xi,\eta\rangle$$

$$= \sum_{x\in\Gamma}\sum_{h,k\in H} \overline{g(xk)}f(xkh)\langle S_h\xi,\eta\rangle$$

$$= \sum_{x\in\Gamma}\sum_{h,k\in H} \overline{g(xk)}f(xh)\langle S_{k^{-1}h}\xi,\eta\rangle$$

$$= \sum_{x\in\Gamma}\sum_{h,k\in H} \overline{g(xk)}f(xh)\langle S_h\xi,S_k\eta\rangle$$

$$= \sum_{x\in\Gamma} \langle F(f \otimes \xi)(x), F(g \otimes \eta)(x)\rangle$$

$$= \langle F(f \otimes \xi), F(g \otimes \eta)\rangle_{X(T)} . \qquad \ldots(12)$$

From this it follows that $\langle S_{[f,f]_B}\xi,\xi\rangle \geq 0$ for all f in A and ξ in X ; so by Proposition 4.1 S' is positive with respect to p, hence inducible to A via p. Let $W = \mathrm{Ind}^{p}_{B\uparrow A}(S')$; and as in §1 let $f \tilde{\otimes} \xi$ stand for the image of $f \otimes \xi$ in $X(W)$. Since F is onto $X(T)$, (11) implies that $X(W)$ and $X(T)$ are isometrically isomorphic under the bijection $\tilde{F} : X(W) \longrightarrow X(T)$ given by:

$$\tilde{F}(f \tilde{\otimes} \xi) = F(f \otimes \xi) \qquad (f \in A; \xi \in X).$$

One easily verifies that

$$F((g^*f)\otimes \xi) = T'_g(F(f \otimes \xi)) \qquad (f,g \in A; \xi \in X)$$

(where T' is the integrated form of T). Now the last two equations imply that W and T' are unitarily equivalent under $\tilde{F}$. This completes the proof of (a) and (b).

We shall see in Chapter III that not only the induced representations of finite groups, but even the induced representations of arbitrary locally compact groups are special cases of the abstract inducing process defined above. Indeed, there are many other constructions by which a *-representation of one *-algebra is obtained from a *-representation of another *-algebra, which, though apparently quite dissimilar to the classical inducing process on groups, nevertheless emerge as special cases of the above abstract inducing process. Several such examples will appear in §9 and Chapter III.

§5. General properties of abstract induction.

Direct sums.

Let A and B be two *-algebras and $\underset{\sim}{L}$ a B-rigged A-module.

Proposition 5.1. *Let* $\{S^i\}$ $(i \in I)$ *be an indexed collection of* *-*representations of* B *whose Hilbert direct sum* *-*representation* $S = \sum_{i\in I}^{\oplus} S^i$ *can be formed. For* S *to be inducible to* A *via* $\underset{\sim}{L}$ *it is necessary and sufficient that* (a) *each* S^i *be inducible to* A *via* $\underset{\sim}{L}$, *and* (b) *the Hilbert direct sum* *-*representation* $T = \sum_{i\in I}^{\oplus} \mathrm{Ind}_{B\uparrow A}^{\underset{\sim}{L}}(S^i)$ *exist. In that case* $T \cong \mathrm{Ind}_{B\uparrow A}^{\underset{\sim}{L}}(S)$.

Thus the abstract inducing process preserves direct sums.

The proof of this proposition follows from Proposition 1.3.

Tensor products.

Let A , B , C , D be four *-algebras; let L,[ ] be a B-rigged A-module; and let M,[,]' be a D-rigged C-module. Then $L \otimes M$ is a left $(A \otimes C)$-module and a right $(B \otimes D)$-module under the natural module-structures:

$$(a \otimes c)(r \otimes u) = ar \otimes cu \ , \quad (r \otimes u)(b \otimes d) = rb \otimes ud$$

$(a \in A; b \in B; c \in C; d \in D; r \in L; u \in M)$; and the equation

$$[r \otimes u, s \otimes v]'' = [r,s] \otimes [u,v]'$$

$(r,s \in L; u,v \in M)$ defines $L \otimes M, [\ ,\ ]''$ as a $(B \otimes D)$-rigged $(A \otimes C)$-module.

Proposition 5.2. *Let* S *be a* *-*representation of* B *which is inducible to* A *via* L, [,], *and* W *a* *-*representation of* D *which is inducible to* C *via* M, [,]'. *Then the (outer) tensor product* *-*representation* $S \otimes W$ *of* $B \otimes D$ *is inducible to* $A \otimes C$ *via* $L \otimes M$, [,]"; *and*

$$\mathrm{Ind}(S \otimes W) \cong \mathrm{Ind}(S) \otimes \mathrm{Ind}(W).$$

This follows easily from Proposition 1.4.

Thus the abstract inducing operation preserves (outer) tensor products.

Inducing in stages.

From the next proposition it will follow that, under suitable conditions, the result of iterating the abstract inducing process is the same as the result of a single abstract inducing process.

Let A and B be two $*$-algebras, and $\underset{\sim}{L} = (L,[\ ,\ ])$ a B-rigged A-module. Further, let V be an operator inner product on a linear space K; and let K be also a bounded Hermitian left B-module with respect to V (see §3). Thus by §3 we can form the $*$-representation S of B deduced from K, V. Putting $P = L \otimes K$ and $X = X(V)$, we can also form the map $\Gamma : P \times P \longrightarrow \underset{\sim}{O}(X)$ given by

$$\Gamma(r \otimes k, s \otimes m) = V([s,r]k,m) \quad (r,s \in L; k,m \in K),$$

which is linear in the first variable and conjugate-linear in the second. Notice that P is a natural left A-module:

$$a(r \otimes k) = ar \otimes k \quad (a \in A; r \in L; k \in K).$$

As a left A-module, P is Hermitian with respect to Γ in view of §4(5).

Proposition 5.3. (I) *S is positive with respect to $\underset{\sim}{L}$ if and only if Γ is an operator inner product on P.* (II) *S is inducible to A via $\underset{\sim}{L}$ if and only if Γ is an operator inner product and P, as a left A-module, is Γ-bounded.* (III) *If the conditions of (II) hold, $\mathrm{Ind}_{B\uparrow A}^{\underset{\sim}{L}}(S)$ is unitarily equivalent to the $*$-representation of A*

deduced from P and Γ.

Proof. We define conjugate-bilinear forms $\langle\ ,\ \rangle_0$ and $\langle\ ,\ \rangle_{00}$ on $P \otimes X$ and $L \otimes X(V)$ respectively as follows:

$$\langle r \otimes k \otimes \xi, s \otimes m \otimes \eta\rangle_0 = \langle \Gamma(r \otimes k, s \otimes m)\xi, \eta\rangle_X ,$$

$$\langle r \otimes \zeta, s \otimes \zeta'\rangle_{00} = \langle S_{[s,r]}\zeta, \zeta'\rangle_{X(V)}$$

($r, s \in L; m \in K; \xi, \eta \in X; \zeta, \zeta' \in X(V)$). Then, for all such r, s, k, m, ξ, η,

$$\begin{aligned}\langle r \otimes k \otimes \xi, s \otimes m \otimes \eta\rangle_0 &= \langle V([s,r]k, m)\xi, \eta\rangle_X \\ &= \langle [s,r]k \tilde{\otimes} \xi, m \tilde{\otimes} \eta\rangle_{X(V)} \\ &= \langle S_{[s,r]}(k \tilde{\otimes} \xi), m \tilde{\otimes} \eta\rangle_{X(V)} \\ &= \langle r \otimes (k \tilde{\otimes} \xi), s \otimes (m \tilde{\otimes} \eta)\rangle_{00} . \qquad \ldots (1)\end{aligned}$$

Now Γ is an operator inner product if and only if $\langle\ ,\ \rangle_0$ is positive. By (1) this happens if and only if

$$\sum_{i,j=1}^{n} \langle r_i \otimes (k_i \tilde{\otimes} \xi_i), r_j \otimes (k_j \tilde{\otimes} \xi_j)\rangle_{00} \geq 0$$

for all n and all r_i in L, k_i in K, ξ_i in X. Since $K \tilde{\otimes} X$ is dense in $X(V)$, this happens if and only if

$$\sum_{i=1}^{n} \langle r_i \otimes \zeta_i, r_j \otimes \zeta_j\rangle_{00} \geq 0 \qquad \ldots (2)$$

for all n and all r_i in L and ζ_i in $X(V)$. But (2) asserts that S is positive with respect to L. So (I) is proved.

We observe from (1) that, if the conditions of (I) hold, then

$\underset{\sim}{X}(\Gamma)$ is isometrically isomorphic with $\underset{\sim}{X}(S)$ under the linear map

$$\beta : (r \otimes k) \tilde{\otimes} \xi \longmapsto r \tilde{\otimes} (k \tilde{\otimes} \xi). \qquad \ldots(3)$$

(Here by $\underset{\sim}{X}(S)$ we mean $\underset{\sim}{X}(W)$, where $W_{r,s} = S_{[s,r]}$.)

To prove (II), suppose that S is inducible to A. This means that for each a in A there exists a bounded linear operator T_a on $\underset{\sim}{X}(S)$ such that $T_a(r \tilde{\otimes} \zeta) = ar \tilde{\otimes} \zeta$ $(r \in L; \zeta \in X(S))$. Passing to $\underset{\sim}{X}(\Gamma)$ via the isometry β of (3) we conclude that there exists a bounded linear operator T'_a on $\underset{\sim}{X}(\Gamma)$ satisfying

$$T'_a((r \otimes k) \tilde{\otimes} \xi) = (ar \otimes k) \tilde{\otimes} \xi \quad (r \in L; k \in K; \xi \in X).$$

So P, as a left A-module, is Γ-bounded. Since the argument works equally well in the opposite direction, statement (II) is proved. The unitary equivalence required for statement (III) is just the β of (3). □

The following special case of Proposition 5.3 describes a genuine iteration of inducing processes.

Let A, B, C be three *-algebras. Let $\underset{\sim}{L} = (L, [\ ,\ ])$ be a B-rigged A-module, and let $\underset{\sim}{K} = (K, [\ ,\ ]')$ be a C-rigged B-module. Thus $P = L \otimes K$ is a natural left A-module and right C-module:

$$a(r \otimes k) = ar \otimes k, \quad (r \otimes k)c = r \otimes kc.$$

Defining $[\ ,\ ]'' : P \times P \longrightarrow C$ by means of the relation

$$[r \otimes k, s \otimes m]'' = [[s,r]k, m]' \quad (r,s \in L; k,m \in K) \qquad \ldots(4)$$

one verifies without difficulty that $\underset{\sim}{P} = (P, [\ ,\ ]'')$ is a C-rigged A-module. We call $\underset{\sim}{P}$ the <u>composition</u> of $\underset{\sim}{L}$ and $\underset{\sim}{K}$. In this context we have:

<u>Corollary 5.4</u>. <u>Let</u> W <u>be a</u> *-<u>representation of</u> C <u>which is inducible to</u> B <u>via</u> $\underset{\sim}{K}$, <u>and set</u> $S = \mathrm{Ind}_{C\uparrow B}^{\underset{\sim}{K}}(W)$. <u>Then the following two conditions are equivalent</u>: (i) S <u>is inducible to</u> A <u>via</u> $\underset{\sim}{L}$; (ii) W <u>is inducible to</u> A <u>via</u> $\underset{\sim}{P}$. <u>If</u> (i) <u>and</u> (ii) <u>hold, then</u>

$$\mathrm{Ind}_{B\uparrow A}^{\underset{\sim}{L}}(S) \cong \mathrm{Ind}_{C\uparrow A}^{\underset{\sim}{P}}(W).$$

This follows immediately from Proposition 5.3.

This Corollary should be compared with Rieffel [1], Thm. 5.9.

<u>Remark</u>. In all interesting cases the above two-sided modules L and K will satisfy the associative law:

$$(ar)b = a(rb), \quad (bk)c = b(kc) \qquad \ldots(5)$$

($r \in L; k \in K; a \in A; b \in B; c \in C$). Let us for the moment assume (5). Let N denote the linear span in P of $\{rb \otimes k - r \otimes bk : r \in L, k \in K, b \in B\}$, and P_B the quotient space P/N ; and let $r \otimes_B k$ be the image of $r \otimes k$ under the quotient map $P \longrightarrow P_B$. In view of (5) P_B inherits a left A-module and a right C-module structure from P :

$$a(r \otimes_B k) = ar \otimes_B k, \quad (r \otimes_B k)c = r \otimes_B kc.$$

Likewise, (4) can be lifted to a C-rigging $[\ ,\ ]_B''$ of P_B :

$$[r \otimes_B k, s \otimes_B m]_B'' = [[s,r]k,m]' ; \qquad \ldots(6)$$

and $\underset{\sim}{P}_B = (P_B, [\ ,\ ]_B'')$ is a C-rigged A-module. We call it the <u>reduced composition</u> of $\underset{\sim}{L}$ and $\underset{\sim}{K}$. Now it is quite clear that the two processes of inducing from C to A via $\underset{\sim}{P}$ and via $\underset{\sim}{P}_B$ are identical. Hence

Corollary 5.4 holds when $\underset{\sim}{P}$ is replaced (in both its occurrences) by $\underset{\sim}{P}_B$.

In the context of induced representations of finite groups, the application of Corollary 5.4 with $\underset{\sim}{P}$ replaced by $\underset{\sim}{P}_B$ gives just the classical statement of the theorem on inducing in stages.

§6. The regional topology of representations and the continuity of the inducing process.

We intend to show that the abstract inducing process is continuous. To do this one must introduce a topology into the space of all (not necessarily irreducible) *-representations of a given *-algebra. Most of this section deals with the definition and elementary properties of one such topology -- the so-called regional topology.

For comparisons of this topology with other such topologies that have appeared in the literature, see the end of this section.

Let A be a fixed *-algebra, and $\underset{\sim}{T}(A)$ the class of all (concrete) *-representations of A . As usual, $X(T)$ is the Hilbert space on which the element T of $\underset{\sim}{T}(A)$ acts.

Consider an element T of $\underset{\sim}{T}(A)$; and let ε be a positive number, $\xi_1,\dots,\xi_p$ a finite sequence of vectors in $X(T)$, and F a finite non-void subset of A . Then we shall define

$$U = U(T;\varepsilon;\xi_1,\dots,\xi_p;F) \qquad \dots(1)$$

to be the family of all those T' in $\underset{\sim}{T}(A)$ such that there exist vectors $\xi_1',\dots,\xi_p'$ in $X(T')$ satisfying:

$$|\langle\xi_i',\xi_j'\rangle-\langle\xi_i,\xi_j\rangle| < \varepsilon \ , \qquad \dots(2)$$

$$|\langle T_a'\xi_i',\xi_j'\rangle - \langle T_a\xi_i,\xi_j\rangle| < \varepsilon \qquad \ldots(3)$$

for all a in F and all $i,j = 1,\cdots,p$. The collection of all such families $U(T;\varepsilon;\{\xi_i\};F)$, with T fixed and ε, p, $\{\xi_i\}$, F varying, will be denoted by $\underset{\sim}{U}(T)$.

Now the following properties of $\underset{\sim}{U}(T)$ are either obvious or easily verified:

(I) $T \in U$ for every U in $\underset{\sim}{U}(T)$.

(II) If U and V are in $\underset{\sim}{U}(T)$, there is a family W in $\underset{\sim}{U}(T)$ such that $W \subset U \cap V$.

(III) If $U \in \underset{\sim}{U}(T)$, there is a family V in $\underset{\sim}{U}(T)$ such that, for each S in V, we have $W \subset U$ for some W in $\underset{\sim}{U}(S)$.

These three properties assure us that there is a unique topology for $\underset{\sim}{T}(A)$ relative to which, for each T in $\underset{\sim}{T}(A)$, $\underset{\sim}{U}(T)$ is a basis of neighborhoods of T. (See Kelley [1], Problem B, p. 56.)

<u>Definition</u>. The topology of $\underset{\sim}{T}(A)$ constructed above, in which $\underset{\sim}{U}(T)$ is a basis of neighborhoods of T for each T in $\underset{\sim}{T}(A)$, will be called the <u>regional topology (of *-representations of</u> A).

For brevity, a basis of neighborhoods of T in the regional topology will be called simply a <u>regional neighborhood basis</u> of T; sets which are closed in the regional topology will be called <u>regionally closed</u>; and so forth.

<u>Remark</u>. It can be shown without much difficulty (though we shall not need it) that the definition of the regional topology is not altered if we replace (2) by

$$\langle \xi_i', \xi_j' \rangle = \langle \xi_i, \xi_j \rangle \quad (i,j = 1,\ldots,p).$$

Remark. Notice that any subrepresentation of a *-representation T of A belongs to the regional closure of $\{T\}$.

The following proposition asserts that there are regional neighborhood bases which are smaller and easier to work with than $\underset{\sim}{U}(T)$.

Proposition 6.1. *Let T be a *-representation of A; and let Z be any subset of $X(T)$ such that the smallest closed T-stable subspace of $X(T)$ containing Z is all of $X(T)$. Then the set of all $U(T;\varepsilon;\xi_1,\ldots,\xi_p;F)$, where $\varepsilon > 0$, F is a finite subset of A, and $\xi_1,\ldots,\xi_p$ is a finite sequence of vectors in Z, is a regional neighborhood basis of T.*

Proof. To begin with, the reader will observe that this is true if Z is a dense subset X' of $X(T)$.

Next, suppose that Z is a subset X'' of $X(T)$ whose linear span X' is dense in $X(T)$. If $\xi_1,\ldots,\xi_p$ is a given finite sequence of vectors in X', we can represent each ξ_i as a finite linear combination of vectors $\eta_1,\ldots,\eta_q$ in X'':

$$\xi_i = \sum_{j=1}^{q} \lambda_{ij}\eta_j \quad (\lambda_{ij} \in \mathbb{C}; i = 1,\ldots,p).$$

For any $\varepsilon > 0$ and any finite subset F of A, putting $m = \max\{|\lambda_{ij}|: i = 1,\ldots,p; j = 1,\ldots,q\}$ and $\delta = \varepsilon q^{-2}m^{-2}$, we see by an easy calculation that

$$U(T;\delta;\eta_1,\ldots,\eta_q;F) \subset U(T;\varepsilon;\xi_1,\ldots,\xi_p;F).$$

This combined with the first sentence of the proof gives the required conclusion in case $Z = X''$.

Finally, let Z be any subset of $X(T)$ with the property assumed in the proposition. Then $X'' = Z \cup \{T_b\xi : b \in A, \xi \in Z\}$ has dense linear span in $X(T)$. If η_1, η_2 are any two vectors in X'' and $a \in A$, one verifies that $\langle \eta_1, \eta_2 \rangle$ and $\langle T_a\eta_1, \eta_2 \rangle$ are each of the form either $\langle \xi_1, \xi_2 \rangle$ or $\langle T_b\xi_1, \xi_2 \rangle$, where $\xi_1, \xi_2 \in Z$ and $b \in A$. It follows that for any $\varepsilon > 0$, any finite sequence $\eta_1, \ldots, \eta_q$ of vectors in X'', and any finite subset F of A, there is a finite sequence $\xi_1, \ldots, \xi_r$ of vectors in Z and a finite subset G of A such that

$$U(T;\varepsilon;\xi_1,\ldots,\xi_r;G) \subset U(T;\varepsilon;\eta_1,\ldots,\eta_q;F).$$

Combining this with the preceding paragraph, we obtain the conclusion of the proposition. □

<u>Corollary 6.2</u>. <u>Let</u> T <u>be a cyclic</u> $*$-<u>representation of</u> A, <u>with cyclic vector</u> ξ. <u>Then the family of all</u> $U(T;\varepsilon;\xi;F)$, <u>where</u> $\varepsilon > 0$ <u>and</u> F <u>runs over all finite subsets of</u> A, <u>is a regional neighborhood basis of</u> T.

Notice that the regional topology does not distinguish between unitarily equivalent $*$-representations of A. Conversely, it can be shown that if S and T are two <u>finite-dimensional</u> $*$-representations of A which are not distinguished by the regional topology (i.e. an open subset of $\underset{\sim}{T}(A)$ contains S if and only if it contains T), then S and T are unitarily equivalent. However, it is perfectly possible for two <u>infinite-dimensional</u> $*$-representations of A to be unitarily inequivalent and yet fail to be distinguished by the regional topology.

Identifying unitarily equivalent elements of $\underset{\sim}{T}(A)$ we shall speak of the <u>regional topology</u> of the space $\underset{\sim}{\tilde{T}}(A)$ of all unitary equivalence classes of elements of $\underset{\sim}{T}(A)$. By the preceding paragraph the regional topology of $\underset{\sim}{\tilde{T}}(A)$ does not have even the T_0 separation property. Relativized to the subspace of finite-dimensional elements of $\underset{\sim}{\tilde{T}}(A)$, it has the T_0 property (by the result quoted in the preceding paragraph), but not the T_1 property (in view of the Remark preceding Prop. 6.1).

Let $\underset{\sim}{\tilde{T}}_1(A)$ be the set of all *-homomorphisms of A into $\mathbb{C}$; thus $\underset{\sim}{\tilde{T}}_1(A)$ can be identified with the space of all one-dimensional elements of $\underset{\sim}{\tilde{T}}(A)$. In view of Cor. 6.2 we have:

<u>Corollary 6.3</u>. <u>The regional topology relativized to</u> $\underset{\sim}{\tilde{T}}_1(A)$ <u>coincides with the topology of pointwise convergence on</u> A.

It is often useful to formulate the definition of the regional topology in terms of convergence. The following statements are easy consequences of the definition of the regional topology and Prop. 6.1:

Let T be an element of $\underset{\sim}{T}(A)$; and let Z be a subset of $X(T)$ with the property in Prop. 6.1. Suppose now that $\{T^{\nu}\}$ is a net of elements of $\underset{\sim}{T}(A)$ with the following property: For any finite sequence $\xi_1,\ldots,\xi_p$ of vectors in Z, we can find vectors $\xi_1^{\nu},\ldots,\xi_p^{\nu}$ in $X(T^{\nu})$ for each ν such that

$$\langle \xi_i^{\nu},\xi_j^{\nu}\rangle \underset{\nu}{\longrightarrow} \langle \xi_i,\xi_j\rangle \qquad \ldots(4)$$

and

$$\langle T_a^{\nu}\xi_i^{\nu},\xi_j^{\nu}\rangle \xrightarrow[\nu]{} \langle T_a\xi_i,\xi_j\rangle \qquad \ldots(5)$$

for each $i,j = 1,\ldots,p$ and each a in A. Then $T^{\nu} \xrightarrow[\nu]{} T$ in the regional topology.

Conversely, suppose that $\{T^{\nu}\}$ is a net converging to T in $\underset{\sim}{T}(A)$ in the regional topology. Then every subnet of $\{T^{\nu}\}$ has a subnet $\{T'^{\mu}\}$ with the following property: For every ξ in $X(T)$ and every μ there is a vector ξ_{μ} in $X(T'^{\mu})$ such that

$$\langle \xi_{\mu},\eta_{\mu}\rangle \xrightarrow[\mu]{} \langle \xi,\eta\rangle \qquad \ldots(6)$$

and

$$\langle T'^{\mu}_a\xi_{\mu},\eta_{\mu}\rangle \xrightarrow[\mu]{} \langle T_a\xi,\eta\rangle \qquad \ldots(7)$$

for all ξ, η in $X(T)$ and all a in A.

<u>*-Subalgebras and quotient *-algebras</u>.

We shall now briefly mention the behavior of the regional topology under passage to *-subalgebras and quotient *-algebras.

Let B be a *-subalgebra of the *-algebra A, and $\Phi : \underset{\sim}{T}(A) \to \underset{\sim}{T}(B)$ the <u>restriction map</u> sending T into $T|B$. It is clear that Φ is continuous in the regional topologies. In fact, if A is a Banach *-algebra and B is dense, then Φ is a homeomorphism.

<u>Proposition 6.4</u>. <u>Suppose that A is a Banach *-algebra and B is a dense *-subalgebra of A. Then the restriction map $\Phi : \underset{\sim}{T}(A) \to \underset{\sim}{T}(B)$ is a regional homeomorphism</u>.

<u>Proof</u>. An element T of $T(A)$ is determined by $T|B$ (since T is norm-decreasing and B is dense); so Φ is one-to-one. We have

already mentioned that Φ is continuous.

Let $T \in \underset{\sim}{T}(A)$; and consider the regional neighborhood $U = U(T;\varepsilon;\xi_1,\ldots,\xi_p;F)$ of T (see (1)), where $\varepsilon < 1$, $F = \{a_1,\ldots,a_n\}$, and $\|\xi_j\| \leq 1$ for all j. For each $i = 1,\ldots,n$ let b_i be an element of B such that $\|a_i - b_i\| < 6^{-1}\varepsilon$; and put $G = \{b_1,\ldots,b_n\}$. An easy calculation shows that if T' is any element of $\underset{\sim}{T}(A)$ such that $T'|B \in U(T|B;6^{-1}\varepsilon;\xi_1,\ldots,\xi_p;G)$, then $T' \in U$. This proves the continuity of Φ^{-1}. □

As regards quotient *-algebras, we have:

Proposition 6.5. Let I be a *-ideal of the *-algebra A, and $\pi : A \longrightarrow A/I$ the quotient *-homomorphism. Then the composition map $T \longmapsto T\circ\pi$ is a regional homeomorphism of $\underset{\sim}{T}(A/I)$ onto the regionally closed subset $\{S \in \underset{\sim}{T}(A): I \subset \mathrm{Ker}(S)\}$ of $\underset{\sim}{T}(A)$.

The simple proof of this is left to the reader.

Now let A be a Banach *-algebra and A_c its C*-completion (see Dixmier [1], p. 41). It is well known that there is a natural one-to-one correspondence between *-representations of A and of A_c. By Props. 6.4 and 6.5 this correspondence is a regional homeomorphism. Thus, as regards the regional topology, it makes no difference whether we look at the *-representations of A or the corresponding *-representations of A_c.

The regional topology and the norm.

Proposition 6.6. Let A be a Banach *-algebra. For each a in A the real function $T \longmapsto \|T_a\|$ is lower semicontinuous on $\underset{\sim}{T}(A)$.

Proof. Given T in $\underset{\sim}{T}(A)$ and $\varepsilon > 0$, choose unit vectors ξ and η

in $X(T)$ so that

$$|\langle T_a\xi,\eta\rangle| > \|T_a\| - \varepsilon . \qquad \ldots(8)$$

Then there is a regional neighborhood U of T such that, for every T' in U, there are vectors ξ' and η' in $X(T')$ satisfying

$$\|\xi'\| < 1+\varepsilon , \; \|\eta'\| < 1+\varepsilon$$

$$|\langle T_a'\xi',\eta'\rangle| > |\langle T_a\xi,\eta\rangle| - \varepsilon .$$

These inequalities combined with (8) imply

$$\|T_a'\| > (1+\varepsilon)^{-2}(\|T_a\|-2\varepsilon)$$

for all T' in U. □

Remark. The "norm-functions" $T \longmapsto \|T_a\|$ are far from being continuous on $\underset{\sim}{T}(A)$.

Corollary 6.7. Let A be a Banach *-algebra. If T belongs to the regional closure of a subset $\underset{\sim}{S}$ of $\underset{\sim}{T}(A)$, then

$$\|T_a\| \le \sup\{\|S_a\| : S \in \underset{\sim}{S}\} \quad (a \in A).$$

Weak containment.

If $\underset{\sim}{S}$ is any subfamily of $\underset{\sim}{T}(A)$, let us denote by $\underset{\sim}{S}_f$ the family of all Hilbert direct sums of finitely many elements of $\underset{\sim}{S}$ (repetitions being allowed).

Definition. If $T \in \underset{\sim}{T}(A)$ and $\underset{\sim}{S} \subset \underset{\sim}{T}(A)$, we say that T is weakly contained in $\underset{\sim}{S}$ if T belongs to the regional closure of $\underset{\sim}{S}_f$.

Remark. For non-degenerate *-representations of C^*-algebras this definition coincides with that given in Fell [8], p. 368 (see also Dixmier [1], 3·4·5). This follows easily from Theorem 1.2 of Fell [8] (see also Dixmier [1], 3·4·4) together with Prop. 6.1 of the present work.

Remark. If A is a Banach *-algebra and A_c its C^*-completion, one deduces from the paragraph following Prop. 6.5 that the relation of weak containment is the same for *-representations of A as for the corresponding *-representations of A_c.

Proposition 6.8. *Let A be a Banach *-algebra and A_0 a dense subset of A. Let T be a non-degenerate *-representation of A and $\underset{\sim}{S}$ a family of non-degenerate *-representations of A. Then the following two conditions are equivalent: (I) T is weakly contained in $\underset{\sim}{S}$. (II) We have*

$$\|T_a\| \leq \sup\{\|S_a\|: \quad S \in \underset{\sim}{S}\} \qquad \dots(9)$$

for every a in A_0.

Proof. The implication (I)⇒(II) follows from Cor. 6.7 together with the evident fact that the right side of (9) is unchanged on replacing $\underset{\sim}{S}$ by $\underset{\sim}{S}_f$.

Assume (II); and let A_c, $\| \|_c$ be the C^*-completion of A. We shall denote corresponding *-representations of A and of A_c by the same letters. Now since A_0 is dense in A it is dense in A_c. Hence one deduces easily that (9) holds for all a in A_c. It follows from this that

$$\cap\{\mathrm{Ker}(S) : S \in \underset{\sim}{S}\} \subset \mathrm{Ker}(T), \qquad \ldots(10)$$

where the kernels in (10) are the kernels in A_c . Combining (10) with Theorem 1.2 of Fell [8] (or Dixmier [1], 3·4·4), we obtain (I). □

Continuity of the inducing construction.

We shall now prove the main theorem of this section -- that the inducing construction of §4 is continuous.

Let A and B be two *-algebras, and $\underset{\sim}{L} = (L,[\ ,\ ])$ a B-rigged A-module. We shall denote by $\underset{\sim}{S}$ the set of all those *-representations of B which are inducible to A via $\underset{\sim}{L}$.

Theorem 6.9. The inducing map $\Psi : S \mapsto \mathrm{Ind}_{B\uparrow A}^{\underset{\sim}{L}}(S)$ is continuous on $\underset{\sim}{S}$ with respect to the regional topologies of $\underset{\sim}{T}(B)$ and $\underset{\sim}{T}(A)$.

Proof. Take a *-representation S in $\underset{\sim}{S}$, with space X ; and let $T = \mathrm{Ind}_{B\uparrow A}(S)$, $Y = X(T)$. Let $Z = \{r \tilde{\otimes} \xi : r \in L; \xi \in X\}$ (where as usual $r \tilde{\otimes} \xi$ is the image of $r \otimes \xi$ in Y). Thus Z has dense linear span in Y ; and so by Prop. 6.1 the set of all

$$U^0 = U(T;\varepsilon;\zeta_1,\ldots,\zeta_n;F) \qquad \ldots(11)$$

where $\varepsilon > 0$, $\zeta_1,\ldots,\zeta_n$ are in Z , and F is a finite subset of A , is a basis of regional neighborhoods of T . Thus, to prove the continuity of Ψ , we need only fix a set U^0 of the form (11) with $\zeta_i = r_i \tilde{\otimes} \xi_i$ $(r_i \in L; \xi_i \in X)$, and exhibit a regional neighborhood U of S such that $\Psi(U \cap \underset{\sim}{S}) \subset U^0$.

Set

$$U = U(S;\varepsilon;\xi_1,\ldots,\xi_n;G), \qquad \ldots(12)$$

where G is the (finite) subset of B consisting of all $[r_j,r_i]$ and all $[r_j, ar_i]$ $(i,j=1,\ldots,n; a \in F)$. This is a regional neighborhood of S; and I claim that $\Psi(\mathcal{U} \cap \underset{\sim}{S}) \subset U^0$. Indeed: Let $S' \in \mathcal{U} \cap \underset{\sim}{S}$; and put $X' = X(S')$, $T' = \mathrm{Ind}_{B\uparrow A}(S')$, $Y' = X(T')$. Thus by (12) there exist $\xi_1', \ldots, \xi_n'$ in X' such that

$$|\langle S'_{[r_j, r_i]}\xi_i', \xi_j'\rangle - \langle S_{[r_j, r_i]}\xi_i, \xi_j\rangle| < \epsilon$$

and

$$|\langle S'_{[r_j, ar_i]}\xi_i', \xi_j'\rangle - \langle S_{[r_j, ar_i]}\xi_i, \xi_j\rangle| < \epsilon$$

for all i, j and all a in F. If we set $\zeta_i' = r_i \tilde{\otimes} \xi_i' \in Y'$, these inequalities become

$$|\langle \zeta_i', \zeta_j'\rangle_{Y'} - \langle \zeta_i, \zeta_j\rangle_Y| < \epsilon$$

and

$$|\langle T_a'\zeta_i', \zeta_j'\rangle_{Y'} - \langle T_a\zeta_i, \zeta_j\rangle_Y| < \epsilon$$

for all i, j and all a in F. This says that $T' \in U^0$. So the claim is proved and the proposition established. □

<u>Comparison of the regional and other topologies</u>.

We conclude this section with some comments on the relationship between the regional topology as defined here and other topological notions that have appeared in the literature. In view of the remark following Proposition 6.5 we shall confine our remarks to C^*-algebras.

Let A be a fixed C^*-algebra; and let $\hat{A}$ be the <u>structure space</u> of A, that is, the space of all unitary equivalence classes of irreducible $*$-representations of A. It follows from our Corollary

6.2, together with Theorem 3.4.10 of Dixmier [1], that <u>the regional topology relativized to $\hat{A}$ coincides with the well-known hull-kernel or Jacobson topology of $\hat{A}$</u> (see Dixmier [1] 3.1.1).

Next, let α be a (large) cardinal number; and let $\underset{\sim}{S}_\alpha$ stand for the family of all unitary equivalence classes of non-degenerate *-representations of A of dimension no greater than α. In Fell [1], p. 225, we introduced a so-called <u>quotient topology</u> into $\underset{\sim}{S}_\alpha$ by fixing a Hilbert space Y of dimension α, topologizing the concrete representations acting on subspaces of Y, and then passing to the quotient topology modulo the relation of unitary equivalence. It is not hard to see that <u>this quotient topology is identical with the relativized regional topology of $\underset{\sim}{S}_\alpha$</u>.

Finally, in our paper Fell [4], we studied the continuity of the inducing process on <u>groups</u> with respect to the so-called <u>inner hull-kernel topology</u> of representations. For the definition of this topology we refer the reader to Fell [4]; it will not be repeated here. If $\underset{\sim}{S}$ stands for the space of all non-degenerate *-representations of the C^*-algebra, it is easy to see that the inner hull-kernel topology of $\underset{\sim}{S}$ is strictly smaller (i.e. has strictly fewer open sets) than the regional topology of $\underset{\sim}{S}$. In fact the following precise relation holds between them: <u>If $T \in \underset{\sim}{S}$ and $\{T^i\}$ is a net of elements of $\underset{\sim}{S}$, then $T^i \rightarrow T$ in the inner hull-kernel topology if and only if $\infty \cdot T^i \rightarrow \infty \cdot T$ regionally</u>. (<u>Here $\infty \cdot T$ and $\infty \cdot T^i$ denote the Hilbert direct sum of countably infinitely many copies of T and T^i respectively</u>.)

§7. Imprimitivity bimodules.

Given any two *-algebras A and B , by an A , B imprimitivity bimodule we mean, very roughly speaking, a B-rigged A-module which is also an A-rigged B-module. With this double structure we shall be able to induce in both directions -- from B to A and also from A to B . In fact, under simple general conditions, the Duality Theorem 2.2 will assert that the inducing map sets up an "isomorphism" between the "*-representation theories" of A and B . This will be the abstract Imprimitivity Theorem.

Fix two *-algebras A and B .

Definition. An A , B imprimitivity bimodule is a system $\underset{\sim}{L} = (L,[\ ,\ ]_A,[\ ,\ ]_B)$, where (i) L is a left A-module and a right B-module; (ii) $[\ ,\ ]_A : L \times L \to A$ is linear in its first variable and conjugate-linear in its second variable, (iii) $[\ ,\ ]_B: L \times L \to B$ is linear in its second variable and conjugate-linear in its first variable; and (iv) the following relations hold for all a in A , b in B , and r , s , t in L :

$$[s,r]_A = [r,s]_A^* , \ [ar,s]_A = a[r,s]_A , \qquad \ldots(1)$$

$$[rb,s]_A = [r,sb^*]_A , \qquad \ldots(2)$$

$$[s,r]_B = [r,s]_B^* , \ [r,sb]_B = [r,s]_B b , \qquad \ldots(3)$$

$$[ar,s]_B = [r,a^*s]_B , \qquad \ldots(4)$$

$$[r,s]_A t = r[s,t]_B . \qquad \ldots(5)$$

$[\ ,\ ]_A$ and $[\ ,\ ]_B$ are called the A-rigging and B-rigging of $\underset{\sim}{L}$ respectively.

In view of relations (3) and (4), $(L,[\ ,\ ]_B)$ is a B-rigged

A-module.

Notice the high degree of symmetry between A and B in the above definition. To make this symmetry precise, suppose that M is any left A-module (the module operation being denoted as usual by $(a,m) \mapsto am$). Then the equation

$$m:a = a^* m \quad (m \in M; a \in A) \qquad \ldots(6)$$

defines a right A-module structure : for $\bar{M}$. (The replacement of M by $\bar{M}$ is necessary in order to make m:a linear in a.) We call $\bar{M}$, : the right A-module complex-conjugate to M. Similarly, if M is a right A-module, the equation

$$a:m = ma^* \quad (a \in A; m \in M) \qquad \ldots(7)$$

defines a left A-module structure : for $\bar{M}$; and $\bar{M}$, : is the left A-module complex-conjugate to M.

Now suppose that $\underset{\sim}{L} = (L,[\ ,\ ]_A,[\ ,\ ]_B)$ is an A, B imprimitivity bimodule. Equipping $\bar{L}$ with the complex-conjugate right A-module and left B-module structures defined in (6) and (7), one easily checks that $(\bar{L},[\ ,\ ]_B,[\ ,\ ]_A)$ becomes a B, A imprimitivity bimodule. (To see this, recall §4(1) and observe that (1) implies the analogous relation $[r,a^*s]_A = [r,s]_A a$ for $[\ ,\ ]_A$.) We call $(\bar{L},[\ ,\ ]_B,[\ ,\ ]_A)$ the imprimitivity bimodule complex-conjugate to $\underset{\sim}{L}$, and denote it by $\underset{\sim}{\bar{L}}$. In particular $(\bar{L},[\ ,\ ]_A)$ is an A-rigged B-module.

Definition. The A, B imprimitivity bimodule $(L,[\ ,\ ]_A,[\ ,\ ]_B)$ is said to be strict if the linear span of $\{[r,s]_A : r,s \in L\}$ is all of A, and the linear span of $\{[r,s]_B : r,s \in L\}$ is all of B.

Even without strictness, it follows from (1) and (3) that these linear spans are automatically $*$-ideals of A and B respectively.

Example. As a first example of an imprimitivity bimodule, suppose we have a $*$-algebra D, two $*$-subalgebras A and B of D, and a linear subspace L of D with the following properties:

$$AL \subset L , \quad LB \subset L , \qquad \ldots(8)$$

$$LL^* \subset A , \quad L^*L \subset B . \qquad \ldots(9)$$

By (8) L is a left A-module and a right B-module under multiplication. By (9) the equations

$$[r,s]_A = rs^* , \quad [r,s]_B = r^*s \quad (r,s \in L) \qquad \ldots(10)$$

define maps $[\ ,\ ]_A$ and $[\ ,\ ]_B$ of $L \times L$ into A and B respectively; and the algebraic properties of D oblige $(L,[\ ,\ ]_A,[\ ,\ ]_B)$ to be an A, B imprimitivity bimodule.

An interesting special case of the preceding example is that in which $A = D$ and B is a so-called block $*$-subalgebra of A; that is, B is a $*$-subalgebra of A satisfying

$$a \in A , \ b,c \in B \Rightarrow bac \in B . \qquad \ldots(11)$$

In that case let us take L to be the left ideal of A generated by B (i.e., L is the linear span of $B \cup AB$). Condition (11) then assures us that (8) and (9) hold; so (10) defines an A, B imprimitivity bimodule. We shall return to this example in §9.

The imprimitivity $*$-algebra.

One might ask whether, given any B-rigged A-module $L,[\ ,\ ]_B$, it would be possible to adjoin a map $[\ ,\ ]_A : L \times L \longrightarrow A$ so as to make $(L,[\ ,\ ]_A,[\ ,\ ]_B)$ an A , B imprimitivity bimodule. The answer is 'no'. However, the following less ambitious program is feasible: Let us start with a single *-algebra B and a B-rigged space $(L,[\ ,\ ]_B)$. It turns out that there is a canonical way to construct from $(L,[\ ,\ ]_B)$ a new *-algebra E , a left E-module structure for L , and a rigging $[\ ,\ ]_E$ such that $(L,[\ ,\ ]_E,[\ ,\ ]_B)$ is an E , B imprimitivity bimodule. The construction of E is as follows: We form the tensor product $L \otimes \bar{L}$, and introduce into it the following bilinear multiplication:

$$(r \otimes s)(r' \otimes s') = r[s,r']_B \otimes s' . \qquad \ldots(12)$$

Notice that the right side of (12) is linear in each of its four variables when r , r' are considered as belonging to L and s , s' to $\bar{L}$. The multiplication (12) is associative. Indeed, if $r,r',r'' \in L$ and $s,s',s'' \in \bar{L}$,

$$\begin{aligned}
&((r \otimes s)(r' \otimes s'))(r'' \otimes s'') \\
&= (r[s,r']_B \otimes s')(r'' \otimes s'') \\
&= r[s,r']_B[s',r'']_B \otimes s'' \\
&= r[s,r'[s',r'']_B]_B \otimes s'' \\
&= (r \otimes s)(r'[s',r'']_B \otimes s'') \\
&= (r \otimes s)((r' \otimes s')(r'' \otimes s'')).
\end{aligned}$$

Thus $L \otimes \bar{L}$ has become an (associative) algebra. Let I be the linear span of

$$\{rb \otimes s - r \otimes sb^* : r \in L,\ s \in \bar{L},\ b \in B\}$$

in $L \otimes \bar{L}$. One checks without difficulty that I is a two-sided ideal of $L \otimes \bar{L}$. We now define E to be the quotient algebra $(L \otimes \bar{L})/I$; and denote by $r \tilde{\otimes} s$ the image of $r \otimes s$ in E $(r \in L; s \in \bar{L})$. We have by (12)

$$(r \tilde{\otimes} s)(r' \tilde{\otimes} s') = r[s,r']_B \tilde{\otimes} s' ; \qquad \ldots(13)$$

and, in view of the definition of I ,

$$rb \tilde{\otimes} s = r \tilde{\otimes} sb^* \quad (r \in L; s \in \bar{L}; b \in B). \qquad \ldots(14)$$

We next observe that the equation

$$(r \otimes s)^* = s \otimes r \quad (r,s \in L)$$

defines $*$ as a conjugate-linear mapping of order 2 of $L \otimes \bar{L}$ onto itself. Since $*$ leaves I stable, it generates a conjugate-linear mapping (also called $*$) of E onto itself of order 2:

$$(r \tilde{\otimes} s)^* = s \tilde{\otimes} r \quad (r,s \in L). \qquad \ldots(15)$$

If $r,r' \in L$ and $s,s' \in \bar{L}$, we have by (13), (14), and (15):

$$\begin{aligned} ((r \tilde{\otimes} s)(r' \tilde{\otimes} s'))^* &= s' \tilde{\otimes} r[s,r']_B \\ &= s'[r',s]_B \tilde{\otimes} r \\ &= (r' \tilde{\otimes} s')^*(r \tilde{\otimes} s)^* . \end{aligned}$$

Thus E , with the operations (13) and (15), is a $*$-algebra.

Definition. This E is called _the imprimitivity *-algebra constructed from_ B _and_ $L,[\ ,\ ]_B$.

It is easy to give to L the structure of a left E-module.

First observe that the equation

$$(r \otimes s)t = r[s,t]_B \quad (r,t \in L; s \in \bar{L}) \qquad \ldots(16)$$

defines a left $(L \otimes \bar{L})$-module structure for L. (Indeed, $\{(r \otimes s)(r' \otimes s')\}t = (r[s,r']_B \otimes s')t = r[s,r']_B[s',t]_B$ $= r[s,r'[s',t]_B]_B = (r \otimes s)\{r'[s',t]_B\} = (r \otimes s)\{(r' \otimes s')t\}$.) By §4(1) $\zeta t = 0$ whenever $\zeta \in I$ and $t \in L$. Hence the action (16) generates a left E-module structure for L:

$$(r \tilde{\otimes} s)t = r[s,t]_B \quad (r,t \in L; s \in \bar{L}). \qquad \ldots(17)$$

As a left E-module, L satisfies:

$$(\zeta t)b = \zeta(tb) \quad (\zeta \in E; t \in L; b \in B). \qquad \ldots(18)$$

We shall now write $[r,s]_E$ for $r \tilde{\otimes} s$ $(r,s \in L)$. Notice that $[\ ,\ ]_E$, as a function on $L \times L$ to E, is linear in the first variable and conjugate-linear in the second.

<u>Proposition 7.1</u>. $(L,[\ ,\ ]_E,[\ ,\ ]_B)$ <u>is an</u> E, B <u>imprimitivity bimodule</u>.

<u>Proof</u>. It remains to observe the following relations:

$$[r,s]_E^* = [s,r]_E, \qquad \ldots(19)$$

$$[\zeta r,s]_E = \zeta[r,s]_E, \qquad \ldots(20)$$

$$[rb,s]_E = [r,sb^*]_E, \qquad \ldots(21)$$

$$[\zeta r,s]_B = [r,\zeta^* s]_B, \qquad \ldots(22)$$

$$[r,s]_E t = r[s,t]_B, \qquad \ldots(23)$$

$(r,s,t \in L; \zeta \in E; b \in B)$. To prove (20) and (22) it is enough to suppose that $\zeta = r' \tilde{\otimes} s'$ $(r',s' \in L)$. The five relations then follow from (13), (14), (15), (17), and §4(1). □

<u>The example of finite groups</u>.

Let H be a subgroup of the finite group G (with unit e); let A and B be the convolution group $*$-algebras of G and H respectively; and as in §4(11) let

$$[f,g]_B = p(f^* * g) \qquad (f,g \in A),$$

where $p : f \mapsto f|H$ is the restriction mapping of A onto B. Since p is a conditional expectation (see §4), $(A,[\ ,\ ]_B)$ is a B-rigged space. What is the imprimitivity $*$-algebra E of $(A,[\ ,\ ]_B)$?

We are going to show that E essentially coincides with the so-called G, G/H <u>transformation</u> $*$-<u>algebra</u> F defined as follows: F is the linear space of all complex functions on $G \times G/H$; and multiplication $*$ and involution * in F are given by

$$(f*g)(x,m) = \sum_{y \in G} f(y,m)g(y^{-1}x,y^{-1}m), \qquad \ldots(24)$$

$$f^*(x,m) = \overline{f(x^{-1},x^{-1}m)} \qquad \ldots(25)$$

$(f,g \in F; x \in G; m \in G/H)$. One checks that F is a $*$-algebra under the operations (24) and (25). Notice that A and $\underset{\sim}{C}(G/H)$ can be regarded as $*$-subalgebras of F. More precisely, if we put

$$f_\varphi(x,m) = \varphi(x) \qquad (\varphi \in A; x \in G; m \in G/H), \qquad \ldots(26)$$

$$g_a(x,m) = \delta_{xe}a(m) \qquad (a \in \underset{\sim}{C}(G/H); x \in G; m \in G/H), \ldots(27)$$

then $\varphi \mapsto f_\varphi$ and $a \mapsto g_a$ are *-isomorphisms of A and $\underset{\sim}{C}(G/H)$ onto *-subalgebras A' and C' respectively of F ; and F is generated by A' and C' . For later use we observe the commutation relation:

$$f_{\delta_u} * g_a * f_{\delta_{u^{-1}}} = g_{ua} \qquad (u \in G; a \in \underset{\sim}{C}(G/H)), \qquad \ldots(28)$$

where $(ua)(m) = a(u^{-1}m)$ $(m \in G/H)$.

We shall now show that $E \cong F$. For this we define a linear map $\Phi : A \otimes \bar{A} \longrightarrow F$ as follows:

$$\Phi(\varphi \otimes \psi)(x, yH) = \sum_{h \in H} \varphi(yh)\psi^*(h^{-1}y^{-1}x) \qquad \ldots(29)$$

$(\varphi, \psi \in A; x, y \in G)$. Evidently the right side of (29) depends on y only through yH ; so (29) makes sense. One easily verifies that

$$\Phi((\varphi * b) \otimes \psi) = \Phi(\varphi \otimes (\psi * b^*))$$

$(\varphi, \psi \in A; b \in B)$. Thus Φ vanishes on the ideal I used in the definition of the imprimitivity *-algebra, and so gives rise to a linear map $\tilde{\Phi} : E \longrightarrow F$:

$$\tilde{\Phi}(\varphi \tilde{\otimes} \psi) = \Phi(\varphi \otimes \psi) \qquad (\varphi, \psi \in A).$$

The fact that $E \cong F$ now finds precise expression in the following proposition:

<u>Proposition 7.2</u>. $\tilde{\Phi}$ <u>is a</u> *-<u>isomorphism of</u> E <u>onto</u> F .

<u>Proof</u>. A routine calculation shows that $\tilde{\Phi}$ is a *-homomorphism.

Notice that any function in $\underset{\sim}{C}(G \times G)$ belongs to the linear span of the set of functions of the form $(x,y) \mapsto \varphi(y)\psi^*(y^{-1}x)$, where $\varphi, \psi \in A$. On the other hand every function in F is of the form $(x,yH) \mapsto \sum_{h\in H}\gamma(x,yh)$ for some γ in $\underset{\sim}{C}(G\times G)$. These two facts together imply that Φ, and hence also $\tilde{\Phi}$, is onto F.

Finally, we must show that $\operatorname{Ker}(\Phi) = I$. If n and p are the orders of G and H respectively, $A \otimes \bar{A}$ (= domain(Φ)) and F (= range(Φ)) have dimensions n^2 and $p^{-1}n^2$; so $\operatorname{Ker}(\Phi)$ has dimension $n^2 - p^{-1}n^2$. On the other hand, if Γ is a transversal for G/H, the $(p^{-1}n)n(p-1)$ elements $\delta_{xh} \otimes \delta_y - \delta_x \otimes \delta_{yh^{-1}}$, where x, y and h run over Γ, G and $H-\{e\}$ respectively, are linearly independent and belong to I. Thus I has dimension at least $(p^{-1}n)n(p-1) = n^2 - p^{-1}n^2$. Since $I \subset \operatorname{Ker}(\Phi)$, these facts imply that $I = \operatorname{Ker}(\Phi)$. □

<u>Remark</u>. We have seen that A is a left E-module under the action defined by (17). Thus, by the preceding proposition, the equation

$$f:\varphi = (\tilde{\Phi}^{-1}(f))\varphi \qquad \ldots(30)$$

defines a left F-module structure $:$ for A. An easy calculation gives the following explicit formula for this structure:

$$(f:\varphi)(x) = \sum_{y\in G} f(y,xH)\varphi(y^{-1}x) \quad (f \in F; \varphi \in A; x \in G). \ \ldots(31)$$

With this left F-module structure, and with the F-rigging $[\ ,\ ]_F$ given by

$$[\varphi,\psi]_F(x,yH) = \sum_{h\in H} \varphi(yh)\psi^*(h^{-1}y^{-1}x) \quad (\varphi,\psi \in A; x,y \in G) \ \ldots(32)$$

(see (29)), $(A,[\ ,\ ]_F,[\ ,\ ]_B)$ <u>becomes an</u> F, B <u>imprimitivity bimodule</u>.

§8. The abstract Inprimitivity Theorem.

Fix two *-algebras A and B, and an A, B imprimitivity bimodule $\underset{\sim}{L} = (L,[\ ,\ ]_A,[\ ,\ ]_B)$. A *-representation S of B will be called *positive with respect to* $\underset{\sim}{L}$ (or $\underset{\sim}{L}$*-positive*) if it is positive with respect to the B-rigged A-module $(L,[\ ,\ ]_B)$, that is, if $S_{[r,r]_B}$ is a positive operator for all r in L; it will be called *inducible to* A *via* $\underset{\sim}{L}$ if it is inducible to A via $(L,[\ ,\ ]_B)$ (see §4). If S is inducible to A via $\underset{\sim}{L}$, we write $\mathrm{Ind}^{\underset{\sim}{L}}_{B\uparrow A}(S)$ for the induced *-representation of A. Similarly, a *-representation T of A is *positive with respect to* $\underset{\sim}{L}$ [*inducible to* B *via* $\underset{\sim}{L}$] if it is positive [inducible] with respect to the A-rigged B-module $(\bar{L},[\ ,\ ]_A)$ (see §7). If it is inducible to B via $\underset{\sim}{L}$, we denote by $\mathrm{Ind}^{\underset{\sim}{L}}_{A\uparrow B}(T)$ the induced *-representation of B.

We shall now show that, if $\underset{\sim}{L}$ is strict, positivity with respect to $\underset{\sim}{L}$ automatically implies inducibility.

Proposition 8.1. Assume that $\underset{\sim}{L}$ *is strict; and let* S *be a *-representation of* B *which is positive with respect to* $\underset{\sim}{L}$. *Then* S *is inducible to* A *via* $\underset{\sim}{L}$; *and the induced *-representation* $T = \mathrm{Ind}^{\underset{\sim}{L}}_{B\uparrow A}(S)$ *is non-degenerate and is given by*

$$T_{[r,s]_A} = W_{s,r} \quad (r,s \in L), \qquad \ldots(1)$$

where W *is the operator inner product on* $\bar{L}$ *deduced (as in* §2) *from the operator inner product* $V : (r,s) \longmapsto S_{[s,r]_B}$ *(see Proposition 4.1).*

Proof. We denote $X(S)$ by X, and by Y the deduced Hilbert space $\underset{\sim}{X}(V)$ (see §1). As usual $r \tilde{\otimes} \xi$ $(r \in L; \xi \in X)$ is the image of $r \otimes \xi$

in Y . Now S will induce a *-representation T of A provided that for each a in A there is a bounded linear operator T_a on Y such that

$$T_a(r \tilde{\otimes} \xi) = ar \tilde{\otimes} \xi \quad (r \in L; \xi \in X).$$

Since by strictness the range of $[\ ,\ ]_A$ spans A ; it is enough to show the existence of such a T_a when $a = [s,t]_A$ $(s,t \in L)$. To do this, and at the same time to establish (1), it is enough to verify that

$$W_{t,s}(r \tilde{\otimes} \xi) = ([s,t]_A r) \tilde{\otimes} \xi \qquad \ldots(2)$$

for all r , s , t in L and ξ in X . But we have

$$\begin{aligned} ([s,t]_A r) \tilde{\otimes} \xi &= (s[t,r]_B) \tilde{\otimes} \xi && \text{(by §7(5))} \\ &= s \tilde{\otimes} S_{[t,r]_B}\xi && \text{(by §4(5))} \\ &= s \tilde{\otimes} V_{r,t}\xi && \\ &= W_{t,s}(r \tilde{\otimes} \xi) && \text{(by §2(4))}, \end{aligned}$$

proving (2).

To show that T is non-degenerate we have only to recall from §2 that W is non-degenerate, and to observe from (1) that range (T) contains all the $W_{r,s}$. □

Applying 8.1 with $\underset{\sim}{L}$ replaced by its complex conjugate $\bar{\underset{\sim}{L}}$ (see §7) we conclude that any *-representation T of A which is positive with respect to $\underset{\sim}{L}$ is inducible to B via $\underset{\sim}{L}$.

The following crucial result is a specialization of Prop. 2.1 to the present setting:

<u>Theorem 8.2</u>. <u>Let</u> $\underset{\sim}{L}$ <u>be a strict</u> A , B <u>imprimitivity bimodule, and</u> S <u>a</u> *-<u>representation of</u> B <u>which is positive with respect to</u> $\underset{\sim}{L}$. <u>Then</u> $T = \mathrm{Ind}_{B\uparrow A}^{\underset{\sim}{L}}(S)$ (<u>which exists by Prop</u>. 8.1) <u>is positive with respect to</u> $\underset{\sim}{L}$, <u>so that we can form</u> $\tilde{S} = \mathrm{Ind}_{A\uparrow B}^{\underset{\sim}{L}}(T)$. <u>If</u> S <u>is non-degenerate, then</u> $\tilde{S} \cong S$.

<u>Proof</u>. Let W be the operator inner product on $\bar{L}$ deduced from the operator inner product $V : (r,s) \mapsto S_{[s,r]_B}$. The positivity of T then follows from (1), and the first statement is proved.

Now assume that S is non-degenerate, so that by strictness V is likewise non-degenerate. Let $\tilde{V}$ be the operator inner product on L deduced from W . By (1) applied to $\underset{\sim}{\bar{L}}$

$$\tilde{S}_{[r,s]_B} = \tilde{V}_{s,r} \quad (r,s \in L). \qquad \ldots(3)$$

Since V is non-degenerate, Prop. 2.1 says that V and $\tilde{V}$ are unitarily equivalent under an isometry F . By (3), the definition of V , and the strictness of $\underset{\sim}{L}$, F must intertwine S and $\tilde{S}$. □

Theorem 8.2 gives rise immediately to the following abstract Imprimitivity Theorem, which is the crux of the present chapter.

<u>Abstract Imprimitivity Theorem 8.3</u>. <u>Let</u> A , B <u>be two</u> *-<u>algebras, and</u> $\underset{\sim}{L} = (L,[\ ,\ ]_A,[\ ,\ ]_B)$ <u>a strict</u> A , B <u>imprimitivity bimodule</u>. <u>Let</u> $\underset{\sim}{S}$ <u>and</u> $\underset{\sim}{T}$ <u>be the collections of all unitary equivalence classes of non-degenerate</u> *-<u>representations of</u> B <u>and</u> A <u>respectively which are positive with respect to</u> $\underset{\sim}{L}$. <u>Then</u> $\Phi : S \mapsto \mathrm{Ind}_{B\uparrow A}^{\underset{\sim}{L}}(S)$ (<u>considered as a map of unitary equivalence classes</u>) <u>is a bijection of</u> $\underset{\sim}{S}$ <u>onto</u> $\underset{\sim}{T}$. <u>Its inverse</u> Φ^{-1} <u>is the map</u> $T \mapsto \mathrm{Ind}_{A\uparrow B}^{\underset{\sim}{L}}(T)$.

Let $\underset{\sim}{L}$, S , T , V and W be as in Theorem 8.2 and its proof.

The strictness of $\underset{\sim}{L}$ implies that the commuting algebras of S and V coincide, and likewise the commuting algebras of T and W. Hence Theorem 2.5 gives the following very important result, which should be considered as an integral part of the abstract Imprimitivity Theorem:

Theorem 8.4. The mapping Φ of the preceding theorem preserves the *-isomorphism class of the commuting algebras. More precisely, let S be a non-degenerate *-representation of B which is positive with respect to $\underset{\sim}{L}$, and let $T = \mathrm{Ind}^{\underset{\sim}{L}}_{B\uparrow A}(S)$. Then the commuting algebras $\underset{\sim}{I}(S)$ and $\underset{\sim}{I}(T)$ are *-isomorphic under the mapping $F \longmapsto \tilde{F}$ defined as in Theorem 2.5:

$$\tilde{F}(r \tilde{\otimes} \xi) = r \tilde{\otimes} F\xi \quad (r \in L; \xi \in X(S)).$$

In particular T is irreducible if and only if S is irreducible.

Remark 1. Suppose that in the preceding context the algebra A has a unit $\mathbf{1}$ satisfying $\mathbf{1}r = r$ for all r in L; and suppose also that we can find finitely many elements $s_1, \cdots, s_n$ of L such that

$$\sum_{i=1}^{n} [s_i, s_i]_A = \mathbf{1}.$$

Then for $r \in L$ we have

$$\begin{aligned} [r,r]_B &= [r, \mathbf{1}r]_B \\ &= \sum_{i=1}^{n} [r,[s_i,s_i]_A r]_B \\ &= \sum_i [r, s_i[s_i,r]_B]_B \qquad \text{(by §7(5))} \\ &= \sum_i b_i^* b_i \qquad \text{(by §7(3))} \qquad \ldots(4) \end{aligned}$$

where $b_i = [s_i, r]_B$. If S is any $*$-representation of B we get from (4)

$$S_{[r,r]_B} = \sum_i S_{b_i{}^* b_i} = \sum_i (S_{b_i})^* (S_{b_i}) \geq 0 .$$

Therefore, under these hypotheses, _every_ $*$-_representation_ _of_ B _is_ _positive_ _with_ _respect_ _to_ $\underset{\sim}{L}$.

Similarly, suppose that B has a unit $\mathfrak{1}'$ satisfying $r\mathfrak{1}' = r$ $(r \in L)$, and that there are elements $t_1, \ldots, t_m$ of L such that

$$\sum_{i=1}^{m} [t_i, t_i]_B = \mathfrak{1}' .$$

Then _every_ $*$-_representation_ _of_ A _is_ _positive_ _with_ _respect_ _to_ $\underset{\sim}{L}$.

Let $\underset{\sim}{L}$ be a strict A, B imprimitivity bimodule. Here are a few properties of $*$-representations which are preserved by the correspondence Φ of Theorem 8.3.

In the first place, it is clear from Prop. 5.1 that Φ preserves direct sums. Next we have:

Proposition 8.5. Φ _is_ _a_ _homeomorphism_ _with_ _respect_ _to_ _the_ _regional_ _topologies_ _of_ $\underset{\sim}{T}(A)$ _and_ $\underset{\sim}{T}(B)$ (_see_ §6).

Proof. By Theorem 8.3 both Φ and Φ^{-1} are inducing processes. So the result follows from Theorem 6.9. □

A $*$-representation S of the $*$-algebra B is called _compact_ if S_b is a compact operator for all b in B. From (1) and Theorem 2.7 we see immediately:

Proposition 8.6. _If_ $S \in \underset{\sim}{S}$ _and_ $T = \Phi(S)$ (_see_ _Theorem_ 8.3), _then_ T

<u>is compact if and only if</u> S <u>is compact</u>.

<u>Remark</u>. Not all properties of *-representations are preserved by Φ . See §9, Remark 6.

<u>Example; finite groups</u>. Let H be a subgroup of the finite group G , and A and B the group *-algebras of G and H respectively. Let F be the G , G/H transformation *-algebra defined in the example of finite groups in §7. It was shown there that F can be identified with the imprimitivity *-algebra E constructed from B and $(A,[\ \ ]_B)$, and hence that equations §7(31),(32) and §4(10) define an F , B imprimitivity bimodule $\underset{\sim}{A} = (A,[\ ,\ ]_F,[\ ,\ ]_B)$. Evidently $\underset{\sim}{A}$ is strict. The Remark following Theorem 8.4 can easily be applied to show that every *-representation of B is inducible to F via $\underset{\sim}{A}$, and that every *-representation of F is inducible to B via $\underset{\sim}{A}$.

Now a non-degenerate *-representation S of B is essentially a unitary representation of H . Further, let T be a non-degenerate *-representation of F . By §7(26),(27),(28) T gives rise to a unitary representation $W : x \mapsto T_{f_{\delta_x}}$ of G and a *-representation $P : a \mapsto T_{g_a}$ of $\underset{\sim}{C}(G/H)$ satisfying:

$$W_x P_a W_x^{-1} = P_{xa} \quad (x \in G; a \in \underset{\sim}{C}(G/H)). \qquad \ldots(5)$$

Such a pair W , P is called a <u>system of imprimitivity for</u> G <u>over</u> G/H (see §30). Conversely, every system of imprimitivity for G over G/H arises in this way from a non-degenerate *-representation T of F . Putting these facts together, we deduce from Theorem 8.3 a natural one-to-one correspondence, implemented by the inducing process, between unitary representations of H and systems of imprimitivity for

G over G/H . This is the classical Imprimitivity Theorem for finite groups.

Actually, for finite groups the proof of the classical Imprimitivity Theorem is a simple and routine matter, and certainly does not require the machinery of imprimitivity bimodules. We have indicated here the application to finite groups only to orient the reader toward our approach to the much less simple case of arbitrary locally compact groups -- in fact, arbitrary Banach *-algebraic bundles over these -- discussed in Chapter III.

§9. Topological versions and examples of the abstract Imprimitivity Theorem

The abstract Imprimitivity Theorem 8.3 as it stands is unsatisfactory for topological contexts. In many interesting situations the A and B of Theorem 8.3 are Banach *-algebras, and the imprimitivity bimodule $\underset{\sim}{L}$ is not strict, but merely topologically strict in the following sense:

Definition. Let A and B be normed *-algebras. An A , B imprimitivity bimodule $(L,[\ ,\ ]_A,[\ ,\ ]_B)$ is topologically strict if the linear spans of range$([\ ,\ ]_A)$ and range$([\ ,\ ]_B)$ are norm-dense in A and B respectively.

In this situation the abstract Imprimitivity Theorem continues to hold.

Theorem 9.1. Let A and B be Banach *-algebras and $\underset{\sim}{L} = (L,[\ ,\ ]_A,[\ ,\ ]_B)$ a topologically strict A , B imprimitivity bimodule. Let $\underset{\sim}{S}$ and $\underset{\sim}{T}$ be the spaces of all unitary equivalence classes of non-degenerate *-representations of B and A respectively *which are positive with respect to* $\underset{\sim}{L}$. *Then*: (I) *Every element* S *of* $\underset{\sim}{S}$ *is inducible* (*via* $(L,[\ ,\ ]_B)$) *to an element* $\Phi(S)$ *of* $\underset{\sim}{T}$; (II) *every element* T *of* $\underset{\sim}{T}$ *is inducible* (*via* $(\bar{L},[\ ,\ ]_A)$) *to an element* $\Psi(T)$ *of* $\underset{\sim}{S}$; *and* (III) $\Phi : S \mapsto \Phi(S)$ (*as a mapping of unitary equivalence classes*) *is a bijection of* $\underset{\sim}{S}$ *onto* $\underset{\sim}{T}$ *whose inverse is* $T \mapsto \Psi(T)$. *Furthermore, if* $S \in \underset{\sim}{S}$, *then* S *and* $\Phi(S)$ *have* *-*isomorphic commuting algebras. In particular* S *is irreducible if and only if* $\Phi(S)$ *is*.

Proof. Let A_0 and B_0 denote the linear spans of the ranges of $[\ ,\ ]_A$ and $[\ ,\ ]_B$ respectively. Thus A_0 and B_0 are dense *-subalgebras (in fact *-ideals) of A and B respectively; and $\underset{\sim}{L}$ can be regarded as a *strict* A_0 , B_0 imprimitivity bimodule. Thus, if $\underset{\sim}{S}_0$ and $\underset{\sim}{T}_0$ are the families of all unitary equivalence classes of non-degenerate *-representations of B_0 and A_0 respectively which are positive with respect to $\underset{\sim}{L}$, Theorem 8.3 says that the inducing map $\Phi_0 : S \mapsto \mathrm{Ind}_{B_0 \uparrow A_0}(S)$ is a bijection of $\underset{\sim}{S}_0$ onto $\underset{\sim}{T}_0$ whose inverse is the inducing map $\Psi_0 : T \mapsto \mathrm{Ind}_{A_0 \uparrow B_0}(T)$.

Now each element of $\underset{\sim}{S}$, being continuous on B , is determined by its restriction to B_0 . So $\underset{\sim}{S}$ can be regarded as a subset of $\underset{\sim}{S}_0$; and similarly $\underset{\sim}{T}$ can be regarded as a subset of $\underset{\sim}{T}_0$. By Prop. 4.2 every element of $\underset{\sim}{S}_0$ is inducible not merely to A_0 but to A . From this it follows immediately that $\Phi_0(\underset{\sim}{S}_0) \subset \underset{\sim}{T} \subset \underset{\sim}{T}_0$. Likewise $\Psi_0(\underset{\sim}{T}_0) \subset \underset{\sim}{S} \subset \underset{\sim}{S}_0$. Since Φ_0 and Ψ_0 are bijections onto $\underset{\sim}{T}_0$ and $\underset{\sim}{S}_0$ respectively, it follows that $\underset{\sim}{S}_0 = \underset{\sim}{S}$ and $\underset{\sim}{T}_0 = \underset{\sim}{T}$. With this identification it is clear that Φ and Ψ coincide with Φ_0 and Ψ_0 respectively. Therefore $\Phi : \underset{\sim}{S} \to \underset{\sim}{T}$ is a bijection, with inverse Ψ . □

<u>Remark 1</u>. In the above proof we have actually shown that $\underset{\sim}{S}_0 = \underset{\sim}{S}$ and $\underset{\sim}{T}_0 = \underset{\sim}{T}$; that is, <u>every *-representation of</u> B_0 [resp. A_0] <u>which is positive with respect to</u> $\underset{\sim}{L}$ <u>is continuous in the norm of</u> B [resp. A] <u>and so extendable to a *-representation of</u> B [resp. A].

<u>Remark 2</u>. The above theorem is stronger than Theorem 8.3 in that only topological strictness of $\underset{\sim}{L}$ is required. On the other hand, A and B must be Banach *-algebras in order that it be applicable.

<u>Remark 3</u>. The Propositions 8.5 and 8.6 apply without any change to the correspondence Φ of Theorem 9.1; and we shall not rewrite them here. This is proved, of course, by applying them as they stand to the correspondence Φ_0 of the proof of Theorem 9.1. In the case of Prop. 8.5 we must in addition recall Prop. 6.4.

<u>Remark 4</u>. It is worth observing that Remark 1 of §8 has a topological analogue. Let A, B, $\underset{\sim}{L}$ be as in the first paragraph of §8; and suppose in addition that B is a normed *-algebra, and that there exists a net $\{a_i\}$ of elements of A such that (i) each a_i is of the form $\sum_{j=1}^{n} [s_j,s_j]_A$, where $s_1,\ldots,s_n \in L$ (and of course n depends on a_i), and (ii) $[r,a_ir]_B \underset{i}{\longrightarrow} [r,r]_B$ in B for each r in L. Then I claim that <u>every norm-continuous *-representation</u> S <u>of</u> B <u>is</u> $\underset{\sim}{L}$<u>-positive</u>. To see this, observe that if $a_i = \sum_{j=1}^{n} [s_j,s_j]_A$, then $[r,a_ir]_B = \sum_j [r,s_j[s_j,r]_B]_B = \sum_j [s_j,r]_B^*[s_j,r]_B$; so

$$S_{[r,a_ir]_B} = \sum_j (S_{[s_j,r]_B})^* S_{[s_j,r]_B} \geq 0 .$$

By assumption (ii) and the norm-continuity of S, this implies that $S_{[r,r]_B} \geq 0$ for all r in L. Thus S is $\underset{\sim}{L}$-positive; and the claim is proved. Replacing $\underset{\sim}{L}$ by $\underset{\sim}{\bar{L}}$, we obtain a similar statement with the roles of A and B reversed.

Examples.

(I) The compact operators. Let L be any non-zero Hilbert space (with inner product $< , >$), B the complex field $\mathbb{C}$, and A the C^*-algebra $\underset{\sim}{O}_C(L)$ of all compact operators on L. Thus L is a natural left A-module and a natural right B-module (under scalar multiplication). We define B- and A-valued riggings on L as follows:

$$[\xi,\eta]_B = <\eta,\xi> , \qquad \ldots(1)$$

$$[\xi,\eta]_A(\zeta) = <\zeta,\eta>\xi \qquad \ldots(2)$$

($\xi,\eta,\zeta \in L$). One verifies easily that $\underset{\sim}{L} = (L,[\ ,\ ]_A,[\ ,\ ]_B)$ is then an A, B imprimitivity bimodule. The range of $[\ ,\ ]_A$ is the *-algebra $\underset{\sim}{O}_F(L)$ of all bounded linear operators on L of finite rank, and so is dense in A; thus $\underset{\sim}{L}$ is topologically strict. Notice that if ξ is a unit vector in L we have $[\xi,\xi]_B \geq 0$ and also

$$[\xi,\xi]_A = p_\xi = p_\xi^* p_\xi , \qquad \ldots(3)$$

where p_ξ is projection onto $\mathbb{C}\xi$. Therefore all *-representations of A and of B are positive with respect to $\underset{\sim}{L}$.

Thus we conclude from Theorem 9.1 the existence of a natural one-to-one (direct-sum-preserving) correspondence Φ between the non-degenerate *-representations of B and of A. Now a non-degenerate *-representation of B ($= \mathbb{C}$) is just a direct sum of copies of the unique trivial one-dimensional irreducible *-representation of B. Hence, applying Φ, we recover the well-known fact that A ($= \underset{\sim}{O}_C(L)$) has to within unitary equivalence exactly one irreducible

*-representation T , and that every non-degenerate *-representation of A is a Hilbert direct sum of copies of T .

Remark 5. The argument of (3) shows that every *-representation of $\mathcal{O}_F(\mathcal{L})$ is positive with respect to $\mathcal{L}$. Hence by the preceding Remark 1 every *-representation of $\mathcal{O}_F(\mathcal{L})$ is continuous with respect to the operator norm, and so extends to a *-representation of $\mathcal{O}_C(\mathcal{L})$.

Remark 6. Assume that $\dim(L) > 1$, and let T be the unique irreducible *-representation of A . Then $T \oplus T$ will be cyclic; but the corresponding *-representation of B (consisting of the scalar operators acting on a two-dimensional space) will not be cyclic. This shows that cyclicity is one property of *-representations which is not preserved by the Φ of Theorems 8.3 and 9.1.

(II) Block *-subalgebras. Let B be a block*-subalgebra of a Banach *-algebra A (see §7(11)); and let L be the linear span in A of $B \cup AB$. We saw in §7(10) that the equations $[r,s]_A = rs^*$, $[r,s]_B = r^*s$ $(r,s \in L)$ define $\mathcal{L} = (L,[\ ,\ ]_A,[\ ,\ ]_B)$ as an A , B imprimitivity bimodule. Of course $\mathcal{L}$ is not in general strict. However, if we denote by A^0 and B^0 the linear spans of range$([\ ,\ ]_A)$ and range$([\ ,\ ]_B)$ respectively, then $\mathcal{L}$ considered as an A^0 , B^0 imprimitivity bimodule is strict. We recall that A^0 and B^0 are *-ideals of A and B respectively. In fact, since $B \subset L$, we have $BB \subset B^0$.

Let $\mathcal{S}^0$ and $\mathcal{T}^0$ be the spaces of all unitary equivalence classes of non-degenerate *-representations of B^0 and A^0 respectively which are positive with respect to $\mathcal{L}$. Further, let $\mathcal{T}$ be the space of all unitary equivalence classes of *-representations

T of A such that $T|A^0$ is non-degenerate. Since an element T of $\underset{\sim}{T}$ is determined by $T|A^0$ and is automatically positive with respect to $\underset{\sim}{L}$, we can identify $\underset{\sim}{T}$ with a subset of $\underset{\sim}{T}^0$. Also let $\underset{\sim}{S}$ be the space of all equivalence classes of non-degenerate *-representations of B which are positive with respect to $\underset{\sim}{L}$. Since $BB \subset B^0$, elements of $\underset{\sim}{S}$ are automatically non-degenerate on the *-ideal B^0; and so $\underset{\sim}{S}$ can be identified with a subset of $\underset{\sim}{S}^0$.

Let $\Phi : \underset{\sim}{S}^0 \longrightarrow \underset{\sim}{T}^0$ be the bijection set up by the inducing process as in Theorem 8.3.

We now make two observations. The first is that, since A is assumed to be a Banach *-algebra, Proposition 4.2 permits us to induce any element S of $\underset{\sim}{S}^0$ via $\underset{\sim}{L}$ to all of A; and by Proposition 8.1 the result is non-degenerate on A^0. From this it follows that

$$\Phi(\underset{\sim}{S}^0) \subset \underset{\sim}{T} . \qquad \ldots(4)$$

The second observation is that any *-representation T of A can be induced via $\underset{\sim}{L}$ to B. Indeed, an easy calculation (which we leave to the reader) shows that $S = \mathrm{Ind}_{A\uparrow B}^{\underset{\sim}{L}}(T)$ is obtained to within unitary equivalence by the following simple procedure: Let Y be the essential space of $T|B$; and define

$$S_b = T_b|Y \quad (b \in B). \qquad \ldots(5)$$

From this it follows in particular that

$$\Phi^{-1}(\underset{\sim}{T}) \subset \underset{\sim}{S} . \qquad \ldots(6)$$

Combining (4) and (6) with the fact that Φ is a bijection of $\underset{\sim}{S}^0$ onto $\underset{\sim}{T}^0$, we conclude that $\underset{\sim}{S} = \underset{\sim}{S}^0$, $\underset{\sim}{T} = \underset{\sim}{T}^0$, and $\Phi : \underset{\sim}{S} \longrightarrow \underset{\sim}{T}$

is a bijection. Reformulating these observations we obtain:

Proposition 9.2. Let B be a block *-subalgebra of a Banach *-algebra A ; and let L be the linear span of $B \cup AB$. Let S be a non-degenerate *-representation of B such that $S_{r^*r} \geq 0$ for all r in L . Then there exists a *-representation T of A , unique to within unitary equivalence, with the following two properties:

(i) If Y denotes the essential space of $T|B$, S is unitarily equivalent to the *-representation $b \mapsto T_b|Y$ of B acting on Y ;

(ii) the smallest closed T-stable subspace of X(T) containing Y is X(T).

Furthermore, the commuting algebras of S and T are *-isomorphic. The mapping $S \mapsto T$ (as a mapping of unitary equivalence classes) is a homeomorphism with respect to the regional topologies.

The last two statements come from Theorem 8.4 and Prop. 6 6.

Remark 7. Consider the case that B is a *-ideal of A , hence certainly a block *-subalgebra of A . Then L = B , and every *-representation of B is automatically positive with respect to L . The above proposition then asserts that every *-representation S of a *-ideal B of a Banach *-algebra A can be extended to a *-representation of A acting in the same space as S . (Notice that B need not be closed in A .)

This last assertion would not in general be true if A were merely a *-algebra, not a Banach *-algebra. Thus the use of Proposition 4.2 in the argument of Proposition 9.2 was essential.

Remark 8. Prop. 9.2 is intimately related to the theory of spherical functions on groups. Let G be a locally compact group with a compact subgroup K ; and let p be a self-adjoint idempotent element of the group algebra of K . If A stands for the $\underset{\sim}{L}_1$ group algebra of G, then $B = p*A*p$ is a block *-subalgebra of A ; and Prop. 9.2 provides us with a correspondence between certain irreducible *-representations S of B and certain irreducible *-representations T of A (i.e. irreducible unitary representations of G). It sometimes happens that S is finite-dimensional even though T is not. In this case the trace of the operators of S provides us with a complex-valued function on G which uniquely determines the equivalence class of T . This function is called the spherical function of the pair T , p . See Godement [3].

Remark 9. Remark 8 indicates the possibility of analyzing the irreducible unitary representations of a group G in terms of those of a compact subgroup K of G , by means of a correspondence which is a special case of the abstract inducing construction. The Mackey normal subgroup analysis accomplishes a similar analysis when K is normal. Could it be that both of these are special cases of a generalized analysis, based on the abstract inducing process, of the representations of a locally compact group G in terms of those of an arbitrary closed subgroup of G ?

Norm estimates for imprimitivity bimodules.

Suppose that A and B are two normed *-algebras and that $\underset{\sim}{L} = (L, [\ ,\ ]_A, [\ ,\ ]_B)$ is a strict A , B imprimitivity bimodule; and let Φ be the correspondence of Theorem 8.3 between the non-degenerate $\underset{\sim}{L}$-positive *-representations of A and those of B . Let A_c and B_c be the Banach *-algebra completions of A and B . If the actions of A and B on L can be extended to actions of A_c and B_c on L in such a way that $\underset{\sim}{L}$ becomes a (necessarily topologically strict) A_c , B_c imprimitivity bimodule, then Theorem 9.1 becomes applicable; and by Remark 1 the *-representations in the domain and range of Φ are all automatically continuous with respect to the norms of A and B . However, it may happen that no such extension is possible. In that case it is useful to know a relationship between the continuity of S and the continuity of $\Phi(S)$. The following proposition provides such a relationship.

By a *-seminorm on a *-algebra D we mean a seminorm σ on D such that $\sigma(c^*) = \sigma(c)$ and $\sigma(cd) \leq \sigma(c)\sigma(d)$ for all c , d in D .

Proposition 9.3. Let A and B be two *-algebras, and σ a *-seminorm on A . Let $\underset{\sim}{L} = (L,[\ ,\ ]_A,[\ ,\ ]_B)$ be a strict A , B imprimitivity bimodule, S an $\underset{\sim}{L}$-positive non-degenerate *-representation of B , and T the induced *-representation $\mathrm{Ind}_{B\uparrow A}^{\underset{\sim}{L}}(S)$ of A . Then the following two conditions are equivalent:

(I) $\|T_a\| \leq \sigma(a)$ for all a in A .

(II) For each pair r , s of elements of L , there is a positive number k such that

$$\|S_{[r,as]_B}\| \leq k\sigma(a)$$

for all a in A .

Proof. Assume (I); and take elements r , s of L . By §8(1) and §2(5) (applied to $\bar{\underset{\sim}{L}}$ instead of $\underset{\sim}{L}$) we get for $a \in A$

$$\|S_{[r,as]_B}\|^2 \leq \|T_{[r,r]_A}\|\|T_{[as,as]_A}\|. \qquad \ldots(7)$$

Now $[as,as]_A = a[s,s]_A a^*$; so by (I) and (7)

$$\|S_{[r,as]_B}\|^2 \leq k^2 \|T_a\|^2 \leq k^2 \sigma(a)^2 ,$$

where $k^2 = \|T_{[r,r]_A}\| \|T_{[s,s]_A}\|$. This establishes (II); and we have shown that (I)⇒(II).

Now assume (II); and let us adopt the notation of Prop. 8.1, so that $X = X(S)$, $Y = X(T)$, $V_{r,s} = S_{[s,r]_B}$, $W_{r,s} = T_{[s,r]_A}$. As usual the image of $r \otimes \xi$ in Y $(r \in L;\ \xi \in X)$ is written $r \tilde{\otimes} \xi$. To prove (I) it is enough (see the Remark following this proof) to fix a non-zero vector ξ in X and an element a of A and to show that

$$\|ar \tilde{\otimes} \xi\|_Y \leq \sigma(a) \|r \tilde{\otimes} \xi\|_Y \quad (r \in L). \qquad \ldots(8)$$

For any s , t in L let us define $\{s,t\} = \langle s \tilde{\otimes} \xi, t \tilde{\otimes} \xi\rangle_Y = \langle S_{[t,s]_B} \xi, \xi\rangle$. Notice that $\{\ ,\ \}$ is conjugate-bilinear, and that

$$\{cs,t\} = \{s,c^* t\}, \qquad \ldots(9)$$

$$\{s,s\} \geq 0 \qquad \ldots(10)$$

$(s,t \in L;\ c \in A)$. Applying the ordinary Schwarz Inequality to $\{\ ,\ \}$ (by (10)), and writing $N(r)$ for $\{r,r\}^{\frac{1}{2}}$, we get by (9)

$$(N(ar))^2 = \{a^* ar,r\} \leq N(a^* ar)N(r). \qquad \ldots(11)$$

Replacing a by $a^* a$ in (11) we obtain $N(a^* ar) \leq N((a^* a)^2 r)^{\frac{1}{2}} N(r)^{\frac{1}{2}}$. Combining this with (11) gives

$$(N(ar))^2 \leq N((a^* a)^2 r)^{\frac{1}{2}} N(r)^{1+\frac{1}{2}} .$$

Iterating this argument, we find for each $n = 1,2,\ldots$ and each r in L :

$$(N(ar))^2 \leq N((a^*a)^{2^n}r)^{2^{-n}}(N(r))^{2-2^{-n}} . \qquad \ldots(12)$$

Now fix $r \in L$. By (II) there is a positive number k (depending of course on r) such that for all c in A

$$(N(cr))^2 = (S_{[r,c^*cr]_B}\xi, \xi)$$

$$\leq \|\xi\|^2 k^2 \sigma(c^*c) \leq \|\xi\|^2 k^2 \sigma(c)^2 . \qquad \ldots(13)$$

Substituting (13) in (12) with $c = (a^*a)^{2^n}$, we find

$$(N(ar))^2 \leq (\|\xi\|k)^{2^{-n}}(\sigma((a^*a)^{2^n}))^{2^{-n}}(N(r))^{2-2^{-n}} . \qquad \ldots(14)$$

In view of the seminorm properties of σ,

$$(\sigma((a^*a)^{2^n}))^{2^{-n}} \leq \sigma(a^*a) \leq \sigma(a)^2 .$$

So (14) becomes

$$(N(ar))^2 \leq (\|\xi\|k)^{2^{-n}}(\sigma(a))^2(N(r))^{2-2^{-n}} . \qquad \ldots(15)$$

Now (15) holds for all positive integers n. Passing to the limit $n \to \infty$ in (15) we obtain

$$(N(ar))^2 \leq (\sigma(a))^2(N(r))^2 .$$

By the definition of N this is just (8). So we have shown that (II)$\Rightarrow$(I). □

<u>Remark 10</u>. In the preceding proof, we used the following fact: <u>Let</u> Z <u>be a Hilbert space</u>, $\underset{\sim}{S}$ <u>a collection of closed linear subspaces of</u> Z <u>whose linear span is dense in</u> Z, <u>and</u> D <u>the</u> C^*<u>-subalgebra of</u>

$\mathcal{O}(Z)$ consisting of those γ such that every W in $\mathcal{S}$ is stable under both γ and γ^*. Then for each γ in D we have $\|\gamma\| = \sup\{\|\gamma|W\|: W \in \mathcal{S}\}$. To prove this fact, we have only to observe that $\gamma \mapsto \{\gamma|W\}_{W\in\mathcal{S}}$ is a one-to-one *-homomorphism of D into the C^*-direct product of the $\mathcal{O}(W)$ $(W \in \mathcal{S})$, and hence an isometry.

Corollary 9.4. In Proposition 9.3 suppose that A and B are normed *-algebras, and that, for every pair r, s of elements of L, the linear map $a \mapsto [r,as]_B$ of A into B is continuous. Then, for each $\mathcal{L}$-positive *-representation of B which is norm-continuous on B, the induced *-representation $\mathrm{Ind}^{\mathcal{L}}_{B\uparrow A}(S)$ of A is norm-continuous on A.

Proof. Apply Prop. 9.3 with $\sigma(a) = \|a\|$. □

On replacing $\mathcal{L}$ by $\bar{\mathcal{L}}$ the roles of A and B become reversed. Corollary 9.4 thus gives:

Corollary 9.5. In Proposition 9.3 suppose that A and B are normed *-algebras, and that for each pair r, s of elements of L the two linear maps $a \mapsto [r,as]_B$ of A into B and $b \mapsto [rb,s]_A$ of B into A are continuous. Then, under the correspondence Φ of Theorem 8.3, the norm-continuous elements of $\mathcal{S}$ correspond exactly with the norm-continuous elements of $\mathcal{T}$.

CHAPTER II

BANACH *-ALGEBRAIC BUNDLES

§10. <u>Banach bundles</u>.

We have described at length in the Introduction the role played
by Banach *-algebraic bundles in our investigations of the inducing
process. They form a half-way house, as it were, between the classical
inducing process on groups and the abstract Rieffel inducing process
on rigged modules. They are also of great interest for their own sake.

Every Banach *-algebraic bundle has, underlying it, the structure
of a Banach bundle. Banach bundles are related to Banach *-algebraic
bundles just as Banach spaces are related to Banach *-algebras. The
present section is devoted to Banach bundles. For brevity we shall
omit those proofs that can be found in Fell [6].

It should be remarked that Banach bundles are a special case of
a much more general class of structures studied by Dauns and Hofmann
[1] under the name of 'uniform fields'. These generalize Banach
bundles to somewhat the same extent that topological uniform spaces
generalize Banach spaces. Even the 'uniform fields of Banach spaces'
occurring in Dauns and Hofmann [1] are more general than our Banach
bundles, inasmuch as their norm-functions $b \longmapsto \|b\|$ are not required
to be continuous.

Let X be a fixed Hausdorff topological space.

<u>Definition</u>. A <u>bundle over</u> X is a pair $\underset{\sim}{B} = (B,\pi)$ such that B is
a Hausdorff topological space (called the <u>bundle space</u> of $\underset{\sim}{B}$) and
$\pi : B \longrightarrow X$ is a continuous open surjection (called the <u>bundle
projection</u> of $\underset{\sim}{B}$). We refer to X as the <u>base space</u> of $\underset{\sim}{B}$. If $x \in X$,

we call $\pi^{-1}(x)$ the <u>fiber over</u> x, and denote it by B_x. Sometimes it is convenient to refer to the bundle $\underset{\sim}{B}$ itself as $(B,\{B_x\}_{x\in X})$.

A <u>cross-section</u> of $\underset{\sim}{B}$ is a function $f : X \to B$ such that $f(x) \in B_x$ for each x in X. We say that $\underset{\sim}{B}$ has <u>enough continuous cross-sections</u> if for every b in B there exists a continuous cross-section $f : X \to B$ which passes through b (that is, $f(\pi(b)) = b$).

<u>Definition</u>. A <u>Banach bundle over</u> X is a bundle $\underset{\sim}{B} = (B,\pi)$ over X, together with operations and a norm making each fiber B_x $(x \in X)$ into a (complex) Banach space and satisfying the following conditions:

(i) $b \mapsto \|b\|$ is continuous on B to $\mathbb{R}$.

(ii) The operation $+$ is continuous on $\{(b,c) \in B \times B : \pi(b) = \pi(c)\}$ to B.

(iii) For each λ in $\mathbb{C}$, the map $b \mapsto \lambda\cdot b$ is continuous on B to B.

(iv) If $x \in X$ and $\{b_i\}$ is any net of elements of B such that $\|b_i\| \to 0$ and $\pi(b_i) \to x$, then $b_i \to 0_x$.

<u>Remark</u>. Here and in the sequel we write $+$, $\cdot$, $\|\ \|$ for the operations of addition, scalar multiplication and norm in each fiber. In connection with (ii) notice that $b+c$ is defined only if b and c are in the same fiber. As usual we write λb instead of $\lambda\cdot b$ for scalar multiplication. The symbol 0_x in (iv) stands for the zero element of the Banach space B_x.

<u>Definition</u>. A Banach bundle in which each fiber is a Hilbert space is called a <u>Hilbert bundle</u>.

The simplest Banach bundles are the so-called trivial ones. Let A be any Banach space; and put $B = A \times X$, $\pi(a,x) = x$; then (B,π) is a bundle over X. If we equip each fiber $A \times \{x\}$ with the obvious Banach space structure making $a \mapsto (a,x)$ an isometric isomorphism, then (B,π) becomes a Banach bundle; it is called the <u>trivial bundle with constant fiber</u> A. Cross-sections of this (B,π) will often be deliberately confused with functions $f: X \to A$.

Let $\underset{\sim}{B} = (B,\pi)$ be a Banach bundle over X. If Y is a topological subspace of X and $B_Y = \pi^{-1}(Y)$, then $(B_Y, \pi|B_Y)$ is clearly a Banach bundle over Y; we call it the <u>reduction of</u> $\underset{\sim}{B}$ <u>to</u> Y, and denote it by $\underset{\sim}{B}_Y$.

More generally, let X and Y be any two Hausdorff spaces, and $\varphi: Y \to X$ a continuous map; and let $\underset{\sim}{B} = (B,\pi)$ again be a Banach bundle over X. Let C be the topological subspace $\{(y,b): y \in Y, b \in B, \varphi(y) = \pi(b)\}$ of $Y \times B$; and define $\rho: C \to Y$ by $\rho(y,b) = y$. Clearly ρ is a continuous surjection; and the openness of π implies that ρ is open. So (C,ρ) is a bundle over Y. For each y in Y we make $C_y = \rho^{-1}(y)$ into a Banach space in such a way that the bijection $b \mapsto (y,b)$ of $B_{\varphi(y)}$ onto C_y becomes a linear isometry. One then verifies immediately that (C,ρ) is a Banach bundle; it is called the <u>Banach bundle retraction of</u> $\underset{\sim}{B}$ <u>by</u> φ.

If Y is a topological subspace of X and $\varphi: Y \to X$ is the identity map, the Banach bundle retraction of $\underset{\sim}{B}$ by φ is essentially just $\underset{\sim}{B}_Y$.

Here are a few elementary properties of Banach bundles. We fix a Banach bundle $\underset{\sim}{B} = (B,\pi)$ over X.

Proposition 10.1. *The scalar multiplication map* $(\lambda,b) \mapsto \lambda b$ *is (jointly) continuous on* $\mathbb{C} \times B$ *to* B .

This is Prop. 1.1 of Fell [6].

Proposition 10.2. *The topology of* B *relativized to any fiber* B_x *is the norm-topology of* B_x .

This is Prop. 1.3 of Fell [6].

Proposition 10.3. *Let* b *be an element of* B *and* $\{b_i\}$ *a net of elements of* B *such that* $\pi(b_i) \to \pi(b)$. *Suppose further that for each* $\varepsilon > 0$ *there is a continuous cross-section* f *of* $\underset{\sim}{B}$ *such that* $\|b-f(\pi(b))\| < \varepsilon$ *and* $\|b_i-f(\pi(b_i))\| < \varepsilon$ *for all large enough* i . *Then* $b_i \to b$ *in* B .

This is essentially Prop. 1.4 of Fell [6].

Corollary. *Suppose* f *is a continuous cross-section of* $\underset{\sim}{B}$, $x \in X$, *and* $f(x) = c$. *For each positive number* ε *and each open* X-*neighborhood* U *of* x , *put*

$$V(U,\varepsilon) = \{b \in B : \pi(b) \in U, \|b-f(\pi(b))\| < \varepsilon\}.$$

The $V(U,\varepsilon)$ *are open subsets of* B ; *and in fact the set of all* $V(U,\varepsilon)$ $(U,\varepsilon$ *varying) is a basis of neighborhoods of* c .

A very useful method of constructing specific Banach bundles is available when we start with a family of cross-sections which we would like to make continuous. It is expressed in the following result:

Proposition 10.4. *Let* X *be a Hausdorff space,* B *an (untopologized) set, and* $\pi : B \to X$ *a surjection; and suppose that for each* x *in*

X *the set* $B_x = \pi^{-1}(x)$ *has a Banach space structure* $+ , \cdot , \| \|$. *Let* Γ *be a linear space of cross-sections of* (B,π) *such that* (i) *for each* f *in* Γ *the numerical function* $x \mapsto \|f(x)\|$ *is continuous on* X, *and* (ii) *for each* x *in* X, $\{f(x): f \in \Gamma\}$ *is dense in* B_x. *Then there exists a unique topology for* B *making* (B,π) *a Banach bundle such that all the cross-sections in* Γ *are continuous on* X.

This is Proposition 1.6 of Fell [6].

Now almost all the Banach bundles encountered in this work will have locally compact base spaces. In this connection the following remarkable result has been proved (though not yet published) by A. Douady and L. dal Soglio-Hérault [1]:

Theorem 10.5. *If* X *is either paracompact or locally compact, every Banach bundle over* X *has enough continuous cross-sections.*

The reader will find the proof of this theorem in the Appendix. It will follow from this that all the specific Banach bundles encountered in this work automatically have enough continuous cross-sections.

For the rest of this section X is a *locally compact* Hausdorff space, and $\underset{\sim}{B} = (B,\pi)$ is a fixed Banach bundle over X.

We shall denote by $\underset{\sim}{C}(\underset{\sim}{B})$ the linear space of all continuous cross-sections of $\underset{\sim}{B}$, and by $\underset{\sim}{L}(\underset{\sim}{B})$ the subspace of $\underset{\sim}{C}(\underset{\sim}{B})$ consisting of those f which vanish outside some compact set (that is, there is a compact subset C of X, depending on f, such that $f(x) = 0_x$ for all $x \in X-C$). If $f \in \underset{\sim}{L}(\underset{\sim}{B})$, the (compact) closure of $\{x : f(x) \neq 0_x\}$ is the *support* of f, written $\text{supp}(f)$.

Since scalar multiplication is jointly continuous (Prop. 10.1), $\underset{\sim}{C}(\underset{\sim}{B})$ is a $\underset{\sim}{C}(X)$-module; that is, if $\varphi \in \underset{\sim}{C}(X)$ and $f \in \underset{\sim}{C}(\underset{\sim}{B})$, then

$\varphi f\colon x \mapsto \varphi(x)f(x)$ is in $\mathcal{C}(B)$. From this and Theorem 10.5 it follows that the elements of $\mathcal{L}(B)$ pass through every point of B.

The most useful topology of $\mathcal{C}(B)$ is the topology of uniform convergence on compact sets, in which a net $\{f_i\}$ converges to f if and only if $\|f_i(x)-f(x)\| \to 0$ uniformly in x on each compact subset of X. For this topology we have the following important density result:

Proposition 10.6. *Let Γ be a linear subspace of $\mathcal{C}(B)$ with the following two properties:*

(i) *If* $\varphi \in \mathcal{C}(X)$ *and* $f \in \Gamma$, *then* $\varphi f \in \Gamma$;

(ii) $\{f(x)\colon f \in \Gamma\}$ *is dense in* B_x *for each* x *in* X. *Then* Γ *is uniformly-on-compacta dense in* $\mathcal{C}(B)$.

This is Prop. 1.7 of Fell [6].

As an application of this we shall deduce a version of Tietze's Extension Theorem for Banach bundles (see Godement [2], Prop. 7, p. 82):

Theorem 10.7. *Let Y be any closed subset of X. Then for any g in $\mathcal{L}(B_Y)$ there exists an element f of $\mathcal{L}(B)$ such that $f|Y = g$.*

Proof. Let $g \in \mathcal{L}(B_Y)$; and choose an open subset U of X such that $\mathrm{supp}(g) \subset U$ and $\bar{U}$ is compact. By the Tietze Extension Theorem for numerical functions, the family $S = \{f|Y : f \in \mathcal{L}(B)\}$ satisfies both hypotheses of Prop. 10.6. Therefore by Prop. 10.6 there is a sequence $\{f_n\}$ of elements of $\mathcal{L}(B)$ such that

$$f_n|Y \to g \text{ uniformly on } \bar{U} \cap Y. \qquad \ldots(1)$$

Multiplying the f_n by a fixed element of $\mathcal{L}(X)$ which is 1 on $\mathrm{supp}(g)$ and vanishes outside U, we may assume that

$$\text{all } f_n \text{ vanish outside } U . \qquad \ldots(2)$$

Passing to a subsequence of $\{f_n\}$, we may assume by (1) that, for each $n \geq 2$,

$$\sup\{\|f_n(y) - f_{n-1}(y)\|: \ y \in \bar{U} \cap Y\} < 2^{-n} . \qquad \ldots(3)$$

Now put $h_n' = f_n - f_{n-1}$ $(n \geq 2)$ and define

$$h_n(x) = \begin{cases} h_n'(x) & \text{if } \|h_n'(x)\| \leq 2^{-n} , \\ \|h_n'(x)\|^{-1} 2^{-n} h_n'(x) & \text{if } \|h_n'(x)\| \geq 2^{-n} . \end{cases}$$

Evidently $h_n \in \underset{\sim}{L}(\underset{\sim}{B})$ and $\|h_n(x)\| \leq 2^{-n}$ for all x. Thus

$$f(x) = f_1(x) + \sum_{n=2}^{\infty} h_n(x) \qquad (x \in X) \qquad \ldots(4)$$

converges absolutely uniformly on X; and the cross-section f defined by (4) is continuous by Prop. 10.3. By (2) f vanishes outside U, and so belongs to $\underset{\sim}{L}(\underset{\sim}{B})$. By (3) $h_n(y) = f_n(y) - f_{n-1}(y)$ for $y \in \bar{U} \cap Y$, whence by (1) and (4) $f(y) = g(y)$ for $y \in \bar{U} \cap Y$. Since f and g both vanish on $Y - \bar{U}$, this implies $f|Y = g$, completing the proof. □

The inductive limit topology.

A very useful topology for $\underset{\sim}{L}(\underset{\sim}{B})$ is the inductive limit topology, defined as follows: Consider a compact subset K of X, and put $\underset{\sim}{L}^K(\underset{\sim}{B}) = \{f \in \underset{\sim}{L}(\underset{\sim}{B}): \ \text{supp}(f) \subset K\}$. Equipped with the supremum norm:

$$\|f\|_\infty = \sup\{\|f(x)\|: \ x \in K\},$$

$L^K(B)$ becomes a normed linear space (in fact a Banach space). Let $i_K : L^K(B) \longrightarrow L(B)$ be the identity map. Just as for $L(X)$ (see for example Schaefer [1], Chap. II, §6), one proves:

<u>Proposition 10.8</u>. <u>There is a unique topology T for $L(B)$ which makes $L(B)$ a (Hausdorff) locally convex linear space and has the following universal property</u> (P): <u>For any locally convex linear space</u> M <u>and any linear map</u> $F : L(B) \longrightarrow M$, F <u>is continuous if and only if</u> $F \cdot i_K : L^K(B) \longrightarrow M$ <u>is continuous for every compact subset</u> K <u>of</u> X.

<u>The injections</u> $i_K : L^K(B) \longrightarrow L(B)$ <u>are homeomorphisms with respect to</u> T.

<u>Definition</u>. The topology T of this proposition is called the <u>inductive limit topology</u> of $L(B)$.

<u>Remark</u>. If B is the trivial bundle with one-dimensional fiber $\mathbb{C}$, this reduces to the usual definition of the inductive limit topology of $L(X)$. The inductive limit topology for $L(B)$ when B is a trivial bundle is studied by Bourbaki [3], Chap. 3, §1.

<u>Remark</u>. It follows immediately from the universal property (P) that the multiplication map $(\varphi, f) \longmapsto \varphi f$ of $L(X) \times L(B)$ into $L(B)$ is <u>separately</u> continuous in its two variables with respect to the inductive limit topologies. In general, however, it is <u>not jointly continuous</u>.

The following proposition is now an easy corollary of Prop. 10.6:

<u>Proposition 10.9</u>. <u>Let</u> Γ <u>be a linear subspace of</u> $L(B)$ <u>with the following two properties</u>:

(i) <u>If</u> $\varphi \in L(X)$ <u>and</u> $f \in \Gamma$, <u>then</u> $\varphi f \in \Gamma$;

(ii) $\{f(x): f \in \Gamma\}$ *is dense in* B_x *for each* x *in* X. *Then* Γ *is dense in* $\mathcal{L}(B)$ *in the inductive limit topology.*

In this work we shall not make much use of countability or separability properties. However, a few remarks on them are in order.

Proposition 10.10. *Suppose that the bundle space* B *is second-countable (i.e., has a countable base of open sets). Then* $\mathcal{L}(B)$ *is separable (i.e., has a dense countable subset) with respect to the inductive limit topology.*

Proof. Since π is open, the second countability of B implies that of X. Now it is well known that, when X is second-countable, $\mathcal{L}(X)$ is separable in the inductive limit topology. So we may choose a countable subset D of $\mathcal{L}(X)$ which is inductive-limit dense. Furthermore, by Prop. 1.8 of Fell [6] there is a countable subset F of $\mathcal{C}(B)$ such that, for each x in X, $\{f(x): f \in F\}$ is dense in B_x. Now let E be the countable family of all finite sums of products of the form φf, where $\varphi \in D$ and $f \in F$. Thus the inductive limit closure $\bar{E}$ of E contains the set Γ of all finite sums of products φf, where $\varphi \in \mathcal{L}(X)$ and $f \in F$. Now this Γ is a linear subspace of $\mathcal{L}(B)$ satisfying hypotheses (i) and (ii) of Prop. 10.9. So by Prop. 10.9 Γ is dense in $\mathcal{L}(B)$ in the inductive limit topology. Hence the same is true of E. □

Remark. If B is second-countable, then, as we have observed, X is second-countable, and also each fiber B_x is separable (see Prop. 10.2). One might conjecture the converse -- that if X is second-countable and each fiber is separable the bundle space B must be second-countable. An interesting example of Maréchal [1] shows that this is false.

Integration in Banach bundles.

We conclude this section with a summary of the main facts about integration in Banach bundles over locally compact spaces, as given in §2 of Fell [6]. (Integration theory in Banach bundles is originally due to Godement [1], [2].)

Fix a regular non-negativeBorel measure μ on X. (We regard μ as defined on the σ-ring of those Borel sets which are σ-bounded, i.e., contained in a countable union of compact sets. Thus any μ-summable numerical function on X vanishes outside some σ-bounded set.) A cross-section f of $\underset{\sim}{B}$ is locally μ-measurable if for each compact subset K of X there is a sequence $\{g_n\}$ of elements of $\underset{\sim}{C}(\underset{\sim}{B})$ such that

$$g_n(x) \longrightarrow f(x) \quad \text{in} \quad B_x \quad \text{for} \quad \mu\text{-almost all} \quad x \quad \text{in} \quad K .$$

If f is locally μ-measurable, then of course $x \longmapsto \|f(x)\|$ is locally μ-measurable also.

Let $1 \leq p < \infty$. A cross-section f of $\underset{\sim}{B}$ is p^{th}-power summable if it is locally μ-measurable and

$$\|f\|_p = \left(\int_X \|f(x)\|^p \, d\mu x\right)^{\frac{1}{p}} < \infty . \qquad \ldots(5)$$

The space of all p^{th} power summable cross-sections is called the cross-sectional $\underset{\sim}{L}_p$-space of $\underset{\sim}{B}$ with respect to μ, and is denoted by $\underset{\sim}{L}_p(\underset{\sim}{B};\mu)$. It is a Banach space under the norm $\| \ \|_p$ defined by (5).

If $\underset{\sim}{B}$ happens to be the trivial Banach bundle over X with constant fiber A, we write $\underset{\sim}{L}_p(X,A;\mu)$ or $\underset{\sim}{L}_p(A;\mu)$ instead of $\underset{\sim}{L}_p(\underset{\sim}{B};\mu)$.

Now obviously $\mathcal{L}(\mathcal{B}) \subset \mathcal{L}_p(\mathcal{B};\mu)$. In fact we have:

Proposition 10.11. $\mathcal{L}(\mathcal{B})$ is dense in $\mathcal{L}_p(\mathcal{B};\mu)$ for all $1 \leq p < \infty$.

Note. This was the definition of $\mathcal{L}_p(\mathcal{B};\mu)$ in Fell [6], §2.

Since the $\mathcal{L}_p$-topology of $\mathcal{L}(\mathcal{B})$ is contained in the inductive limit topology, Props. 10.11 and 10.9 imply the following density result:

Proposition 10.12. Let Γ be a linear subspace of $\mathcal{L}(\mathcal{B})$ with the following two properties:

(i) If $\varphi \in \mathcal{L}(X)$ and $f \in \Gamma$, then $\varphi f \in \Gamma$;

(ii) $\{f(x): f \in \Gamma\}$ is dense in B_x for each x in X. Then Γ is dense in $\mathcal{L}_p(\mathcal{B};\mu)$ for each $1 \leq p < \infty$.

Suppose that $\mathcal{B}$ is a Hilbert bundle, the inner product in each fiber being $\langle\ ,\ \rangle$. Then $\mathcal{L}_2(\mathcal{B};\mu)$ is a Hilbert space, with inner product

$$\langle f,g\rangle = \int_X \langle f(x),g(x)\rangle d\mu x .$$

The Bochner integral.

We now consider the special case that $p = 1$ and $\mathcal{B}$ is the trivial Banach bundle over X with constant fiber A. In that case there exists a unique continuous linear map $I : \mathcal{L}_1(X,A;\mu) \to A$ such that

$$I(Ch_W a) = \mu(W)a \qquad \ldots(6)$$

for every a in A and every σ-bounded Borel subset W of X with

$\mu(W) < \infty$. We call $I(f)$ $(f \in \underset{\sim}{L}_1(X,A;\mu))$ the A-valued Bochner integral of f with respect to μ, and denote it by $\int_X f(x)d\mu x$ or $\int f\,d\mu$. Functions in $\underset{\sim}{L}_1(X,A;\mu)$ are called μ-summable. If $A = \mathbb{C}$, this definition reduces of course to the usual Lebesgue integral of a summable complex function.

It is a simple but important fact that the Bochner integral commutes with continuous linear maps:

Proposition 10.13. Let A_1 and A_2 be Banach spaces and $\varphi : A_1 \to A_2$ a continuous linear map. If $f \in \underset{\sim}{L}_1(X,A_1;\mu)$, then $\varphi \circ f \in \underset{\sim}{L}_1(X,A_2;\mu)$ and

$$\int_X (\varphi \circ f)\,d\mu = \varphi\left(\int_X f\,d\mu\right).$$

This is Prop. 2.5 of Fell [6].

We shall now state an important continuity lemma for Bochner integrals. Take another locally compact Hausdorff space Y and a regular Borel measure ν on Y; let $p : X \times Y \to X$ be the surjection $(x,y) \mapsto x$; and form the Banach bundle retraction $\underset{\sim}{C} = (C,\rho)$ of $\underset{\sim}{B}$ by p. Thus $\underset{\sim}{C}$ is a Banach bundle over $X \times Y$ whose bundle space C can be identified with $B \times Y$, the bundle projection then being $\rho : (b,y) \mapsto (\pi(b),y)$. For each x in X, $\underset{\sim}{C}_{\{x\}\times Y}$ is thus the trivial bundle with constant fiber B_x. So, given $h \in \underset{\sim}{L}(\underset{\sim}{C})$, for each x in X we can form the Bochner integral $\int_Y h(x,y)d\nu y$, and this will belong to B_x.

Lemma 10.14. In this context, for each h in $\underset{\sim}{L}(\underset{\sim}{C})$ the map $x \mapsto \int_Y h(x,y)d\nu y$ $(x \in X)$ is a continuous cross-section of $\underset{\sim}{B}$.

This is a reformulation of Lemma 2.3 of Fell [6].

Here is the Bochner form of the Fubini Theorem.

Fubini Theorem 10.15. *Let X and Y be two locally compact Hausdorff spaces; let μ and ν be regular Borel measures on X and Y respectively; and let A be any Banach space. Suppose that $f : X\times Y \to A$ belongs to $\mathcal{L}_1(X\times Y, A; \mu\times\nu)$ ($\mu\times\nu$ being the regular Borel product measure). Then (I) $x \mapsto f(x,y)$ belongs to $\mathcal{L}_1(X,A;\mu)$ for ν almost all y in Y; (II) $y \mapsto \int f(x,y)d\mu x$ belongs to $\mathcal{L}_1(Y,A;\nu)$; and (III)*

$$\int_Y\int_X f(x,y)d\mu x d\nu y = \int_{X\times Y} f(x,y)d(\mu\times\nu)(x,y).$$

This is Prop. 2.12 of Fell [6].

Remark. It is often useful to discuss Bochner integrals of functions whose values lie in an arbitrary locally convex space. We shall sketch this generalization very briefly, leaving the reader to fill in the details.

Let μ be a regular Borel measure on a locally compact Hausdorff space X, A an arbitrary locally convex space, and P a family of continuous seminorms on A which generate the topology of A. Each p in P gives rise to a Banach space A_p, obtained by factoring from A the null-space of p and completing with respect to p; let $\rho_p : A \to A_p$ be the natural (continuous) quotient map. We now define a function $f : X \to A$ to be μ-*summable* if, for every p in P, $\rho_p \circ f : X \to A_p$ is μ-summable in the earlier sense. If $f : X \to A$ is μ-summable, and if there is an element a of A such that

$\int_X(\rho_p \circ f)d\mu = \rho_p(a)$ for all p in P, then a is called the X-valued Bochner integral of f with respect to μ; and we write

$$a = \int_X f \, d\mu .$$

If f is μ-summable and A is complete, then $\int f \, d\mu$ exists.

It is easy to see that Prop. 10.13 continues to hold in this generalized context:

Proposition 10.16. *If A_1 and A_2 are locally convex spaces, $\varphi : A_1 \to A_2$ is a continuous linear map, and $f : X \to A_1$ is μ-summable, then $\varphi \circ f : X \to A_2$ is μ-summable. If in addition $a = \int_X f \, d\mu$ exists, then*

$$\varphi(a) = \int_X (\varphi \circ f) \, d\mu .$$

§11. Banach *-algebraic bundles.

In this section we begin the study of Banach *-algebraic bundles. A Banach *-algebraic bundle is, among other things, a Banach bundle whose base space is a topological group. We shall therefore fix a (Hausdorff) topological group G, with unit e.

Definition. A *Banach *-algebraic bundle over* G is a Banach bundle (B,π) over G, together with a binary operation $\cdot$ (called *multiplication*) and a unary operation $*$ (called *involution*) on B having the following properties:

(i) $\pi(b \cdot c) = \pi(b)\,\pi(c)$ for all b,c in B.

(Equivalently, $B_x \cdot B_y \subset B_{xy}$ for all x,y in G.)

(ii) For each pair of elements x,y of G, the map $(b,c) \mapsto b \cdot c$

is bilinear from $B_x \times B_y$ to B_{xy} .

(iii) The associative law $b\cdot(c\cdot d) = (b\cdot c)\cdot d$ holds $(b,c,d \in B)$.

(iv) $\|b\cdot c\| \le \|b\|\,\|c\|$ $\quad (b,c \in B)$.

(v) The multiplication map $\cdot$ is (jointly) continuous on $B \times B$ to B .

(vi) $\pi(b^*) = (\pi(b))^{-1}$ $\quad (b\in B)$. (Equivalently, $(B_x)^* \subset B_{x^{-1}}$ for all x in G .)

(vii) For each x in G the map $b \mapsto b^*$ is conjugate-linear on B_x to $B_{x^{-1}}$.

(viii) $(b\cdot c)^* = c^*\cdot b^*$ $\quad (b,c \in B)$.

(ix) $b^{**} = b$ $\quad (b\in B)$.

(x) $\|b^*\| = \|b\|$ $\quad (b\in B)$.

(xi) $b \mapsto b^*$ is continuous on B to B .

<u>Remarks</u>. As usual B_x is the fiber $\pi^{-1}(x)$. The elements of B_x are said to be <u>of order</u> x in G . 0_x will be the zero element of B_x . Usually we write bc instead of $b\cdot c$.

The above postulates are just the postulates of a Banach *-algebra in so far as the latter make sense in a system in which addition is not universally defined.

Notice that

$$b0_x = 0_{\pi(b)x}\,, \quad 0_x b = 0_{x\pi(b)} \qquad \ldots(1)$$

for all x in G and b in B .

In view of (vi) and (ix) we have

$$(B_x)^* = B_{x^{-1}} \quad (x\in G) \,. \qquad \ldots(2)$$

In a Banach *-algebra the postulates (v) and (xi) follow from the remaining ones (especially (iv) and (x)). This is not true for Banach *-algebraic bundles.

Suppose that there is a family Γ of continuous cross-sections of (B,π) such that $\{f(x) : f \in \Gamma\}$ is dense in B_x for each x in G. Then condition (v) of the preceding definition can be replaced (without altering the definition) by the following condition:

(v′) For each f,g in Γ, the map $(x,y) \mapsto f(x)\, g(y)$ is continuous on $G \times G$ to B.

Likewise, condition (xi) can be replaced by:

(xi′) For each f in Γ, the map $x \mapsto (f(x))^*$ is continuous on G to B.

(See Fell [6], §3, Remarks 3,4.)

Let $\underset{\sim}{B} = (B,\pi,\cdot,^*)$ be a Banach *-algebraic bundle over G; and let H be another topological group and $\varphi : H \to G$ a continuous homomorphism. One verifies that the Banach bundle retraction (C, ρ) of $\underset{\sim}{B}$ by φ (defined in §10) is a Banach *-algebraic bundle when multiplication and involution are defined in C by

$$(y,b)\,(y',b') = (yy',\, bb'), \qquad (y,b)^* = (y^{-1},b^*) \qquad \ldots(3)$$

$((y,b),\,(y',b') \in C)$. With these operations (C, ρ) becomes the <u>Banach *-algebraic bundle retraction of</u> $\underset{\sim}{B}$ by φ.

In particular, if H is a topological subgroup of G, the <u>reduction</u> $\underset{\sim}{B}_H$ of $\underset{\sim}{B}$ to H is a Banach *-algebraic bundle under the restrictions to $\underset{\sim}{B}_H$ of the operations of $\underset{\sim}{B}$.

If we take H to be the one-element subgroup $\{e\}$, $\underset{\sim}{B}_H$ becomes the fiber B_e; and postulates (i)-(xi) assert that B_e is a Banach

*-algebra under the restrictions to B_e of the operations of $\underset{\sim}{B}$. We call B_e the <u>unit fiber *-algebra</u> of $\underset{\sim}{B}$. Notice that the other fibers B_x $(x \neq e)$ are not algebras, since they are not closed under multiplication or involution.

We observe that any Banach *-algebra can be regarded as a Banach *-algebraic bundle over the one-element group. Thus any definition or theorem given for Banach *-algebraic bundles applies in particular to Banach *-algebras.

<u>Definition</u>. Let $\underset{\sim}{B} = (B,\pi,\cdot,^*)$ and $\underset{\sim}{B}' = (B',\pi',\cdot,^*)$ be two Banach *-algebraic bundles over the same topological group G. By an <u>isomorphism</u> of $\underset{\sim}{B}$ and $\underset{\sim}{B}'$ we mean a homeomorphism φ of B onto B' such that (i) $\pi'\circ\varphi = \pi$ (i.e. $\varphi(B_x) = B'_x$ for each x in G), (ii) $\varphi|B_x : B_x \to B'_x$ is an isomorphism of linear spaces for each x, and (iii) φ preserves multiplication and involution. If such a φ exists, $\underset{\sim}{B}$ and $\underset{\sim}{B}'$ are <u>isomorphic</u>. If in addition φ is an isometry (i.e. $\|\varphi(b)\| = \|b\|$ for all b in B), then $\underset{\sim}{B}$ and $\underset{\sim}{B}'$ are <u>isometrically isomorphic</u>.

Now fix a Banach *-algebraic bundle $\underset{\sim}{B} = (B,\pi,\cdot,^*)$ over G.

A very important property of Banach *-algebraic bundles is that of saturation.

<u>Definition</u>. The Banach *-algebraic bundle $\underset{\sim}{B}$ is <u>saturated</u> if, for every pair x,y of elements of G, the linear span $[B_x B_y]$ of $B_x B_y$ in B_{xy} is dense in B_{xy}.

Observe that $\underset{\sim}{B}$ is saturated if and only if for every x in G (i) $[B_e B_x]$ is dense in B_x, and (ii) $[B_x B_{x^{-1}}]$ is dense in B_e. Indeed: To see that (i) and (ii) imply saturation, notice that $[B_x B_y] \supset [B_x[B_{x^{-1}} B_{xy}]] = [[B_x B_{x^{-1}}] B_{xy}]$. By (ii) and the continuity of the product the last expression is dense in $[B_e B_{xy}]$; and by (i) this is dense in B_{xy} .

Remark. The concept of saturation assures us that every fiber of $\underset{\sim}{B}$ plays its "full role" in the multiplicative structure of $\underset{\sim}{B}$. It thus rules out such trivial situations as the following: Let G be a discrete group and A a Banach *-algebra, and set $B_e = A$, $B_x = \{0\}$ for $x \neq e$.

Clearly any retraction of a saturated Banach *-algebraic bundle is likewise saturated.

Approximate units

A unit element of $\underset{\sim}{B}$ is of course an element $\mathbb{1}$ of B such that $\mathbb{1}b = b\mathbb{1} = b$ for all b in B . Notice that $\mathbb{1}$ is necessarily in B_e .

Definition. An approximate unit of $\underset{\sim}{B}$ is a net $\{u_i\}$ of elements of B_e such that (i) there is a positive constant k such that $\|u_i\| \leq k$ for all i , and (ii) $\|u_i b - b\| \to 0$ and $\|b u_i - b\| \to 0$ for all b in B .

Obviously a unit element is an approximate unit in a natural sense.

If $\underset{\sim}{B}$ has an approximate unit, then so does any Banach *-algebraic bundle retraction of $\underset{\sim}{B}$.

It is well known that if $\{u_i\}$ is an approximate unit of a Banach algebra A , then $\|u_i a - a\| \to 0$ and $\|a u_i - a\| \to 0$ uniformly ~~in~~ a on norm-compact subsets of A . Our next goal is to prove the same for Banach *-algebraic bundles over locally compact groups.

Proposition 11.1. Suppose that G is locally compact; and let $\{u_i\}$ be an approximate unit of $\underset{\sim}{B}$. Then $\|u_i b - b\| \to 0$ and $\|b u_i - b\|$

$\to 0$ uniformly in b on each compact subset of B.

Proof. Let Γ be the linear span in $\underset{\sim}{C}(\underset{\sim}{B})$ of the set of all cross-sections of $\underset{\sim}{B}$ of the form

$$x \mapsto af(x)b\,, \qquad \ldots(4)$$

where $f \in \underset{\sim}{C}(\underset{\sim}{B})$ and $a,b \in B_e$. By Theorem 10.5 and the fact that $\underset{\sim}{B}$ has an approximate unit,

$$\{g(x) : g\in\Gamma\} \text{ is dense in } B_x \text{ for each } x\,. \qquad \ldots(5)$$

Furthermore, I claim that, for each g in Γ, we have $\|u_i g(x) - g(x)\| \to 0$ and $\|g(x)\,u_i - g(x)\| \to 0$ uniformly in x on any compact subset K of G. Indeed: We need only to prove this when g is of the form (4). But then

$$\|u_i\; g(x) - g(x)\| \le \|u_i a - a\|\;\|b\|\;\sup\,\{\|f(x)\| : x\in K\}\,,$$

$$\|g(x)\,u_i - g(x)\| \le \|bu_i - b\|\;\|a\|\;\sup\,\{\|f(x)\| : x\in K\}\,,$$

and the claim now follows from the fact that $\{u_i\}$ is an approximate unit.

Let K be any compact subset of B, ε a positive number, and k a positive number such that $\|u_i\| \le k$ for all i. For each f in Γ the set $V(f) = \{b\in B : \|b - f(\pi(b))\| < \varepsilon\}$ is open in B. By (5) these sets cover B; so we can find finitely many $f_1,\cdots,f_n$ in Γ such that

$$\bigcup_{j=1}^{n} V(f_j) \supset K\,. \qquad \ldots(6)$$

Now by the above claim there is an index i_0 such that

$$\|u_i\; f_j(x) - f_j(x)\| < \varepsilon\,, \quad \|f_j(x)\,u_i - f_j(x)\| < \varepsilon \qquad \ldots(7)$$

for all $i \succ i_0$, all x in the compact set $\pi(K)$, and all $j = 1,\cdots,n$. Thus, if b is any element of K, we have by (6) $b \in V(f_j)$ for some j; so by (7), putting $x = \pi(b)$ $(\in \pi(K))$,

$$\|u_i b - b\| \leq \|u_i(b - f_j(x))\| + \|u_i f_j(x) - f_j(x)\| + \|b - f_j(x)\|$$

$$\leq k\varepsilon + \varepsilon + \varepsilon = (k+2)\varepsilon, \qquad \ldots(8)$$

and similarly

$$\|bu_i - b\| < (k+2)\varepsilon, \qquad \ldots(9)$$

for all $i \succ i_0$. Since this i_0 does not depend on b, (8) and (9) complete the proof of the proposition. □

Remark. In the terminology of Fell [6], this says that, if G is locally compact, every approximate unit of $\underset{\sim}{B}$ is a strong approximate unit.

Corollary 11.2. *Suppose that G is locally compact and $\{u_i\}$ is an approximate unit of $\underset{\sim}{B}$. Then, for any f in $\underset{\sim}{C}(\underset{\sim}{B})$,*

$$\|u_i f(x) - f(x)\| \to 0 \quad \text{and} \quad \|f(x)u_i - f(x)\| \to 0$$

uniformly in x on any compact subset of G.

Corollary 11.3. *Suppose that G is locally compact, $\{u_i\}$ is an approximate unit of $\underset{\sim}{B}$, and $\{b_i\}$ is a net of elements of B (defined on the same directed set as $\{u_i\}$) converging to b in B. Then $u_i b_i \to b$ and $b_i u_i \to b$ in B.*

Proof. By Theorem 10.5 there is an element f of $\underset{\sim}{C}(\underset{\sim}{B})$ such that

$f(\pi(b)) = b$. So, given $\varepsilon>0$, we have

$$\|b_i - f(\pi(b_i))\| < \varepsilon \quad \text{for all large enough } i . \quad \ldots(10)$$

At the same time, by Cor. 11.2, given a compact neighborhood U of $\pi(b)$, we can find an index i_0 such that

$$i \succ i_0 \Rightarrow \|u_i f(x) - f(x)\| < \varepsilon \quad \text{for all } x \text{ in } U . \ldots(11)$$

Thus, by (10) and (11),

$$\|u_i b_i - f(\pi(b_i))\|$$

$$\leq \|u_i (b_i - f(\pi(b_i)))\| + \|u_i f(\pi(b_i)) - f(\pi(b_i))\|$$

$$\leq k\varepsilon + \varepsilon = (k+1)\varepsilon$$

for all large enough i . (Here k is a constant such that $\|u_i\| \leq k$ for all i .) Therefore by Prop. 10.3 $u_i b_i \rightarrow f(\pi(b)) = b$. Similarly $b_i u_i \rightarrow b$. □

Remark. We do not know whether or not Prop. 11.1 and its corollaries fail when G is not locally compact.

Multipliers.

The notion of a multiplier of $\underset{\sim}{B}$ will be very useful in this work, though not as central as in Fell [6].

Consider a mapping $\lambda:B \rightarrow B$ and an element x of G . We say that λ is of left order x [of right order x] if $\lambda(B_y) \subset B_{xy}$ [$\lambda(B_y) \subset B_{yx}$] for all y in G . If λ is of left (or right) order x , it is called quasi-linear if, for each y in G , $\lambda|B_y$ is linear on B_y to B_{xy} (or B_{yx}). Also, λ is bounded if for some non-negative constant k we have $\|\lambda(b)\| \leq k\|b\|$ for all b in B ; the smallest such k is then called $\|\lambda\|$.

We now make the following definition.

<u>Definition</u>. A <u>multiplier of $\underset{\sim}{B}$ of order</u> x is a pair (λ,μ) where λ and μ are continuous bounded quasi-linear maps of B into B , λ is of left order x , μ is of right order x , and the identities

$$b\,\lambda(c) = \mu(b)c\ ,\quad \lambda(bc) = \lambda(b)c\ ,\quad \mu(bc) = b\,\mu(c) \qquad \ldots(12)$$

hold for all b,c in B .

We call λ and μ the <u>left</u> and <u>right actions</u> of the multiplier (λ,μ) .

As a matter of notation, if $u = (\lambda,\mu)$ is a multiplier we shall usually write ub and bu instead of $\lambda(b)$ and $\mu(b)$. Thus (12) has the form of associative laws:

$$b(uc) = (bu)c\ ,\quad u(bc) = (ub)c\ ,\quad (bc)u = b(cu)\ . \qquad \ldots(13)$$

Let $\underset{\sim}{W}_x = \underset{\sim}{W}_x(\underset{\sim}{B})$ denote the collection of all multipliers of $\underset{\sim}{B}$ of order x . Then $\underset{\sim}{W}(\underset{\sim}{B}) = \cup_{x\in G}\,\underset{\sim}{W}_x(\underset{\sim}{B})$ has much of the structure of a Banach *-algebraic bundle over G . Indeed, each $\underset{\sim}{W}_x$ is an obvious linear space. Setting $\|u\|_0 = \max(\|\lambda\|, \|\mu\|)$ whenever $u = (\lambda,\mu) \in \underset{\sim}{W}(\underset{\sim}{B})$, one verifies that each $\underset{\sim}{W}_x$ is a Banach space under $\|\ \|_0$. If $u \in \underset{\sim}{W}_x$ and $v \in \underset{\sim}{W}_y$ $(x,y \in G)$, the product uv defined by

$$(uv)b = u(vb)\ ,\quad b(uv) = (bu)v \quad (b\in B) \qquad \ldots(14)$$

belongs to $\underset{\sim}{W}_{xy}$; and the involution u^* given by

$$u^*b = (b^*u)^*\ ,\quad bu^* = (ub^*)^* \quad (b\in B) \qquad \ldots(15)$$

belongs to $\underset{\sim}{W}_{x^{-1}}$. These operations on multipliers satisfy conditions (i)-(iv) and (vi)-(x) of the definition of a Banach *-algebraic bundle. Thus $\underset{\sim}{W}(\underset{\sim}{B})$ lacks only a suitable topology in order to become a Banach *-algebraic bundle over G . In general we do not know how to find such a topology. In spite of this it is convenient to refer to $\underset{\sim}{W}(\underset{\sim}{B})$ as the <u>multiplier bundle</u> of $\underset{\sim}{B}$.

It should be noted that $\underset{\sim}{W}(\underset{\sim}{B})$ does have one useful topology, namely the <u>strong multiplier topology</u>, in which a net $\{w_i\}$ of elements of $\underset{\sim}{W}(\underset{\sim}{B})$ converges to $w \in \underset{\sim}{W}(\underset{\sim}{B})$ if and only if $w_i b \to wb$ and $bw_i \to bw$ in B for every b in B .

The multiplier whose left and right actions are the identity map on B is the unit element of $\underset{\sim}{W}(\underset{\sim}{B})$.

Each element b of B gives rise to a multiplier u_b of $\underset{\sim}{B}$ whose order is the same as that of b :

$$u_b c = bc , \quad c u_b = cb \quad (c \in B) \quad . \qquad \ldots(16)$$

The map $b \mapsto u_b$ preserves addition (on each fiber), multiplication, and involution; and $\|u_b\|_0 \le \|b\|$.

<u>Remark</u>. In particular, if A is a Banach *-algebra (considered as usual as a Banach *-algebraic bundle over the one-element group), then $\underset{\sim}{W}(A)$ (with the operations (14), (15) and the norm $\| \ \|_0$) is a Banach *-algebra, called the <u>multiplier *-algebra</u> of A . The map $b \mapsto u_b$ $(b \in A)$ is then a *-homomorphism of A into $\underset{\sim}{W}(A)$. In fact the range of this *-homomorphism is a *-ideal of A ; for one easily verifies the relations (valid for any Banach *-algebraic bundle):

$$v u_b = u_{vb} , \quad u_b v = u_{bv} \quad (b \in A;\ v \in \underset{\sim}{W}(A)) \quad . \qquad \ldots(17)$$

Remark. Assume that B has no annihilators, that is, if $0_{\pi(b)} \neq b \in B$ there exist a and c in B satisfying $ab \neq 0_{\pi(ab)}$, $bc \neq 0_{\pi(bc)}$. Then various stronger statements can be made. In the first place, the last two of the three identities (12) can be removed from the definition of a multiplier; secondly, a multiplier is determined by its left action, and also by its right action; and thirdly, the map $b \mapsto u_b$ is one-to-one, so that B can be identified with a subset of $W(B)$. If B has an approximate unit it of course has no annihilators. If B has a unit 1, then $b \mapsto u_b$ is not only one-to-one but onto $W(B)$; and for any w in $W(B)$ we have $w = u_b$, where $b = w1 = 1w$. The proofs of these facts are left to the reader; they are the same as those well known in the purely algebraic case (see Hochschild [1]).

Proposition 11.4. *Suppose that B has an approximate unit $\{u_i\}$ and that, for every x in G, there is a multiplier w of B of order x having an inverse (that is, a multiplier w^{-1} of B of order x^{-1} such that $w^{-1}w = ww^{-1} = 1$). Then B is saturated.*

Proof. Given $x \in G$, let w be an invertible multiplier of order x. For each a in B_e we have

$$a = \lim_i au_i = \lim_i [(aw)(w^{-1}u_i)] ,$$

where $aw \in B_x$ and $w^{-1}u_i \in B_{x^{-1}}$. So $B_xB_{x^{-1}}$ is dense in B_e. By the remark following the definition of saturation this proves that B is saturated. □

Let 1 be the unit element of $W(B)$. A multiplier u of B will be called unitary if $u^*u = uu^* = 1$ and $\|u\|_0 \leq 1$. We denote the set of all unitary multipliers of B by $U(B)$. Thus $U(B)$ is a group under the multiplication operation of $\underset{\sim}{W}(\underset{\sim}{B})$, the inverse in $\underset{\sim}{U}(\underset{\sim}{B})$ being involution. The map $\pi_0 : \underset{\sim}{U}(\underset{\sim}{B}) \to G$ sending each unitary multiplier into its order is a group homomorphism.

Definition. We say that $\underset{\sim}{B}$ has enough unitary multipliers if for every x in G there exists a unitary multiplier of $\underset{\sim}{B}$ of order x -- that is, if the group homomorphism π_0 of the last paragraph is onto G .

Let u be a unitary multiplier of $\underset{\sim}{B}$ of order x . The left action of u is norm-decreasing and its inverse is the norm-decreasing left action of u^* . Therefore the left action of u is an isometry of B onto B , and carries each fiber B_y linearly and isometrically onto B_{xy} . Likewise the right action of u carries each fiber B_y linearly and isometrically onto B_{yx} .

It follows that, if $\underset{\sim}{B}$ has enough unitary multipliers, the fibers of $\underset{\sim}{B}$ are all isometrically isomorphic with each other.

Proposition 11.5. If $\underset{\sim}{B}$ has an approximate unit and enough unitary multipliers, it is saturated.

Proof. By Prop. 11.4. □

Remark. If $\underset{\sim}{B}$ has a unit element 1 , a unitary multiplier of $\underset{\sim}{B}$ is simply an element u of B which is "unitary" in the sense that $u^* u = uu^* = 1$ and $\|ub\| \le \|b\|$, $\|bu\| \le b$ for all b .

Remark. The property of having enough unitary multipliers is somewhat weaker than the property of homogeneity which was central in Fell [6]. The definition of homogeneity in Fell [6], §6, included a condition on the so-called strong topology of $\underset{\sim}{U}(\underset{\sim}{B})$.

C*-algebraic bundles.

A very important role in functional analysis is played by those Banach *-algebras A which are C*-algebras, that is, which satisfy the additional identity $\|a^*a\| = \|a\|^2$ $(a \in A)$. Analogously we will define a special class of Banach *-algebraic bundles called C*-algebraic bundles.

Definition. A C*-algebraic bundle over G is a Banach *-algebraic bundle $\mathcal{B} = (B, \pi, \cdot, ^*)$ over G satisfying (i) $\|b^*b\| = \|b\|^2$ for all b in B, and (ii) b^*b is positive in B_e for all b in B.

Remark. In view of condition (i), the unit fiber *-algebra B_e of $\mathcal{B}$ is a C*-algebra; and the positivity of b^*b referred to in (ii) is to be understood as ordinary positivity of elements of a C*-algebra (see for example Dixmier [1], 1.6).

Remark. Postulate (ii) in the preceding definition is independent of postulate (i). This is illustrated by Example (IV) of the next section, which satisfies (i) but not (ii).

Remark. An isomorphism between two C*-algebraic bundles is automatically an isometric isomorphism. This follows from (i) and the corresponding fact about C*-algebras.

Generalizing the fact that a C*-algebra has an approximate unit, we have:

Proposition 11.6. *Every C*-algebraic bundle has an approximate unit.*

Proof. Since B_e is a C*-algebra, it has an approximate unit $\{u_i\}$ satisfying $u_i^* = u_i$, $\|u_i\| \le 1$ (see Dixmier [1], 1.7.2). Given any b in B we have:

$$\|bu_i - b\|^2 = \|(bu_i-b)^*(bu_i-b)\|$$

$$= \|u_ib^*bu_i - b^*bu_i - u_ib^*b + b^*b\|$$

$$\leq \|u_i\| \, \|b^*bu_i - b^*b\| + \|b^*bu_i - b^*b\| . \qquad ...(18)$$

Since $b^*b \in B_e$ (so that $b^*bu_i \to b^*b$) and $\|u_i\| \leq 1$, (18) implies that $\|bu_i-b\| \to 0$, or $bu_i \to b$, for all b in B. Applying involution, we find in addition that $u_ib^* \to b^*$ for all b. So $\{u_i\}$ is an approximate unit of $\underset{\sim}{B}$. □

§12. Examples.

In this section we shall illustrate the concepts of the preceding section with various examples of Banach *-algebraic bundles.

(I) The most useful and frequently encountered general class of examples is the class of semidirect products.

Fix a topological group G with unit e, a Banach *-algebra A, and a homomorphism ι of G into the group of isometric *-automorphisms of A. We assume that ι is strongly continuous, that is, the map $x \mapsto \iota_x(a)$ is continuous on G to A for each a in A. From this and the isometric character of the ι_x it follows that $(a,x) \mapsto \iota_x(a)$ is (jointly) continuous.

Let (B,π) be the trivial Banach bundle over G whose constant fiber is the Banach space underlying A; thus $B = A \times G$, $\pi(a,x) = x$ (see §10). We now define multiplication $\cdot$ and involution * in B as follows:

$$(a,x)\cdot(b,y) = (a\,\iota_x(b),\, xy), \qquad ...(1)$$

$$(a,x)^* = (\iota_{x^{-1}}(a^*),\, x^{-1}) \qquad ...(2)$$

$(x,y \in G;\ a,b \in A)$. Using the assumed properties of ι one checks easily that $\underset{\sim}{B} = (B,\pi,\cdot,*)$ is a Banach $*$-algebraic bundle over G . For example, the isometric character of $\iota_{x^{-1}}$ gives

$$\|(a,x)^*\| = \|\iota_{x^{-1}}(a^*)\| = \|a^*\| = \|a\| = \|(a,x)\| .$$

Furthermore

$$\begin{aligned}((a,x)\cdot(b,y))^* &= (a\,\iota_x(b),\ xy)^* \\ &= (\iota_{y^{-1}x^{-1}}(a\,\iota_x(b))^*,\ y^{-1}x^{-1}) \\ &= (\iota_{y^{-1}x^{-1}}(\iota_x(b^*)a^*),\ y^{-1}x^{-1}) \\ &= (\iota_{y^{-1}}(b^*)\ \iota_{y^{-1}x^{-1}}(a^*),\ y^{-1}x^{-1}) \\ &= (\iota_{y^{-1}}(b^*),\ y^{-1})\cdot(\iota_{x^{-1}}(a^*),\ x^{-1}) \\ &= (b,y)^*\cdot(a,x)^* ,\end{aligned}$$

proving postulate (viii) of a Banach $*$-algebraic bundle. The other postulates are similarly verified.

Definition. This $\underset{\sim}{B}$ is called the ι-semidirect product of A and G . It may be denoted by $A \times_\iota G$.

Remark. We shall usually write simply $(a,x)(b,y)$ in (1), omitting explicit mention of $\cdot$.

Remark. Notice the strong likeness between the above construction and that of the semidirect products of groups.

If A is saturated (i.e. $[AA]$ is dense in A), then the semidirect product $\underset{\sim}{B}=A \times_{\iota} G$ is also saturated. If A has an approximate unit $\{u_i\}$, then $\{(u_i,e)\}$ is easily seen to be an approximate unit of $\underset{\sim}{B}$. If A is a C*-algebra, $\underset{\sim}{B}$ is a C*-algebraic bundle.

Each x in G gives rise to a unitary multiplier m_x of $\underset{\sim}{B}$, whose left and right actions are given by:

$$\left.\begin{aligned} m_x(a,y) &= (\iota_x(a),xy), \\ (a,y)m_x &= (a,yx) \end{aligned}\right\} \qquad \ldots(3)$$

$((a,y) \in B)$. One verifies that $x \mapsto m_x$ is a homomorphism of G into the group of unitary multipliers, and is continuous with respect to the strong multiplier topology. Thus the semidirect product bundles always have enough unitary multipliers.

Take the special case that $\iota_x(a) = a$ for all x in G and a in A. Then formulae (1), (2) become:

$$(a,x)(b,y) = (ab,xy), \quad (a,x)^* = (a^*,x^{-1}), \qquad \ldots(4)$$

and $A \times_{\iota} G$ is then called the <u>direct product of</u> A <u>with</u> G.

The direct product of $\mathbb{C}$ with G is called the <u>group bundle of</u> G; it is a saturated C*-algebraic bundle with unit.

As we shall repeatedly see later on, it is through the group bundle of G that the theory of Banach *-algebraic bundles can be regarded as a generalization of the theory of topological groups.

Here is a useful special class of semidirect product bundles. Let M be a locally compact Hausdorff space, on which the topological group G acts to the left as a topological transformation group; and let A denote the commutative C*-algebra $\underset{\sim}{C}_0(M)$. The action of G on M gives rise to an action ι of G on A:

$$(\iota_x f)(m) = f(x^{-1}m) \qquad (f\in A;\ x\in G;\ m\in M) .$$

Evidently this ι satisfies the hypotheses on ι laid down above; so we can form $A \times_\iota G$. This semidirect product bundle is called the G,M <u>transformation bundle</u>. It is a special case of the more general transformation bundles to be studied in §29.

(II) Of importance in the theory of projective representations of groups are the so-called <u>circle extension bundles</u>.

Let G be a topological group (with unit e), and $\gamma: \mathbb{E} \underset{i}{\longrightarrow} H \underset{j}{\longrightarrow} G$ a central extension of the circle group $\mathbb{E}$ by G . From γ we shall now construct a C^*-algebraic bundle $\underset{\sim}{B}_\gamma$ over G with one-dimensional fibers. Note first that $\mathbb{E}$ acts as a topological transformation group on $\mathbb{C} \times H$:

$$u(\lambda,h) = (u^{-1}\lambda,\ i(u)h) \qquad (u\in\mathbb{E};\ \lambda\in\mathbb{C};\ h\in H) .$$

Let B be the space of all orbits in $\mathbb{C} \times H$ under this action, with the quotient topology; and let $\rho:\mathbb{C} \times H \to B$ be the (continuous open) quotient map. One verifies that B is Hausdorff, and that the equation

$$\pi(\rho(\lambda,h)) = j(h) \qquad (\lambda\in\mathbb{C};\ h\in H)$$

defines a continuous open surjection $\pi:B \to G$. For fixed h in H we shall transfer the Banach space structure of $\mathbb{C}$ to $\pi^{-1}(j(h)) = \{\rho(\lambda,h) : \lambda\in\mathbb{C}\}$ via the bijection $\lambda \mapsto \rho(\lambda,h)$; this Banach space structure depends only on j(h) (since $\lambda \mapsto u^{-1}\lambda$ is a Banach space isometry of $\mathbb{C}$ onto itself). When the fibers $\pi^{-1}(x)$ $(x\in G)$ are thus made into Banach spaces, (B,π) becomes a Banach bundle over G with one-dimensional fibers. Defining multiplication and involution in B in analogy with the product formulae (4):

$$\rho(\lambda,h)\cdot\rho(\mu,k) = \rho(\lambda\mu,hk), \quad (\rho(\lambda,h))^* = \rho(\bar{\lambda},h^{-1})$$

$(\lambda,\mu\in\mathbb{E};\ h,k\in H)$, we check that $\underset{\sim}{B}_\gamma = (B,\pi,\cdot,{}^*)$ becomes a Banach *-algebraic bundle over G . It is called the circle extension bundle over G corresponding to the extension γ .

If γ is the trivial extension of $\mathbb{E}$ by G , $\underset{\sim}{B}_\gamma$ becomes just the group bundle of G .

Notice that $\underset{\sim}{B}_\gamma$ is a C*-algebraic bundle. It has a unit element $\rho(1,i(1))$; and $\rho(1,h)$ is a unitary element of $\underset{\sim}{B}_\gamma$ for every h in H . So $\underset{\sim}{B}_\gamma$ has enough unitary multipliers and is saturated.

One can prove without much difficulty (though we shall not do so) the following converse fact.

Proposition 12.1. Let G be any topological group, and $\underset{\sim}{B}$ any saturated C*-algebraic bundle over G whose fibers are all one-dimensional. Then there is a central extension γ of $\mathbb{E}$ by G such that $\underset{\sim}{B}$ and $\underset{\sim}{B}_\gamma$ are isometrically isomorphic.

(III) Another important class of Banach *-algebraic bundles is that of the group extension bundles. They are defined in Fell [6], §10, Example 3; but we shall not repeat their definition here, since they will not be needed in what follows.

Remark. Semidirect product bundles, circle extension bundles, and group extension bundles are all homogeneous in the sense of Fell [6], §6. The most general homogeneous Banach *-algebraic bundle can be obtained by a construction using group extensions, in a manner generalizing the construction of circle extension bundles in (II); see Fell [6], §9.

(IV) In preparation for the next, very specific examples, we make brief mention of arbitrary Banach *-algebraic bundles over <u>discrete</u> groups.

Let E be a Banach *-algebra and G a discrete group; and for each x in G let a closed linear subspace E_x of E be given such that:

$$E_x E_y \subset E_{xy}, \quad (E_x)^* \subset E_{x^{-1}} \ (x,y \in G) . \qquad \ldots(5)$$

This immediately gives rise to a Banach *-algebraic bundle $\underset{\sim}{B} = (B,\pi,\cdot,^*)$ over G whose fiber B_x over each x is just E_x, whose bundle space B is the disjoint union of the E_x $(x \in G)$, and whose multiplication and involution are the restrictions to B of the corresponding operations in E. We shall observe in §13 that every Banach *-algebraic bundle over a discrete group is obtained in this way.

If E is a C^*-algebra, then $\underset{\sim}{B}$ is a C^*-algebraic bundle.

(V) As a specific instance of (IV), let us consider the two-dimensional Banach *-algebra E whose underlying linear space is $\mathbb{C}^2$, and whose operations and norm are:

$$(\lambda,\mu)(\lambda',\mu') = (\lambda\lambda',\mu\mu'),$$

$$(\lambda,\mu)^* = (\bar{\mu},\bar{\lambda})$$

$$\|(\lambda,\mu)\| = \max(|\lambda|,|\mu|)$$

$(\lambda,\mu \in \mathbb{C})$. Thus $E_1 = \{(\lambda,\lambda) : \lambda \in \mathbb{C}\}$ and $E_{-1} = \{(\lambda,-\lambda) : \lambda \in \mathbb{C}\}$ are one-dimensional subspaces of E satisfying $E_1 E_{\pm 1} = E_{\pm 1} = E_{\pm 1} E_1$, $E_{-1}E_{-1} = E_1$, $E_{\pm 1}^* = E_{\pm 1}$. By (IV) this gives us a Banach *-algebraic bundle $\underset{\sim}{B}$ over the multiplicative two-element group $G = \{1,-1\}$, with fibers E_1 and E_{-1}. Notice that $\underset{\sim}{B}$ has a unit $(1,1)$ and that both E_1 and E_{-1} have invertible elements, but that E_{-1} has no unitary elements. Thus $\underset{\sim}{B}$ is saturated but does not have enough unitary multipliers.

(VI) Here is another example. Let D be the unit disk $\{z \in \mathbb{C} : |z| \leq 1\}$, and $\mathbb{E}$ as usual its boundary $\{z \in \mathbb{C} : |z| = 1\}$. Let C be the commutative C^*-algebra $\underset{\sim}{C}(D)$ of all continuous complex functions on D, with the pointwise operations. Let G be the two-element group $\{1,-1\}$; and define the following closed linear subspaces C_1 and C_{-1} of C:

$$C_1 = \{f \in C : f(-z) = f(z) \text{ for all } z \text{ in } \mathbb{E}\},$$

$$C_{-1} = \{f \in C : f(-z) = -f(z) \text{ for all } z \text{ in } \mathbb{E}\}.$$

Evidently C_1 and C_{-1} satisfy (5), and so form the fibers of a C^*-algebraic bundle $\underset{\sim}{B}$ over G. $\underset{\sim}{B}$ has a unit element; and the Stone-Weierstrass Theorem shows that the linear span of $C_{-1}C_{-1}$ is dense in C_1. Therefore $\underset{\sim}{B}$ is saturated.

However C_{-1} contains no invertible elements. To see this it is enough to show that every element of C_{-1} vanishes at some point of D. Let $f \in C_{-1}$, and assume that f never vanishes on D. Then, for each z in $\mathbb{E}$ we can denote by $\alpha(z)$ the angle subtended at 0 by the image under f of the straight line path from z to $-z$. Since $f(-z) = -f(z)$, we have $\alpha(z) = \pi + 2\pi n_z$ for some integer n_z. By the continuity of f, α is continuous on $\mathbb{E}$. But $\alpha(z)$ must reverse its sign as z passes continuously in $\mathbb{E}$ from, say, 1 to -1. The last three facts give a contradiction. So f vanishes somewhere on D.

Thus C_{-1} contains no invertible elements, and hence in particular no unitary elements.

(VII) We conclude this section with an interesting class of finite-dimensional saturated bundles.

Let $X_1,\cdots,X_p$ be a finite sequence of finite-dimensional non-zero Hilbert spaces, and $X = \sum_{m=1}^{\oplus p} X_m$ their direct sum. Let G be a finite group (with unit e) acting (to the left) as a transformation group on $M = \{1,2,3,\cdots,p\}$; and for each x in G let E_x be the linear subspace of the C*-algebra $\mathcal{O}(X)$ consisting of those b such that $b(X_m) \subset X_{xm}$ for all m in M and all x in G. One verifies easily that relations (5) hold, so that the E_x form the fibers of a C*-algebraic bundle $\mathcal{B}$ over G. The unit fiber *-algebra E_e of $\mathcal{B}$ is the *-subalgebra of $\mathcal{O}(X)$ consisting of those b which leave stable each $X_m (m = 1,\cdots,p)$; thus E_e is the most general finite-dimensional C*-algebra. Notice that $\mathcal{B}$ is saturated; this follows from the fact that each X_m is non-zero.

We notice also that, for $x \in G$,

$$\dim(E_x) = \sum_{m=1}^{p} \dim(X_m)\dim(X_{xm}) .$$

By Schwarz's Inequality this implies

$$\dim(E_x) \leq \dim(E_e) = \sum_{m=1}^{p} (\dim(X_m))^2 . \qquad \ldots(6)$$

With a suitable choice of G and the $\dim(X_m)$, one can easily ensure that strict inequality holds in (6) for certain x. This shows that in saturated Banach *-algebraic bundles, unlike Banach *-algebraic bundles with enough unitary multipliers, the fibers need not be isomorphic as linear spaces.

In §33 we shall construct a continuous generalization of this example.

§13. The cross-sectional algebra.

As we shall now see, every Banach *-algebraic bundle over a locally compact group gives rise to a cross-sectional algebra analogous to the $\underset{\sim}{L}_1$ group algebra of a locally compact group.

We suppose G is a locally compact group with unit e, left Haar measure λ, and modular function Δ, and that $\underset{\sim}{B} = (B, \pi, \cdot, ^*)$ is a fixed Banach *-algebraic bundle over G. By Theorem 10.5 $\underset{\sim}{B}$ automatically has enough continuous cross-sections. One can then form the cross-sectional $\underset{\sim}{L}_1$-space $\underset{\sim}{L}_1(\underset{\sim}{B};\lambda)$ of the underlying Banach bundle (B,π) as in §10; and we know from Fell [6], §8, that the multiplication and involution in $\underset{\sim}{B}$ induce a multiplication $*$ and an involution * in $\underset{\sim}{L}_1(\underset{\sim}{B};\lambda)$, making the latter a Banach *-algebra. By Prop. 8.1 of Fell [6], these operations are given by:

$$(f*g)(x) = \int_G f(y)\, g(y^{-1}x)\, d\lambda y\,, \qquad \ldots(1)$$

$$f^*(x) = \Delta(x^{-1})(f(x^{-1}))^* \qquad \ldots(2)$$

$(f,g \in \underset{\sim}{L}_1(\underset{\sim}{B};\lambda);\ x \in G)$.

Remark 1. Notice that the integrand $f(y)\, g(y^{-1}x)$ in (1) belongs to B_x for all y. Equation (1) asserts more precisely: (i) For any f,g in $\underset{\sim}{L}_1(\underset{\sim}{B};\lambda)$, the function $y \mapsto f(y)\, g(y^{-1}x)$ belongs to $\underset{\sim}{L}_1(G;B_x;\lambda)$ for λ-almost all x; and (ii) for λ-almost all x (1) holds (the right side of (1) existing as a B_x-valued Bochner integral for λ-almost all x in virtue of (i)).

Definition. $\underset{\sim}{L}_1(\underset{\sim}{B};\lambda)$, with the operations (1) and (2), is called the $\underset{\sim}{L}_1$ cross-sectional algebra of $\underset{\sim}{B}$.

If $f,g \in \underset{\sim}{L}(\underset{\sim}{B})$, it follows from (1) and Lemma 10.14 that $f*g$ is continuous and hence in $\underset{\sim}{L}(\underset{\sim}{B})$. Thus (see Prop. 10.11) $\underset{\sim}{L}(\underset{\sim}{B})$ is a dense *-subalgebra of $\underset{\sim}{L}_1(\underset{\sim}{B};\lambda)$.

Definition. We call $\underset{\sim}{L}(\underset{\sim}{B})$ the compacted cross-sectional algebra of $\underset{\sim}{B}$.

Remark 2. If $\underset{\sim}{B}$ is the group bundle of G , $\underset{\sim}{L}_1(\underset{\sim}{B};\lambda)$ is just the usual $\underset{\sim}{L}_1$ group algebra of G , consisting of all complex-valued λ-summable functions on G with the operations of convolution (1) and involution (2). In the general case, the operation (1) is called generalized convolution.

Remark 3. Suppose that G is discrete. Then $\underset{\sim}{L}_1(\underset{\sim}{B};\lambda)$ consists of all cross-sections f of $\underset{\sim}{B}$ such that $\Sigma_{x\in G}\|f(x)\| < \infty$. In that case each fiber B_x can be identified with a closed linear subspace of $\underset{\sim}{L}_1(\underset{\sim}{B};\lambda)$; and the multiplication and convolution in B are just the restrictions to B of the corresponding operations of $\underset{\sim}{L}_1(\underset{\sim}{B};\lambda)$. This, incidentally, verifies the remark made in §12 (IV) -- that every Banach *-algebraic bundle over a discrete group is obtained by the construction of §12 (IV).

Semidirect product bundles.

The form of the operations on $\underset{\sim}{L}_1(\underset{\sim}{B};\lambda)$ in the semidirect product context deserves special mention.

Let A and ι be as in §12 (I), and form the semidirect product $\underset{\sim}{B} = A \times_{\iota} G$. As usual we identify cross-sections of $\underset{\sim}{B}$ (which as a Banach bundle is trivial) with functions on G to A , so that

$$\underset{\sim}{L}_1(\underset{\sim}{B};\lambda) \cong \underset{\sim}{L}_1(G, A;\lambda) \ . \qquad \ldots(3)$$

With this identification, the product formulae (1) and §12(1) combine to give

$$(f*g)(x) = \int_G f(y) \ \iota_y(g(y^{-1}x))d\lambda y \ , \qquad \ldots(4)$$

while (2) and §12(2) together give

$$f^*(x) = \Delta(x^{-1})\iota_x[(f(x^{-1}))^*] \ . \qquad \ldots(5)$$

The algebraic operations on the right of (4) and (5) are of course performed in A .

These cross-sectional algebras of semidirect product bundles are called <u>covariance algebras</u> by Doplicher, Kastler, and Robinson [1]. They have important applications to physics.

Especially important among the semidirect product bundles are the transformation bundles. Let M be a locally compact Hausdorff space, on which G acts to the left as a topological transformation group; let $A = \underset{\sim}{C}_0(M)$; and let $\underset{\sim}{B} = A \times_{\iota} G$ be the G,M transformation bundle of §12(I). In applying (4) and (5) to describe $\underset{\sim}{L}_1(\underset{\sim}{B};\lambda)$, it is useful to begin with the linear space $E = \underset{\sim}{L}(G \times M)$. Identifying an element f of E with the function $x \mapsto f(x,\cdot)$ on G to A (and noting that this function belongs to $\underset{\sim}{L}(\underset{\sim}{B})$), we can and will consider E as a linear subspace of $\underset{\sim}{L}(\underset{\sim}{B})$. It follows from Prop. 10.12 that E is in fact dense in $\underset{\sim}{L}_1(\underset{\sim}{B};\lambda)$. Applying formulae (4) and (5) to elements of E , we get

$$(f*g)(x,m) = \int_G f(y,m) \ g(y^{-1}x, \ y^{-1}m)d\lambda y \ ,$$

$$f^*(x,m) = \Delta(x^{-1}) \ \overline{f(x^{-1}, \ x^{-1}m)}$$

$(f,g \in E; x \in G; m \in M)$. These operations clearly leave E stable.
Hence with these operations E is a dense *-subalgebra of $\underset{\sim}{L}_1(\underset{\sim}{B};\lambda)$.
The norm of $\underset{\sim}{L}_1(\underset{\sim}{B};\lambda)$ is given on E by the equation

$$\|f\| = \int_G \sup\{|f(x,m)| : m \in M\}\,d\lambda x \quad .$$

We may now regard $\underset{\sim}{L}_1(\underset{\sim}{B};\lambda)$ as the Banach *-algebra completion of E
with respect to this norm.

We call E the <u>compacted transformation algebra</u> of G,M , and
$\underset{\sim}{L}_1(\underset{\sim}{B};\lambda)$ the $\underset{\sim}{L}_1$ <u>transformation algebra</u> of G,M . This construction
will be considerably generalized in §29.

These transformation algebras were first constructed by Glimm [1],
and later studied by several writers including Effros and Hahn [1].

<u>Remark 4</u>. Suppose that the group G is finite. The transformation
algebra E of G,G (G being considered as acting on itself by left
multiplication) then consists of all complex functions on $G \times G$; and
the preceding formulae for the algebraic operations on E become:

$$(f*g)(x,y) = \sum_{z \in G} f(z,y)\; g(z^{-1}x,\; z^{-1}y) \;,$$

$$f^*(x,y) = \overline{f(x^{-1},\; x^{-1}y)}$$

$(f,g \in E; x,y \in G)$. For each u,v in G let e_{uv} be the element
of E given by

$$e_{uv}(x,y) = \delta_{u^{-1}v,x}\; \delta_{u^{-1},y} \quad (x,y \in G) \;.$$

Under multiplication and involution the e_{uv} behave as follows:

$$e_{uv} * e_{rs} = \delta_{vr}\, e_{us} \;,$$

$$(e_{uv})^* = e_{vu} \;;$$

that is, the e_{uv} are canonical basis vectors of a total matrix *-algebra.

Thus the transformation algebra of G,G *is *-isomorphic to the* $n \times n$ *total matrix *-algebra, where* n *is the order of* G.

Approximate units in $\mathcal{L}_1(\mathcal{B};\lambda)$.

Returning to the case of an arbitrary Banach *-algebraic bundle $\mathcal{B}$ over G, we ask whether $\mathcal{L}_1(\mathcal{B};\lambda)$ has an approximate unit. The answer is as simple as one could expect:

Theorem 13.1. *If* $\mathcal{B}$ *has an approximate unit, then* $\mathcal{L}_1(\mathcal{B};\lambda)$ *has an approximate unit.*

Proof. Combine Prop. 8.2 of Fell [6] with the Remark following Prop. 11.1 of the present work. □

Remark 5. Assume that $\mathcal{B}$ has an approximate unit. Examining the proof of Fell [6], Prop. 8.2, we find that $\mathcal{L}_1(\mathcal{B};\lambda)$ has an approximate unit $\{u_i\}$ consisting of elements u_i of $\mathcal{L}(\mathcal{B})$; and that in fact the u_i can be chosen so that in addition, for all f in $\mathcal{L}(\mathcal{B})$,

$$u_i * f \to f, \quad f * u_i \to f, \qquad \ldots(6)$$

and

$$u_i^* * f * u_i \to f \qquad \ldots(7)$$

in the inductive limit topology.

Integrated forms of multipliers.

As one might expect, multipliers on a Banach *-algebraic bundle give rise to corresponding multipliers on the $\mathcal{L}_1$ cross-sectional algebra.

Let $\underset{\sim}{B}$ be a Banach *-algebraic bundle over G. Let u be a multiplier of $\underset{\sim}{B}$ of order x. Then u has a natural left and right action on arbitrary cross-sections of $\underset{\sim}{B}$. Indeed, if f is any cross-section of $\underset{\sim}{B}$, the equations

$$(uf)(y) = u\, f(x^{-1}y) \,, \qquad \ldots(8)$$

$$(fu)(y) = \Delta(x^{-1})\; f(yx^{-1})u \qquad \ldots(9)$$

define cross-sections uf and fu of $\underset{\sim}{B}$. Since the left and right actions of u on B are continuous, uf and fu will be continuous if f is. Therefore, by the boundedness of the actions of u, uf and fu are locally λ-measurable whenever f is (see §10). Thus, if $f \in \underset{\sim}{L}_1(\underset{\sim}{B};\lambda)$,

$$\|uf\|_1 = \int \|uf(x^{-1}y)\| d\lambda y \;\leq\; \|u\|_0 \int \|f(x^{-1}y)\| d\lambda y = \|u\|_0 \; \|f\|_1 \,, \qquad \ldots(10)$$

$$\|fu\|_1 = \Delta(x^{-1}) \int \|f(yx^{-1})u\| d\lambda y \;\leq\; \Delta(x^{-1})\, \|u\|_0 \int \|f(yx^{-1})\| d\lambda y = \|u\|_0 \; \|f\|_1 \;; \qquad \ldots(11)$$

from which it follows that $f \mapsto uf$ and $f \mapsto fu$ are bounded linear endomorphisms of $\underset{\sim}{L}_1(\underset{\sim}{B};\lambda)$. It is easily verified that the three identities

$$u(f*g) = (uf) * g,\; (f*g)u = f * (gu),\; (fu) * g = f * (ug)$$

hold for all f,g in $\underset{\sim}{L}(\underset{\sim}{B})$, and hence by continuity for all f,g in $\underset{\sim}{L}_1(\underset{\sim}{B};\lambda)$. So the left and right actions $f \mapsto uf$ and $f \mapsto fu$ define a bounded multiplier m_u of $\underset{\sim}{L}_1(\underset{\sim}{B};\lambda)$.

<u>Definition</u>. m_u is called the <u>integrated</u> <u>form</u> of u.

Evidently the map $u \mapsto m_u$ preserves the product and involution of multipliers, and also the linear operations on each "fiber" $W_x(B)$. By (10) and (11) $\|m_u\|_0 \le \|u\|_0$. In fact the reader will verify that

$$\|m_u\|_0 = \|u\|_0 . \qquad \ldots(12)$$

The action of m_u leaves $L(B)$ stable; so m_u can also be considered as a multiplier of the compacted cross-sectional algebra $L(B)$.

In particular, every element u of B_x gives rise via (8) and (9) to a multiplier m_u of $L_1(B;\lambda)$ (or of $L(B)$) satisfying

$$\|m_u\|_0 \le \|u\| ; \qquad \ldots(13)$$

and the map $u \mapsto m_u$ preserves all the *-algebraic structure of B.

Suppose that $f,g \in L(B)$. One easily checks that the map

$$x \mapsto m_{f(x)}g \qquad (x \in G)$$

has compact support and is continuous with respect to the inductive limit topology of $L(B)$. In fact, the reader will verify without difficulty (after recalling the last Remark of §10):

<u>Proposition 13.2</u>. <u>If</u> $f,g \in L(B)$, <u>then</u>

$$f*g = \int_G (m_{f(x)}g)\,d\lambda x , \qquad \ldots(14)$$

<u>the right side of</u> (14) <u>being an</u> $L(B)$<u>-valued Bochner integral with respect to the inductive limit topology of</u> $L(B)$.

§14. *-Representations.

It is now time to introduce the concept of a *-representation of a Banach *-algebraic bundle, generalizing that of a unitary representation of a group.

Again we fix a Banach *-algebraic bundle $\underset{\sim}{B} = (B, \pi, \cdot, ^*)$ over a topological group G, with unit e.

Definition. A *-representation of $\underset{\sim}{B}$ acting in a Hilbert space X is a mapping $T: B \longrightarrow \underset{\sim}{O}(X)$ such that: (i) $T|B_x$ is linear on B_x for each x in G; (ii) $T_b T_c = T_{bc}$ $(b, c \in B)$; (iii) $(T_b)^* = T_{b^*}$ $(b \in B)$; (iv) for each ξ in X the map $b \mapsto T_b \xi$ is continuous from B to X.

We call X the space of T, and denote it by $X(T)$.

If T is a *-representation of $\underset{\sim}{B}$, we have, as in the case of Banach *-algebras,

$$\|T_b\| \leq \|b\| \qquad (b \in B). \qquad \dots(1)$$

Indeed, postulates (i)-(iii) in the preceding definition imply in particular that $T|B_e$ is a *-representation of the Banach *-algebra B_e. Thus (1) holds if $b \in B_e$. For arbitrary b in B we have $b^* b \in B_e$, and hence

$$\|T_b\|^2 = \|T_b^* T_b\| = \|T_{b^*b}\| \leq \|b^* b\| \leq \|b\|^2 .$$

As a result of (1), the condition (iv) in the definition of a *-representation can be weakened. In fact I claim that the definition is unaltered if (iv) is replaced by:

(iv′) There is a dense subset X' of X such that $b \mapsto \langle T_b \xi, \xi \rangle$ is continuous on B to $\mathbb{C}$ for all ξ in X'.

Indeed: Obviously (iv) => (iv′). Assume conditions (i)-(iii) and (iv′). Then (1) holds, since (iv) was not used in the proof of (1). It follows from (1) that the set of ξ for which $b \mapsto \langle T_b \xi, \xi \rangle$ is continuous is closed in X, and hence by (iv′) is equal to X. Thus, if $\xi \in X$ and $b_i \to b$ in X,

$$\|T_{b_i} \xi - T_b \xi\|^2 = \langle T_{b_i^* b_i} \xi, \xi \rangle + \langle T_{b^* b} \xi, \xi \rangle$$
$$- \langle T_{b^* b_i} \xi, \xi \rangle - \langle T_{b_i^* b} \xi, \xi \rangle$$
$$\xrightarrow[i]{} 0 .$$

So (iv) holds, and the claim is proved.

One must of course extend the usual definitions of irreducibility, non-degeneracy, etc., to *-representations of $\underset{\sim}{B}$. This can be done either by repeating these definitions almost verbatim or by reducing them to the case of *-algebras. We prefer the latter course. Let A be the *-algebra of all cross-sections of $\underset{\sim}{B}$ which vanish except at finitely many points, with the operations

$$(f*g)(x) = \sum_{y \in G} f(y)\, g(y^{-1}x) , \qquad \ldots(2)$$

$$f^*(x) = (f(x^{-1}))^* \qquad \ldots(3)$$

$(f, g \in A;\ x \in G)$. (In the sum on the right side of (2) all but finitely many terms are 0_x. Observe that A is just $\underset{\sim}{L}(\underset{\sim}{B}_c)$, where $\underset{\sim}{B}_c$ is the Banach *-algebraic bundle obtained from $\underset{\sim}{B}$ by replacing the topology of G by its discrete topology.) This A is called the <u>discrete cross-sectional *-algebra of</u> $\underset{\sim}{B}$. Clearly every *-representation T of $\underset{\sim}{B}$ gives rise to a *-representation T' of A:

$$T'_f = \sum_{x \in G} T_{f(x)} \qquad (f \in A) \ .$$

We now define <u>non-degeneracy</u>, <u>stable subspaces</u>, <u>irreducibility</u>, <u>unitary equivalence</u>, <u>intertwining operators</u>, <u>subrepresentations</u>, and <u>Hilbert direct sums</u> for *-representations of $\underset{\sim}{B}$ by stipulating that these concepts are preserved by the mapping $T \mapsto T'$.

If $\underset{\sim}{B}$ has a unit element 1 , then T is non-degenerate if and only if T_1 is the identity operator on $X(T)$.

Notice from (1) that the Hilbert direct sum of any family of *-representations of $\underset{\sim}{B}$ exists.

<u>Proposition 14.1</u>. <u>A *-representation</u> T <u>of</u> $\underset{\sim}{B}$ <u>is non-degenerate if and only if</u> $T|B_e$ <u>is non-degenerate</u>.

<u>Proof</u>. Obviously the 'if' part holds. Assume then that $T|B_e$ is degenerate. Thus there is a non-zero vector ξ in $X(T)$ such that $T_a\xi = 0$ for all a in B_e . Hence $\|T_b\xi\|^2 = \langle T_b\xi, T_b\xi\rangle = \langle T_{b^*b}\xi, \xi\rangle = 0$ (since $b^*b \in B_e$) for all b in B . So T is degenerate, and the 'only if' part is proved. □

<u>Unitary representations</u>.

Assume for the moment that $\underset{\sim}{B}$ is the group bundle of G , so that $B = \mathbb{C} \times G$. If U is a unitary representation of G , the equation

$$T_{(\lambda, x)} = \lambda U_x \qquad (\lambda \in \mathbb{C};\ x \in G) \qquad \ldots(4)$$

defines a non-degenerate *-representation of $\underset{\sim}{B}$. Conversely, if T is a non-degenerate *-representation of $\underset{\sim}{B}$, the equation

$$U_x = T_{(1, x)} \qquad (x \in G) \qquad \ldots(5)$$

defines a unitary representation of G. Thus, under the correspondence (4), (5), the unitary representations of G can be identified with the non-degenerate *-representations of the group bundle of G.

Tensor products of representations.

Returning to the case of general $\underset{\sim}{B}$, let us consider a unitary representation U of G and a *-representation T of $\underset{\sim}{B}$. From these one can form a new *-representation $U \otimes T$ of $\underset{\sim}{B}$, called the (inner) tensor product of U and T, as follows: The space of $U \otimes T$ is the Hilbert space tensor product $X(U) \otimes X(T)$, and the operators are given by:

$$(U \otimes T)_b = U_{\pi(b)} \otimes T_b \qquad (b \in B) \qquad \ldots(6)$$

For a brief indication of the meaning and importance of this tensor product operation see Remark 4 of §19.

Projective representations of groups.

Suppose that

$$\gamma : \mathbb{E} \underset{i}{\longrightarrow} H \underset{j}{\longrightarrow} G \qquad \ldots(7)$$

is a central extension of the circle group $\mathbb{E}$ by G, and that S is a unitary representation of H satisfying

$$S_{i(\lambda)} = \lambda \mathbb{1} \qquad (\lambda \in \mathbb{E}). \qquad \ldots(8)$$

Such an S is sometimes called a γ-representation of G.

Remark. Strictly speaking the phrase 'γ-representation of G' is a misnomer, since S is not a representation of G but of H.

<u>Remark</u>. γ-representations are referred to by Mackey [5] and others as <u>cocycle representations.</u> We prefer to avoid dealing with cocycles, since the association between cocycles and central extensions of $\mathbb{E}$ seems to depend on the second countability of G ; and we do not wish to assume second countability anywhere.

<u>Remark</u>. Let X be a Hilbert space, $\underset{\sim}{U}(X)$ the topological group of all unitary operators on X with the strong operator topology, $\underset{\sim}{S}(X)$ the normal subgroup of scalar operators $\lambda 1_X$ where $\lambda \in \mathbb{E}$, $\tilde{\underset{\sim}{U}}(X)$ the quotient topological group $\underset{\sim}{U}(X)/\underset{\sim}{S}(X)$, and $\rho : \underset{\sim}{U}(X) \to \tilde{\underset{\sim}{U}}(X)$ the quotient homomorphism. A continuous homomorphism of G into $\tilde{\underset{\sim}{U}}(X)$ is called a <u>projective representation</u> of G on X . Projective representations occur naturally in quantum mechanics.

If S is a γ-representation of G on X (γ being as in (7)), the equation

$$\tilde{S}_{j(h)} = \rho(S_h) \qquad (h \in H)$$

defines a projective representation $\tilde{S}$ of G on X in virtue of (8). Conversely, it can be easily shown (we omit the proof here) that, given any projective representation T of G on X , there exists a central extension γ of $\mathbb{E}$ by G and a γ-representation S of G on X such that $T = \tilde{S}$, and that this γ is determined by T to within isomorphism.

Thus the study of projective representations depends intimately on γ-representations.

Now let γ (as in (7)) be a central extension of $\mathbb{E}$ by G , and $\underset{\sim}{B}_\gamma$ the circle extension bundle over G corresponding to γ , constructed as in §12(II); and let $\rho : \mathbb{C} \times H \to B$ be the quotient surjection of §12(II). If S is a γ-representation of G, the equation

$$S'_{\rho(\lambda,h)} = \lambda S_h \qquad (\lambda \in \mathbb{C}; h \in H) \quad \ldots(9)$$

(meaningful in virtue of (8)) evidently defines S' as a non-degenerate *-representation of $\mathcal{B}_\gamma$. Conversely, if S' is any non-degenerate *-representation of $\mathcal{B}_\gamma$, the relation

$$S_h = S'_{\rho(1,h)} \qquad (h \in H) \quad \ldots(10)$$

defines S as a unitary representation of H; and the fact that $\rho(1, i(\lambda)) = \rho(\lambda,e) = \lambda\rho(1,e)$ (e being the unit of H) shows that S is a γ-representation of G. Thus <u>(9) and (10) constitute a one-to-one correspondence</u> $S \leftrightarrow S'$ <u>between the set of all</u> γ-<u>representations</u> S <u>of</u> G <u>and the set of all non-degenerate</u> *-<u>representations</u> S' of $\mathcal{B}_\gamma$. This correspondence of course generalizes the correspondence mentioned earlier in this section between unitary representations of G and non-degenerate *-representations of the group bundle of G.

<u>Extension of *-representations to multipliers</u>.

It is often important to be able to extend a *-representation of a Banach *-algebraic bundle to a *-representation of its multiplier bundle. To show that this can be done, we begin with Banach *-algebras.

Let A be any Banach *-algebra. We recall from §11 that there is a canonical *-homomorphism $\sigma : a \mapsto u_a$ (given by §11(16)) of A onto a *-ideal I of the multiplier Banach *-algebra $\mathcal{W}(A)$.

<u>Proposition 14.2</u>. <u>Given any non-degenerate</u> *-<u>representation</u> T <u>of</u> A, <u>there is a unique</u> *-<u>representation</u> T' <u>of</u> $\mathcal{W}(A)$ <u>acting in</u> $X(T)$ <u>such that</u>

$$T'_{u_a} = T_a \quad \text{for all} \quad a \text{ in } A.$$

Proof. Clearly the non-degeneracy of T implies the uniqueness of T' if the latter exists.

If $a \in \mathrm{Ker}(\sigma)$, then $ab = 0$, whence $T_a T_b = 0$, for all b in A; so by the non-degeneracy of T we have $T_a = 0$. Thus $\mathrm{Ker}(\sigma) \subset \mathrm{Ker}(T)$; and T generates a non-degenerate *-representation T^0 of $I \cong A/\mathrm{Ker}(\sigma)$. Extending T^0 to $W(A)$ by the last Remark of §9, we obtain a *-representation T' of $W(A)$ with the required property. □

T' is called (somewhat improperly) the extension of T to $W(A)$.

Let us now deduce the corresponding fact for bundles.

Proposition 14.3. *Let T be a non-degenerate *-representation of the Banach *-algebraic bundle B over G. There is a unique *-representation T' of the multiplier bundle $W(B)$, acting on the same space as T, such that*

$$T'_{u_b} = T_b \quad \text{for all} \quad b \ \text{in} \ B.$$

Remark. Since $W(B)$ has all the structure of a Banach *-algebraic bundle over G except a topology (see §11), it is natural to refer to a mapping $T': W(B) \to O(X)$ as a *-representation of $W(B)$ if it satisfies on $W(B)$ the properties (i)-(iii) of the definition of a *-representation. By the same argument as for (1) we then have:

$$\|T'_w\| \leq \|w\|_0 \qquad (w \in W(B)) . \quad \ldots (11)$$

Proof of Proposition. Let A be the discrete cross-sectional *-algebra of B, and A_c its completion with respect to the norm $\|f\|_1 = \Sigma_{x \in G} \|f(x)\|$ $(f \in A)$. Thus A_c is the L_1 cross-sectional algebra of the Banach *-algebraic bundle B_c obtained by replacing G by its discrete topology. Any multiplier w of B, in virtue of the boundedness of its left and right actions, can be extended to a multiplier w' of A_c:

$$(w'f)(x) = w(f(y^{-1}x)), \quad (fw')(x) = f(xy^{-1})w$$

($f \in A_c$; $x \in G$; y = order of w); and the map $w \mapsto w'$ preserves addition on each fiber as well as multiplication and involution.

Now let T^0 be the (non-degenerate) *-representation of A_c corresponding to T:

$$T^0_f = \sum_{x \in G} T_{f(x)} \qquad (f \in A_c) .$$

By Prop. 14.2 T^0 can be extended to a *-representation T^{00} of $W(A_c)$. The map $T': w \mapsto T^{00}_{w'}$ $(w \in W(B))$ is then the *-representation T' required by the proposition. Its uniqueness is evident. □

We refer to T' as the <u>extension of</u> T <u>to</u> $W(B)$.

<u>Remark</u>. This result was proved in Fell [6], Prop. 7.1, under the unnecessary assumption that B has an approximate unit.

Let $W_1(B)$ denote the "unit cylinder" $\{w \in W(B): \|w\|_0 \le 1\}$ of $W(B)$. The following addendum to Prop. 14.3 is important.

<u>Proposition 14.4</u>. <u>The</u> T' <u>of Prop</u>. 14.3 <u>is continuous on</u> $W_1(B)$ <u>with respect to the strong multiplier topology and the strong operator topology</u>.

<u>Proof</u>. Suppose that $w_i \to w$ strongly in $W(B)$, with $\|w_i\|_0 \le 1$; and denote by L the linear space of those ξ in $X(T)$ for which $T'_{w_i}\xi \to T'_w\xi$. By (11) $\|T'_{w_i}\| \le 1$ for all i; and this clearly implies that L is closed in $X(T)$. On the other hand, if $\eta \in X(T)$ and $b \in B$, we have (by §11(17) and the strong continuity of T)

$T'_{w_i}T_b\eta = T'_{w_i}T'_{u_b}\eta = T_{w_ib}\eta \longrightarrow T_{wb}\eta = T'_wT_b\eta$. Therefore $T_b\eta \in L$ for all b in B and η in $X(T)$. Since T is non-degenerate this implies that L is dense in $X(T)$ and hence equal to $X(T)$. □

It is worth noticing that if T and T' are as in Prop. 14.3, then T and T' have identical commuting algebras. This is an easy consequence of the following simple fact about *-algebras.

Proposition 14.5. If I is a *-ideal of a *-algebra A , and S is a *-representation of A such that $S|I$ is non-degenerate, then the commuting algebras of S and $S|I$ are the same.

Proof. Let B be an element of $\mathcal{O}(X(S))$ such that $BS_b = S_bB$ for all b in I . It must be shown that $BS_a = S_aB$ for all a in A . But for any a in A and c in I we have $(BS_a)S_c = BS_{ac} = S_{ac}B$ (since $ac \in I$) $= S_aS_cB = (S_aB)S_c$; since $S|I$ is non-degenerate this implies $BS_a = S_aB$. □

Application to semidirect product bundles.

As an application of Props. 14.3 and 14.4 we will obtain a simplified description of the *-representations of semidirect product bundles.

Let A be a Banach *-algebra, G a topological group (with unit e), and ι a strongly continuous homomorphism of G into the group of isometric *-automorphisms of A ; and form the semidirect product bundle $\mathcal{B} = A \times_\iota G$ as in §12(I).

Suppose that S is a non-degenerate *-representation of A and V is a unitary representation of G with $X(V) = X(S)$, and that S and V are related by the formula:

$$V_xS_aV_x^{-1} = S_{\iota_x(a)} \qquad (x \in G;\ a \in A). \quad \ldots(12)$$

Then one verifies without difficulty that the equation

$$T_{(a,x)} = S_a V_x \quad (x \in G;\ a \in A) . \quad \ldots(13)$$

defines a (non-degenerate) *-representation of $\underset{\sim}{B}$. Conversely, we have:

<u>Proposition 14.6</u>. <u>Let</u> T <u>be any non-degenerate *-representation of</u> $\underset{\sim}{B}$. <u>Then there is a unique non-degenerate *-representation</u> S <u>of</u> A <u>and a unique unitary representation</u> V <u>of</u> G , <u>both acting in</u> X(T), <u>such that</u> (12) <u>and</u> (13) <u>hold</u>.

<u>Proof</u>. We saw in §12(I) that each x in G gives rise to a unitary multiplier m_x of $\underset{\sim}{B}$, and that $x \mapsto m_x$ is continuous on G to $\underset{\sim}{W}(\underset{\sim}{B})$ with respect to the strong topology of the latter. Now let T' be the extension of T to $\underset{\sim}{W}(\underset{\sim}{B})$ obtained in Prop. 14.3, and put $V_x = T'_{m_x}$ $(x \in G)$. By Prop. 14.4 together with the preceding sentence, V is strongly continuous and hence a unitary representation of G .

Furthermore, let

$$S_a = T_{(a,e)} \quad (a \in A) . \quad \ldots(14)$$

By Prop. 14.1 S is a non-degenerate *-representation of A . From the equations

$$(m_x(a,e))m_{x^{-1}} = (\iota_x(a),e) ,$$

$$(a,e)m_x = (a,x) ,$$

$(x \in G;\ a \in A)$ it follows that (12) and (13) hold. This proves the existence of the required S and V .

To prove the uniqueness of S and V , we notice first that (13) implies $S_a = T_{(a,e)}$ $(a \in A)$, so that T determines S . Since S is non-degenerate, (13) then determines V_x for all x . □

In view of this result, non-degenerate *-representations of $\underset{\sim}{B}$ are essentially just pairs S,V with the properties specified above.

§15. The integrated form of a *-representation.

Let $\underset{\sim}{B} = (B,\pi,\cdot,{}^*)$ be a fixed Banach *-algebraic bundle over a locally compact group G with unit e , left Haar measure λ , and modular function Δ . The $\underset{\sim}{L}_1$ cross-sectional algebra $\underset{\sim}{L}_1(\underset{\sim}{B};\lambda)$ of $\underset{\sim}{B}$, whose operations are given by §13(1),(2), will be denoted simply by $\underset{\sim}{L}_1$.

Now let T be any *-representation of $\underset{\sim}{B}$. Then, for each f in $\underset{\sim}{L}_1$ and ξ in X , the function $x \mapsto T_{f(x)}\xi$ is in $\underset{\sim}{L}_1(G,X(T);\lambda)$; the equation

$$T'_f\xi = \int_G T_{f(x)}\xi \, d\lambda x \qquad \ldots(1)$$

($X(T)$-valued Bochner integral) defines an element $T'_f: \xi \mapsto T'_f\xi$ of $\underset{\sim}{O}(X(T))$; and $T': f \mapsto T'_f$ is a *-representation of $\underset{\sim}{L}_1$.

Definition. This T' is called the integrated form of T . The restriction of T' to the dense *-subalgebra $\underset{\sim}{L}(\underset{\sim}{B})$ of $\underset{\sim}{L}_1(\underset{\sim}{B};\lambda)$ is called the $\underset{\sim}{L}(\underset{\sim}{B})$-integrated form of T .

Remark. These facts are essentially proved in Fell [6], §8. (The assumption, made in Fell [6], §8, that T is non-degenerate is clearly irrelevant to what we have said so far.) Equation (1) was used in Fell [6], §8, only for $f \in \underset{\sim}{L}(\underset{\sim}{B})$; but an easy convergence argument based on the density of $\underset{\sim}{L}(\underset{\sim}{B})$ in $\underset{\sim}{L}_1$ shows that (1) holds for all f in $\underset{\sim}{L}_1$.

Proposition 15.1. T *is non-degenerate if and only if* T' *is non-degenerate*.

Proof. It is shown in Fell [6], §8, that T' is non-degenerate if T is. Assume that T is degenerate. Then there is a non-zero vector ξ in $X(T)$ such that $T_b \xi = 0$ for all b in B. By (1) this implies $T'_f \xi = 0$ for all f in $\mathcal{L}_1$, whence T' is degenerate. □

Proposition 15.2. *A closed subspace of* $X(T)$ *is* T-*stable if and only if it is* T'-*stable*. *In particular* T *is irreducible if and only if* T' *is*.

This is Fell [6], Prop. 8.3.

Proposition 15.3. *The construction of integrated forms preserves Hilbert direct sums. That is, if* $\{T^i\}$ *is an indexed collection of* *-*representations of* $\mathcal{B}$, *and* $(T^i)'$ *is the integrated form of* T^i, *then* $\Sigma_i^{\oplus}(T^i)'$ *is the integrated form of* $\Sigma_i^{\oplus} T^i$.

This is evident.

Proposition 15.4. *If* S *and* T *are two* *-*representations of* $\mathcal{B}$, *with integrated forms* S' *and* T' *respectively, the* S,T *intertwining operators and the* S',T' *intertwining operators are identical. In particular, the commuting algebras of* S *and* S' *are identical*.

Proof. This follows from Props. 15.2 and 15.3 when we recall that the graph of an element B of $\mathcal{O}(X(S),X(T))$ is a closed subspace of $X(S) \oplus X(T)$, and is (i) $(S \oplus T)$-stable if and only if B is S,T intertwining, and (ii) $(S' \oplus T')$-stable if and only if B is S',T' intertwining. □

The important fact of this section is that, conversely, every

*-representation of $\underset{\sim}{L}_1$ is the integrated form of a unique *-representation of $\underset{\sim}{B}$.

__Theorem 15.5__. __Given a *-representation__ S __of__ $\underset{\sim}{L}_1$, __there is a unique *-representation__ T __of__ $\underset{\sim}{B}$ __having__ S __as its integrated form__.

__Proof.__ Since every *-representation is the direct sum of a zero representation and a non-degenerate *-representation, it is enough to assume that S is non-degenerate. In this case the proof coincides with that given for Fell [6], Prop. 8.4. (The assumption made in Fell [6], Prop. 8.4, that $\underset{\sim}{B}$ has a strong approximate unit served to ensure that $\underset{\sim}{L}_1$ would have an approximate unit, so that Prop. 7.1 of Fell [6] would be applicable to it. But we have remarked in connection with Prop. 14.3 of the present work that Fell [6], Prop. 7.1, is true without the assumption of an approximate unit.) □

__Remark__. Let $\underset{\sim}{B}$ be the group bundle of G . Then the non-degenerate *-representations of $\underset{\sim}{B}$ are just the unitary representations of G . So in this case Theorem 15.5 asserts the well-known correspondence between unitary representations of G and non-degenerate *-representations of the $\underset{\sim}{L}_1$ group algebra of G .

__Remark__. More generally, let $\underset{\sim}{B}$ be the semidirect product Banach *-algebraic bundle $A \times_{\iota} G$ as in §12(I). Then $\underset{\sim}{L}_1(\underset{\sim}{B};\lambda)$ is the so-called covariance algebra, which has been studied by various authors beginning with Doplicher, Kastler and Robinson [1]. Combining Theorem 15.5 with Prop. 14.6, we obtain a one-to-one correspondence between non-degenerate *-representations of $\underset{\sim}{L}_1(\underset{\sim}{B};\lambda)$ and pairs S,V , where S is a non-degenerate *-representation of A and V is a unitary representation of G acting in X(S) and satisfying §14(12).

§16. Positive functionals on the cross-sectional algebra.

Fix a Banach *-algebraic bundle $\underset{\sim}{B} = (B,\pi,\cdot,^*)$ over a locally compact group G with unit e, left Haar measure λ, and modular function Δ.

Our first goal is to prove the following important bundle version of the Gelfand-Naimark-Segal construction of *-representations of *-algebras:

Theorem 16.1. *Let p be a positive linear functional on the compacted cross-sectional algebra $\mathcal{L}(\underset{\sim}{B})$ which is continuous in the inductive limit topology (see §10). Then p is admissible and so generates a *-representation S of $\mathcal{L}(\underset{\sim}{B})$. Furthermore S is the $\mathcal{L}(\underset{\sim}{B})$-integrated form of a (unique) *-representation T of $\underset{\sim}{B}$.*

Proof. We have seen in §13 that every element b of B gives rise to a multiplier m_b of $\mathcal{L}_1(\underset{\sim}{B};\lambda)$ as follows:

$$(m_b f)(y) = b\, f(x^{-1}y)\ , \qquad \dots(1)$$

$$(f m_b)(y) = \Delta(x^{-1})\ f(yx^{-1})b \qquad \dots(2)$$

$(f \in \mathcal{L}_1(\underset{\sim}{B};\lambda);\ b \in B;\ y \in G;\ x = \pi(b))$. We also notice that m_b leaves $\mathcal{L}(\underset{\sim}{B})$ stable; that is, $f \in \mathcal{L}(\underset{\sim}{B}) \Rightarrow m_b f,\ f m_b \in \mathcal{L}(\underset{\sim}{B})$. The mapping $b \mapsto m_b$ is linear on each fiber and preserves multiplication and involution. For the rest of this proof we write bf and fb instead of $m_b f$ and $f m_b$.

The reader will also recall from Prop. 13.2 that the convolution in $\mathcal{L}(\underset{\sim}{B})$ can be expressed in terms of these multipliers:

$$f*g = \int_G (f(x)g)\,d\lambda x \qquad (f,g \in \mathcal{L}(\underset{\sim}{B})),\dots(3)$$

the right side being an $\underset{\sim}{L}(\underset{\sim}{B})$-valued Bochner integral with respect to the inductive limit topology of $\underset{\sim}{L}(\underset{\sim}{B})$.

I claim that, for b in B and $f \in \underset{\sim}{L}(\underset{\sim}{B})$,

$$p((bf)^* * bf) \leq \|b\|^2 p(f^* * f) . \qquad \ldots(4)$$

Indeed: By Schwarz's Inequality together with the identity $(b^*f)^* = f^*b$, we have

$$\begin{aligned} p((bf)^* * bf) &= p(f^* * b^*bf) \\ &\leq p(f^* * f)^{\frac{1}{2}} p((b^*bf)^* * (b^*bf))^{\frac{1}{2}} \\ &= p(f^* * f)^{\frac{1}{2}} p(f^* * (b^*b)^2 f)^{\frac{1}{2}} . \end{aligned}$$

Iterating this argument (with b replaced by b^*b, $(b^*b)^2$, etc.) we obtain for each positive integer n

$$\begin{aligned} p((bf)^* * bf) &\leq p(f^* * f)^{\frac{1}{2}+\frac{1}{4}} p(f^* * (b^*b)^4 f)^{\frac{1}{4}} \\ &\leq \cdots\cdot \\ &\leq (p(f^* * f))^{\frac{1}{2}+\frac{1}{4}+\cdots+2^{-n}} (p(f^* * (b^*b)^{2^n} f))^{2^{-n}} . \end{aligned} \qquad \ldots(5)$$

We must now fix f in $\underset{\sim}{L}(\underset{\sim}{B})$ and estimate the size of $p(f^* * af)$ for $a \in B_e$. Let K be the compact support of f ; then $\text{supp}(f^* * af) \subset K^{-1}K$. Hence, since p is continuous in the inductive limit topology, there is a positive constant k such that

$$p(f^* * af) \leq k\|f^* * af\|_\infty \qquad (a \in B_e) .$$

Now an easy calculation shows that

$$\|f^* * af\|_\infty \leq \lambda(K) \ \|f\|_\infty^2 \ \|a\| \qquad (a \in B_e) .$$

Combining the last two facts we get

$$p(f^{*} * af) \leq k\, \lambda(K)\, \|f\|_{\infty}^{2}\, \|a\| \qquad (a \in B_e). \quad \ldots(6)$$

Putting $(b^{*}b)^{2^n}$ for a in (6), and substituting (6) in (5), we obtain

$$p((bf)^{*} * bf) \leq p(f^{*} * f)^{\frac{1}{2}+\frac{1}{4}+\cdots+2^{-n}} (k\, \lambda(K)\, \|f\|_{\infty}^{2})^{2^{-n}} \|b^{*}b\|$$

for all $n = 1,2,\cdots$. Passing to the limit $n \to \infty$ in this inequality, we obtain (4).

Now, let I be the null left ideal $\{a \in \underset{\sim}{L}(\underset{\sim}{B}) : p(a^{*}a) = 0\}$ of p, construct the pre-Hilbert space $\underset{\sim}{L}(\underset{\sim}{B})/I$, and denote its Hilbert space completion by X. Let $\tilde{f}$ be the image in X of the element f of $\underset{\sim}{L}(\underset{\sim}{B})$, so that

$$\langle \tilde{f}, \tilde{g} \rangle = p(g^{*} * f) \qquad (f,g \in \underset{\sim}{L}(\underset{\sim}{B})). \quad \ldots(7)$$

The inequality (4) asserts that for each b in B the equation

$$T_b \tilde{f} = (bf)^{\sim} \qquad (f \in \underset{\sim}{L}(\underset{\sim}{B})) \quad \ldots(8)$$

defines a bounded linear operator T_b on X with $\|T_b\| \leq \|b\|$. Evidently $b \mapsto T_b$ is linear on each fiber, and $T_b T_c = T_{bc}$. Also $\langle T_b \tilde{f}, \tilde{g} \rangle = p(g^{*} * bf) = p((b^{*}g)^{*} * f) = \langle \tilde{f}, T_{b^*} \tilde{g} \rangle$; so $T_{b^*} = (T_b)^{*}$.

Thus, to show that T is a *-representation of $\underset{\sim}{B}$, it is enough (see §14(iv′)) to show that $b \mapsto \langle T_b \tilde{f}, \tilde{f} \rangle = p(f^{*} * bf)$ is continuous on B for each f in $\underset{\sim}{L}(\underset{\sim}{B})$. But this follows immediately from the following two easily verifiable facts: (i) $b \mapsto bf$ is continuous on B to $\underset{\sim}{L}(\underset{\sim}{B})$ with respect to the inductive limit topology of the latter, and (ii) $f * g$ is separately continuous in f and g on $\underset{\sim}{L}(\underset{\sim}{B}) \times \underset{\sim}{L}(\underset{\sim}{B})$ to $\underset{\sim}{L}(\underset{\sim}{B})$. So we have shown that T is a *-representation of $\underset{\sim}{B}$.

Now let S be the integrated form of T. Since S determines T, the proposition will be proved if we show that S is just the *-representation of $\underset{\sim}{L}(\underset{\sim}{B})$ generated by p. For this we must verify that

$$\langle S_h\tilde{f},\tilde{g}\rangle = \langle (h * f)^{\sim},\tilde{g}\rangle \qquad \ldots(9)$$

for all f,g,h in $\underset{\sim}{L}(\underset{\sim}{B})$. Now

$$\langle S_h\tilde{f},\tilde{g}\rangle = \int \langle T_{h(x)}\tilde{f},\tilde{g}\rangle \, d\lambda x$$

$$= \int p(g^{*} * h(x)f)\, d\lambda x \,. \qquad \ldots(10)$$

Also, by (3),

$$h*f = \int (h(x)f)\, d\lambda x \qquad \ldots(11)$$

(Bochner integral with respect to the inductive limit topology of $\underset{\sim}{L}(\underset{\sim}{B})$). Noting that the linear functional $q \mapsto p(g^{*} * q)$ is continuous on $\underset{\sim}{L}(\underset{\sim}{B})$ in the inductive limit topology, and applying it to both sides of (11), we obtain by Prop. 10.16

$$p(g^{*} * h * f) = \int p(g^{*} * h(x)f)\, d\lambda x \;;$$

and this, together with (10), gives (9). □

<u>Definition</u>. The *-representation T of $\underset{\sim}{B}$ constructed in the preceding theorem is said to be <u>generated by</u> the positive linear functional p on $\underset{\sim}{L}(\underset{\sim}{B})$.

Easy examples show that the T generated by p will in general be degenerate. However, we have:

<u>Proposition 16.2</u>. <u>If</u> $\underset{\sim}{B}$ <u>has an approximate unit, the</u> T <u>of Theorem 16.1 will be non-degenerate</u>.

Proof. Let L_0 be the linear span of $\{af: a \in B_e, f \in \underset{\sim}{L}(\underset{\sim}{B})\}$. Then L_0 is closed under multiplication by complex continuous functions on G; and since $\underset{\sim}{B}$ has an approximate unit $\{f(x): f \in L_0\}$ is dense in B_x for every x in G. So by Prop. 10.9 L_0 is dense in $\underset{\sim}{L}(\underset{\sim}{B})$ in the inductive limit topology. Further, the continuity of p implies that the quotient map $f \mapsto \tilde{f}$ is continuous from $\underset{\sim}{L}(\underset{\sim}{B})$ (with the inductive limit topology) to the Hilbert space X. These facts imply that $(L_0)^{\sim}$ is dense in X. On the other hand, $(L_0)^{\sim}$ is contained in the linear span of $\{\text{range}(T_a): a \in B_e\}$. So T is non-degenerate. □

If $\underset{\sim}{B}$ is the group bundle, then in particular $\underset{\sim}{B}$ has a unit; and Theorem 16.1 and Prop. 16.2 become the following classical result on groups:

Proposition 16.3. Let p be any positive linear functional on $\underset{\sim}{L}(G)$ (the compacted group algebra of G) which is continuous in the inductive limit topology. Then p is admissible; and there is a (unique) unitary representation T of G whose integrated form is the *-representation of $\underset{\sim}{L}(G)$ generated by p.

As before, T is said to be generated by p.

As a matter of fact, it is important to observe that the argument of Theorem 16.1 holds in somewhat greater generality. Indeed, suppose that E is any *-subalgebra of $\underset{\sim}{L}(\underset{\sim}{B})$ which is dense in $\underset{\sim}{L}(\underset{\sim}{B})$ in the inductive limit topology and such that

$$b \in B, f \in E \Rightarrow bf \in E. \qquad \ldots(12)$$

Let p be a positive linear functional on E which is continuous with respect to the inductive limit topology (relativized to E). Then the same proofs as in Theorem 16.1 and Prop. 16.2 show that (a) p *is admissible*; (b) *there is a unique *-representation* T *of* $\mathcal{B}$ *such that the integrated form of* T, *when restricted to* E, *is just the *-representation of* E *generated by* p; (c) *if in addition* $\mathcal{B}$ *has an approximate unit,* T *is non-degenerate.*

Our next step is to deduce a much stronger version of Theorem 15.5.

What condition on a *-representation S of $\mathcal{L}(\mathcal{B})$ will guarantee that S is the $\mathcal{L}(\mathcal{B})$-integrated form of a *-representation of $\mathcal{B}$? By Theorem 15.5 $\mathcal{L}_1$-continuity is certainly sufficient; but in fact much weaker conditions will do. In view of the preceding paragraph we may even work with a dense *-subalgebra of $\mathcal{L}(\mathcal{B})$ rather than $\mathcal{L}(\mathcal{B})$ itself.

Theorem 16.4. *Let* E *be a *-subalgebra of* $\mathcal{L}(\mathcal{B})$ *which is dense in* $\mathcal{L}(\mathcal{B})$ *in the inductive limit topology and which satisfies* (12). *Let* S *be a *-representation of* E *and* X_0 *a subset of* $X(S)$ *with the following two properties:* (i) *The smallest closed* S-*stable subspace of* $X(S)$ *containing* X_0 *is* $X(S)$ *itself; and* (ii) *for every* ξ *in* X_0, *the linear functional* $f \mapsto \langle S_f\xi, \xi\rangle$ *is continuous on* E *in the (relativized) inductive limit topology. Then* S *is the restriction to* E *of the integrated form of a unique *-representation* T *of* $\mathcal{B}$. *In particular,* S *must be continuous on* E *in the* $\mathcal{L}_1$ *norm.*

Proof. It is sufficient to assume that S is non-degenerate. Indeed, otherwise we replace S by its non-degenerate part, and X_0 by its projection onto the non-degenerate part.

Take a vector ξ in X_0; and let S^ξ be the subrepresentation of S acting on the closed S-stable subspace X_ξ of $X(S)$ generated by ξ. Then $\{S_f\xi : f \in E\}$ is dense in X_ξ; so the positive functional $p_\xi : f \mapsto \langle S_f\xi, \xi\rangle$ on E generates S^ξ. Since by hypothesis p_ξ is continuous in the inductive limit topology, it follows from Theorem 16.1 that S^ξ is the restriction of the integrated form of a *-representation of $\underset{\sim}{B}$, and so

$$\|S_f^\xi\| \leq \|f\|_1 \qquad (f \in E) . \qquad \ldots(13)$$

Now by hypothesis (i) the linear span of $\{X_\xi : \xi \in X_0\}$ is dense in $X(S)$. Hence (see Remark 10 of §9) (13) implies that

$$\|S_f\| \leq \|f\|_1 \qquad (f \in E) ; \qquad \ldots(14)$$

that is, S is norm-decreasing on E. Now E is dense in $\underset{\sim}{L}(\underset{\sim}{B})$ in the inductive limit topology, hence $\underset{\sim}{L}_1$ dense in $\underset{\sim}{L}_1(\underset{\sim}{B};\lambda)$. Thus (14) permits us to extend S to a *-representation of $\underset{\sim}{L}_1(\underset{\sim}{B};\lambda)$. By Theorem 15.5 this extension is the $\underset{\sim}{L}_1$ integrated form of a *-representation T of $\underset{\sim}{B}$.

The existence of the required T is now proved. Its uniqueness is evident from Theorem 15.5 and the denseness of E in $\underset{\sim}{L}(\underset{\sim}{B})$. □

§17. When do enough *-representations exist?

In this section we attack the question: When does a Banach *-algebraic bundle have enough *-representations to distinguish its points?

Fix a Banach *-algebraic bundle $\underset{\sim}{B} = (B, \pi, \cdot, ^*)$ over a locally compact group G with unit e, left Haar measure λ, and modular function Δ.

We recall that a Banach *-algebra A is reduced if A has enough *-representations to distinguish its points. Our goal in this section is to find conditions under which $\underset{\sim}{L}_1(\underset{\sim}{B};\lambda)$ will be reduced. If $\underset{\sim}{L}_1(\underset{\sim}{B};\lambda)$ is reduced, then $\underset{\sim}{B}$ itself will have enough *-representations (in fact enough irreducible *-representations) to distinguish its points.

If $\underset{\sim}{B}$ were the group bundle, one would proceed by exhibiting the regular representation R of G and showing that the integrated form of R is faithful on the $\underset{\sim}{L}_1$ group algebra. In the general case, we will use Theorem 16.1 to construct generalizations of the regular representation which, taken together, will separate points in $\underset{\sim}{L}_1(\underset{\sim}{B};\lambda)$ under suitable conditions.

Let f be any element of $\underset{\sim}{L}_1(\underset{\sim}{B};\lambda)$. I claim that, for any g in $\underset{\sim}{L}(\underset{\sim}{B})$, the cross-sections $g*f$ and $f*g$ (belonging to $\underset{\sim}{L}_1(\underset{\sim}{B};\lambda)$) are continuous.

Indeed: Choose a sequence $\{f_n\}$ of elements of $\underset{\sim}{L}(\underset{\sim}{B})$ converging to f in $\underset{\sim}{L}_1(\underset{\sim}{B};\lambda)$. Then one verifies easily that $g*f_n \to g*f$ and $f_n*g \to f*g$ uniformly on compact subsets of G. Since the $g*f_n$ and f_n*g are continuous, it follows that $g*f$ and $f*g$ are continuous, as we claimed.

Lemma 17.1. Assume that $\underset{\sim}{B}$ has an approximate unit and is saturated.

<u>If</u> $f \in \underset{\sim}{L}_1(\underset{\sim}{B};\lambda)$ <u>and</u> $(h*f*g)(e) = 0$ <u>for all</u> $g,h \in \underset{\sim}{L}(\underset{\sim}{B})$, <u>then</u> $f = 0$.

<u>Remark</u>. Since $h*f*g$ is continuous by the preceding claim, the expression $(h*f*g)(e)$ makes sense.

<u>Proof</u>. We first observe that, if $x \in G$ and $0_x \neq b \in B_x$, then $cb \neq 0$ for some c in $B_{x^{-1}}$. Indeed: Since $\underset{\sim}{B}$ has an approximate unit, $ab \neq 0$ for some a in B_e. Since $\underset{\sim}{B}$ is saturated, a belongs to the closed linear span of $B_x B_{x^{-1}}$; and so $(dc)b \neq 0$ for some d in B_x and c in $B_{x^{-1}}$. This implies that $cb \neq 0$, proving the observation.

Now let $g,h \in \underset{\sim}{L}(\underset{\sim}{B})$. For any x in G and c in $B_{x^{-1}}$, we have

$$\begin{aligned} c[(h*f*g)(x)] &= [c(h*f*g)](e) \\ &= ((ch)*f*g)(e). \qquad \ldots(1) \end{aligned}$$

(Here c is treated as a multiplier of $\underset{\sim}{L}_1(\underset{\sim}{B};\lambda)$ as in §16(1),(2).) Since $ch \in \underset{\sim}{L}(\underset{\sim}{B})$, by hypothesis the right side of (1) is 0; and so $c[(h*f*g)(x)] = 0$ for all c in $B_{x^{-1}}$. By the preceding observation this implies that $(h*f*g)(x) = 0$ for all x. Thus

$$h*f*g = 0 \quad \text{for all } g,h \text{ in } \underset{\sim}{L}(\underset{\sim}{B}). \qquad \ldots(2)$$

Now in view of the Remark following Theorem 13.1 $\underset{\sim}{L}_1(\underset{\sim}{B};\lambda)$ has an approximate unit $\{g_i\}$ consisting of elements of $\underset{\sim}{L}(\underset{\sim}{B})$. Thus, replacing g in (2) by g_i, we conclude from (2) that

$$h*f = 0 \quad \text{for all } h \text{ in } \underset{\sim}{L}(\underset{\sim}{B}). \qquad \ldots(3)$$

Replacing h in (3) by g_i we conclude from (3) that $f = 0$. □

<u>Proposition 17.2</u>. <u>Assume that</u> $\underset{\sim}{B}$ <u>has enough unitary multipliers</u>. <u>If</u>

q *is a continuous positive linear functional on* B_e *, the equation*

$$p(f) = q(f(e)) \quad (f \in \underset{\sim}{L}(\underset{\sim}{B})) \qquad \ldots(4)$$

defines p *as a positive linear functional on* $\underset{\sim}{L}(\underset{\sim}{B})$ *which is continuous in the inductive limit topology.*

Proof. The inductive limit continuity of p follows from the continuity of q .

Suppose that $b \in B_x$ $(x \in G)$. By hypothesis there is a unitary multiplier u of order x . Thus $b = ua$ where $a = u^*b \in B_e$, whence $b^*b = (a^*u^*)ua = a^*a$. Thus by the positivity of q ,

$$q(b^*b) \geq 0 \quad \text{for all } b \text{ in } B . \qquad \ldots(5)$$

Consequently, if $f \in \underset{\sim}{L}(\underset{\sim}{B})$,

$$\begin{aligned} p(f^* * f) &= q[(f^* * f)(e)] \\ &= q\left[\int f^*(y) f(y^{-1}) d\lambda y\right] \\ &= q\left[\int f(y)^* f(y) d\lambda y\right] \\ &= \int q(f(y)^* f(y)) d\lambda y \qquad \text{(by Prop. 10.13)} \\ &\geq 0 \qquad \text{(by (5))}; \end{aligned}$$

and p is positive. □

Suppose that $\underset{\sim}{B}$, q and p are as in Prop. 17.2. Combining Prop. 17.2 with Theorem 16.1 we see that p generates a *-representation of $\underset{\sim}{B}$, which we shall call R^q . *-Representations of the form R^q will be called *generalized regular representations* of $\underset{\sim}{B}$. (If $\underset{\sim}{B}$ were the group bundle and q the identity functional on $\mathbb{C}$, one verifies that R^q would be the ordinary regular representation of G .)

<u>Theorem 17.3</u>. <u>Suppose that $\underset{\sim}{B}$ has an approximate unit and enough unitary multipliers. Then, if the unit fiber *-algebra B_e is reduced, $\underset{\sim}{L}_1(\underset{\sim}{B};\lambda)$ is reduced.</u>

<u>Proof</u>. We shall show that the integrated forms of the generalized regular representations R^q separate the points of $\underset{\sim}{L}_1(\underset{\sim}{B};\lambda)$.

Let f be an element of $\underset{\sim}{L}_1(\underset{\sim}{B};\lambda)$; and let q be a continuous positive linear functional on B_e. Denoting by S^q the integrated form of R^q, we shall show that

$$\langle S_f^q\,\tilde{g},\tilde{h}\rangle = q[(h^* * f * g)(e)] \qquad \ldots(6)$$

for all g,h in $\underset{\sim}{L}(\underset{\sim}{B})$ ($\tilde{g}$ and $\tilde{h}$ being the images of g and h in $X(R^q)$). Indeed: Fix $g,h \in \underset{\sim}{L}(\underset{\sim}{B})$. If $f \in \underset{\sim}{L}(\underset{\sim}{B})$, (6) holds by the definition of R^q. For arbitrary f in $\underset{\sim}{L}_1(\underset{\sim}{B};\lambda)$, choose a sequence $\{f_n\}$ of elements of $\underset{\sim}{L}(\underset{\sim}{B})$ converging to f in the $\underset{\sim}{L}_1$ norm. Then $h^* * f_n * g \to h^* * f * g$ uniformly on compact subsets of G, and so

$$q[(h^* * f_n * g)(e)] \to q[(h^* * f * g)(e)]\,. \qquad \ldots(7)$$

On the other hand, since S^q is norm-decreasing,

$$\langle S_{f_n}^q\,\tilde{g},\tilde{h}\rangle \to \langle S_f^q\,\tilde{g},\tilde{h}\rangle\,. \qquad \ldots(8)$$

Combining (7) and (8) with the fact that (6) holds for the f_n, we see that (6) holds for f.

Suppose now that f is an element of $\underset{\sim}{L}_1(\underset{\sim}{B};\lambda)$ which is not separated from 0 by the S^q; that is, $S_f^q = 0$ for all continuous positive linear functionals q on B_e. By (6) this means that $q[(h^* * f * g)(e)] = 0$ for all g,h in $\underset{\sim}{L}(\underset{\sim}{B})$ and all continuous positive linear functionals q on B_e. Now by assumption B_e is reduced; so there are enough such q to separate points of B_e. Hence the last equality asserts that

$$(h^* * f * g)(e) = 0 \qquad \dots(9)$$

for all g,h in $L(B)$. Now by Prop. 11.4 the hypotheses of the theorem imply that B is saturated. Therefore, by Lemma 17.1, (9) gives $f = 0$.

Thus the set of all the S^q separates the points of $L_1(B;\lambda)$. Consequently the latter is reduced. □

Remark. The preceding theorem becomes false if the hypothesis that B has enough unitary multipliers is replaced by the hypothesis that B is saturated. A simple example of this is the B of §12(V). This B has a unit element and is saturated, but $E = L_1(B;\lambda)$ is not reduced. Indeed, if T is a *-representation of E, we have

$$0 \le (T_{(1,-1)})^* T_{(1,-1)} = T_{(1,-1)*(1,-1)}$$
$$= T_{(-1,-1)} = -T_{(1,1)} \le 0 ;$$

whence $T_{(1,-1)} = 0$ for all such T.

We shall next show that the conclusion of Theorem 17.3 holds whenever B is a C*-algebraic bundle.

Theorem 17.4. *Let B be a C*-algebraic bundle over G. Then the L_1 cross-sectional algebra of B is reduced.*

Proof. Although B need not have enough unitary multipliers (in fact, it need not even be saturated), the same development that led to Theorem 17.3 also holds here.

To begin with, the conclusion of Lemma 17.1 holds here. Indeed, we have seen in Prop. 11.6 that $\underset{\sim}{B}$ has an approximate unit. The hypothesis of saturation was used in the proof of Lemma 17.1 only to show that $0_x \neq b \in B_x$ implies $cb \neq 0$ for some c in $B_{x^{-1}}$; but in the present context this follows from condition (i) of the definition of a C*-algebraic bundle on taking $c = b^*$.

Further, the conclusion of Prop. 17.2 holds in the present context, since the hypothesis of enough unitary multipliers in Prop. 17.2 served only to show that each b^*b is of the form a^*a for some a in B_e ; and this follows here from condition (ii) of the definition of a C*-algebraic bundle.

Notice that B_e, being a C*-algebra, is automatically reduced. Thus, finally, the proof of Theorem 17.3 remains valid in the present context. □

Corollary 17.5. Assume either that $\underset{\sim}{B}$ is a C*-algebraic bundle, or that $\underset{\sim}{B}$ has an approximate unit and enough unitary multipliers and B_e is reduced. Let $\underset{\sim}{W}$ be the set of all irreducible *-representations of $\underset{\sim}{B}$. Then: (I) The set of all the integrated forms of elements of $\underset{\sim}{W}$ separates the points of $\underset{\sim}{L}_1(\underset{\sim}{B};\lambda)$. (II) If b,c are two distinct elements of B which are not both zero elements (i.e., either $b \neq 0_{\pi(b)}$ or $c \neq 0_{\pi(c)}$), then there is an element T of $\underset{\sim}{W}$ such that $T_b \neq T_c$.

Proof. (I) By Theorems 17.3 and 17.4 $\underset{\sim}{L}_1(\underset{\sim}{B};\lambda)$ is reduced. As is well known, this implies that the points of $\underset{\sim}{L}_1(\underset{\sim}{B};\lambda)$ are separated by the irreducible *-representations of $\underset{\sim}{L}_1(\underset{\sim}{B};\lambda)$. But by Theorem 15.5 and Prop. 15.2 these are just the integrated forms of elements of $\underset{\sim}{W}$. So (I) holds.

(II) Let b,c be as in (II); and let $\{u_i\}$ be an approximate unit of $\underset{\sim}{B}$. By choosing an element f of $\underset{\sim}{L}(\underset{\sim}{B})$ which vanishes outside a small enough neighborhood of e and such that $f(e) = u_i$ for large enough i, we can ensure that $bf \neq cf$. Thus by (I) there is an irreducible *-representation T of $\underset{\sim}{B}$ whose integrated form T' satisfies $T'_{bf} \neq T'_{cf}$. But we have $T'_{bf} = T_b T'_f$ and $T'_{cf} = T_c T'_f$. The last two statements imply that $T_b \neq T_c$. □

<u>Corollary 17.6</u>. <u>Given any central extension</u> γ <u>of</u> $\mathbb{E}$ <u>by</u> G, <u>there exist irreducible</u> γ-<u>representations of</u> G.

<u>Proof</u>. Combine the preceding corollary with the correspondence between γ-representations and *-representations of $\underset{\sim}{B}_\gamma$ established in §14. □

<u>Semidirect product bundles</u>.

Theorem 17.3 has interesting consequences when applied to semidirect product bundles.

Let A be a reduced Banach *-algebra with an approximate unit; and let ι be a strongly continuous homomorphism of G into the group of isometric *-automorphisms of A, as in §12(I). Let $\underset{\sim}{P}$ be the set of all pairs (S,V) where S is a non-degenerate *-representation of A, V is a unitary representation of G acting in the same space as S, and

$$V_x S_a V_x^{-1} = S_{\iota_x(a)} \quad (x \in G;\ a \in A)$$

(see Prop. 14.6).

<u>Corollary 17.7</u>. <u>The set</u> $\underset{\sim}{P}_1$ <u>of all those</u> S <u>which occur as first members of pairs in</u> $\underset{\sim}{P}$ <u>separates the points of</u> A. <u>The set</u> $\underset{\sim}{P}_2$ <u>of all the integrated forms of the</u> V <u>which occur as second members of pairs in</u> $\underset{\sim}{P}$ <u>separates the points of</u> $\underset{\sim}{L}_1(\lambda)$.

Proof. Let $\underset{\sim}{B}$ be the semidirect product bundle $A \times_{\iota} G$. By Cor. 17.5 the set $\underset{\sim}{Q}$ of all non-degenerate *-representations of $\underset{\sim}{B}$ separates the points of $A \cong B_e$. Since the elements of $\underset{\sim}{P}_1$ are just the restrictions to A of elements of $\underset{\sim}{Q}$, this proves the first assertion of the corollary.

To prove the second assertion, we define a linear map $\varphi \mapsto m_{\varphi}$ of $\underset{\sim}{L}_1(\lambda)$ into the multiplier *-algebra of $\underset{\sim}{L}_1(\underset{\sim}{B};\lambda)$ as follows:

$$(m_{\varphi}f)(x) = \int \varphi(y)\, \iota_y\, f(y^{-1}x)\,d\lambda y\ ,$$

$$(fm_{\varphi})(x) = \int \varphi(y)\ \Delta(y^{-1})\ f(xy^{-1})\,d\lambda y$$

$(\varphi \in \underset{\sim}{L}_1(\lambda);\ f \in \underset{\sim}{L}_1(\underset{\sim}{B};\lambda);\ x \in G)$. The reader will easily verify that these equations do indeed define a multiplier m_{φ}, and furthermore that $\varphi \mapsto m_{\varphi}$ is one-to-one. Now fix a non-zero element φ of $\underset{\sim}{L}_1(\lambda)$. Since $m_{\varphi} \neq 0$, there are elements f,g of $\underset{\sim}{L}_1(\underset{\sim}{B};\lambda)$ such that $m_{\varphi}f = g \neq 0$. By Theorem 17.3 $\underset{\sim}{L}_1(\underset{\sim}{B};\lambda)$ is reduced; so there is a non-degenerate *-representation T of $\underset{\sim}{L}_1(\underset{\sim}{B};\lambda)$ satisfying $T_g \neq 0$. Thus, if T' is the extension of T to the multiplier *-algebra of $\underset{\sim}{L}_1(\underset{\sim}{B};\lambda)$ (see Prop. 14.3), we have $T'_{m_{\varphi}}T_f = T_g \neq 0$, whence $T'_{m_{\varphi}} \neq 0$. Now let (S,V) be the pair in $\underset{\sim}{P}$ corresponding to T by Prop. 14.6. The reader will verify without difficulty that $T'_{m_{\varphi}} = V_{\varphi}$ (V and its integrated form being denoted by the same symbol). Thus $V_{\varphi} \neq 0$ and $V \in \underset{\sim}{P}_2$. This proves the second assertion of the corollary. □

Remark. Let $\underset{\sim}{W}$ be any collection of non-degenerate *-representations of a reduced Banach *-algebra C; and consider the following two statements: (i) Every *-representation of C is weakly contained in $\underset{\sim}{W}$; (ii) $\underset{\sim}{W}$ separates the points of C. Since C is assumed to be reduced, (i) clearly implies (ii). So it is natural to ask whether the conclusions of Cor. 17.7 can be strengthened to assert (i) instead of (ii) for A and $\underset{\sim}{P}_1$, and also for $\underset{\sim}{L}_1(\lambda)$ and $\underset{\sim}{P}_2$. The answer to these questions is 'yes' for A and $\underset{\sim}{P}_1$, but 'no in general' for $\underset{\sim}{L}_1(\lambda)$ and $\underset{\sim}{P}_2$. The affirmative answer for A and $\underset{\sim}{P}_1$ will follow from our later Cor. 22.4 and Prop. 27.2. To get a counter-example for $\underset{\sim}{L}_1(\lambda)$ and $\underset{\sim}{P}_2$, let us take $A = \underset{\sim}{C}_0(G)$ (with pointwise multiplication and involution and the supremum norm), with $(\tau_x \varphi)(y) = \varphi(x^{-1}y)$ $(\varphi \in \underset{\sim}{C}_0(G);\ x,y \in G)$. Then a non-degenerate *-representation of $\underset{\sim}{B} = A \underset{\tau}{\times} G$ is identifiable (by Prop. 14.6) with a system of imprimitivity for G over G (see §30); and by the Imprimitivity Theorem (Theorem 32.9) the latter is nothing but a direct sum of copies of the system of imprimitivity associated with the regular representation of G. It follows that in this case the elements of $\underset{\sim}{P}_2$ are just the direct sums of copies of the regular representation of G. So, if G is not amenable, $\underset{\sim}{P}_2$ will not weakly contain all *-representations of $\underset{\sim}{L}_1(\lambda)$.

§18. The bundle C*-completion.

Given any Banach *-algebra A, there is a well known method for constructing a C*-algebraic "photograph" of A: For any a in A one defines $\|a\|_c = \sup\{\|T_a\|: T \text{ is a *-representation of } A\}$. The null-space $N = \{a \in A: \|a\|_c = 0\}$ of $\|\ \|_c$ is then a *-ideal of A; and A/N is a normed *-algebra under the norm $a + N \mapsto \|a\|_c$. The completion of A/N under this norm is a C*-algebra, called the C*-*completion* A^c of A. In view of the construction of A^c, the *-representations of A and of A^c are in natural one-to-one correspondence.

In this section we shall point out the extension of this construction to the bundle context.

Let $\underset{\sim}{B} = (B,\pi,\cdot,^*)$ be a fixed Banach *-algebraic bundle over a locally compact group G.

For each b in B we define

$$\|b\|_c = \sup\{\|T_b\| : T \text{ is a *-representation of } \underset{\sim}{B}\} . \qquad \ldots(1)$$

By §14(1)

$$\|b\|_c \leq \|b\| . \qquad \ldots(2)$$

Clearly

$$\|b^*b\|_c = \|b\|_c^2 . \qquad \ldots(3)$$

Restricted to each fiber B_x, $\| \ \|_c$ is a seminorm. If $N_x = \{b \in B_x : \|b\|_c = 0\}$, then B_x/N_x is a normed linear space (under the norm $b + N_x \mapsto \|b\|_c$), whose completion we denote by $C_x, \| \ \|_c$. Let C stand for the disjoint union of the $\{C_x\}$ $(x \in G)$; let $\pi' : C \to G$ be the surjection given by $\pi'^{-1}(x) = C_x$; and let $\rho : B \to C$ be the "quotient" map $b \mapsto b + N_{\pi(b)}$.

Now $b \mapsto \|b\|_c$ is continuous on B. This follows from the identity (3), the continuity of $b \mapsto b^*b$, and the continuity of $\| \ \|_c$ on B_e (see (2)). Also, B has enough continuous cross-sections by Theorem 10.5. Thus the family of cross-sections of (C,π') of the form

$$x \mapsto \rho(f(x)) \qquad \ldots(4)$$

(where f is a continuous cross-section of $\underset{\sim}{B}$) satisfies the hypotheses of Prop. 10.4; and so by the latter there is a unique topology on C making $\underset{\sim}{C} = (C,\pi')$ a Banach bundle (with the norm $\| \ \|_c$) such that the cross-sections (4) are all continuous.

From the continuity of the cross-sections (4) and the fact that ρ is norm-decreasing by (2), we verify that $\rho : B \to C$ is continuous.

Evidently

$$\|ab\|_c \leq \|a\|_c \|b\|_c, \quad \|b^*\|_c = \|b\|_c \qquad \ldots(5)$$

$(a,b \in B)$. From this we see that the equations

$$\rho(a)\ \rho(b) = \rho(ab), \quad (\rho(b))^* = \rho(b^*) \qquad \ldots(6)$$

$(a,b \in B)$ determine a product $\cdot$ and involution * on C satisfying (3) and (5) for all a,b in C . The continuity of these operations on C follows from the continuity of ρ and from conditions §11(v′), (xi′) applied to the family Γ of cross-sections of the form (4). Thus $\underset{\sim}{C}$ has become a Banach *-algebraic bundle over G .

The *-representation theories of $\underset{\sim}{B}$ and of $\underset{\sim}{C}$ coincide in the following sense: <u>The *-representations of $\underset{\sim}{B}$ are precisely the maps of the form</u>

$$T: b \mapsto T'_{\rho(b)} \qquad (b \in B) , \qquad \ldots(7)$$

<u>where T' is a *-representation of $\underset{\sim}{C}$</u> . The correspondence $T \leftrightarrow T'$ preserves closed stable subspaces, non-degeneracy, irreducibility, intertwining operators, and Hilbert direct sums. The reader will verify this without difficulty.

Now $\underset{\sim}{C}$ is a C^*-algebraic bundle. Indeed: The first condition in the definition of a C^*-algebraic bundle holds by (3). To prove the second, we recall (1) and form the Hilbert direct sum T of enough *-representations of $\underset{\sim}{B}$ so that $\|T_b\| = \|b\|_c$ for all b in B . The *-representation T' of $\underset{\sim}{C}$ corresponding to T by (7) thus satisfies $\|T'_b\| = \|b\|_c$ for all b in C . In particular $S = T'|C_e$ is an isomorphism of the C^*-algebra C_e into the C^*-algebra $\underset{\sim}{O}(X(T))$. If $b \in C$, $S_{b^*b} = (T'_b)^* T'_b$ is a positive operator, and so is positive with respect to range(S) . So b^*b is positive in C_e , and the second condition in the definition of a C^*-algebra holds.

Definition. The C^*-algebraic bundle $\underset{\sim}{C}$ which we have constructed is called the bundle C^*-completion of $\underset{\sim}{B}$. It will in future be denoted by $\underset{\sim}{B}^c$.

As we saw above, the *-representation theories of $\underset{\sim}{B}$ and $\underset{\sim}{B}^c$ coincide just as they do for a Banach *-algebra and its C^*-completion.

Remark 1. If $\underset{\sim}{B}$ is saturated, then $\underset{\sim}{B}^c$ is evidently also saturated.

If $\underset{\sim}{B}$ has enough unitary multipliers, so does $\underset{\sim}{B}^c$. We omit the easy proof of this.

Remark 2. As before let us denote $\underset{\sim}{B}^c$ by $\underset{\sim}{C}$. Although C_e is a C^*-algebra, it differs in general from the C^*-completion $(B_e)^c$ of B_e. Indeed: $(B_e)^c$ is constructed from the following seminorm $\| \|'_c$ on B_e:

$$\|a\|'_c = \sup\{\|S_a\|: S \text{ is a } ^*\text{-representation of } B_e\}.$$

Clearly $\|a\|'_c \geq \|a\|_c$ for all a in B_e. So the identity map on B_e generates a norm-decreasing surjective *-homomorphism $H: (B_e)^c \to C_e$. This H will be one-to-one if and only if $\| \|_c$ and $\| \|'_c$ coincide on B_e; but this, as it turns out, is not in general the case. For another condition that H be one-to-one see Cor. 26.4.

Proposition 18.1. Suppose that $\underset{\sim}{B}$ is itself a C^*-algebraic bundle. Then $\|b\|_c = \|b\|$ for all b in B. Hence $\underset{\sim}{B}$ is its own bundle C^*-completion.

Proof. Using Cor. 17.5 and forming Hilbert direct sums, we can find a *-representation T of $\underset{\sim}{B}$ such that $T|B_e$ is faithful on the

C^*-algebra B_e. So T is an isometry on B_e. Hence, for any b in B,

$$\|T_b\|^2 = \|(T_b)^*T_b\| = \|T_{b^*b}\|$$
$$= \|b^*b\| \quad (\text{since } b^*b \in B_e)$$
$$= \|b\|^2 .$$

So $\|b\| = \|T_b\| \le \|b\|_c$ for all b in B. Combining this with (2) completes the proof. □

Remark. We have thus shown that C^*-algebraic bundles over the locally compact group G are precisely the C^*-completions of Banach $*$-algebraic bundles over G. This generalizes the analogous fact for C^*-algebras.

§19. Bundle structures for C^*-algebras.

In the Introduction, to motivate the definition of a Banach *-algebraic bundle, we started with a *-algebra A and a finite group G, and defined a *-algebraic bundle structure for A over G as a certain collection $\{B_x\}$ $(x \in G)$ of closed linear subspaces (fibers) of A. However, when in §11 we came to define a Banach *-algebraic bundle $\underset{\sim}{B}$ over an arbitrary topological group G, we disregarded the "global" algebra A entirely, and concentrated our attention on the fibers B_x. Only later, in §13 (and this only in the case of locally compact G), were we able to reconstruct a "global" counterpart of A, namely the $\underset{\sim}{L}_1$ cross-sectional algebra of $\underset{\sim}{B}$. And here the fibers B_x are not subspaces of the $\underset{\sim}{L}_1$ cross-sectional algebra unless G is discrete.

Nevertheless, it is worth raising the question: Given a Banach *-algebra A and a locally compact group G, what do we mean by a "Banach *-algebraic bundle structure" for A over G? One's first thought might be to understand by this a Banach *-algebraic bundle $\underset{\sim}{B}$ over G along with a homeomorphic *-isomorphism between A and the $\underset{\sim}{L}_1$ cross-sectional algebra of $\underset{\sim}{B}$. This will in fact be our definition, except that in order to avoid dependence on the accidental nature of the norm in $\underset{\sim}{B}$ and in $\underset{\sim}{L}_1(\underset{\sim}{B};\lambda)$ we shall confine our attention to C^*-algebras and C^*-algebraic bundles.

Fix a locally compact group, with unit e and left Haar measure λ.

<u>Definition</u>. If $\underset{\sim}{B}$ is a Banach *-algebraic bundle over G, we shall refer to the C^*-completion of $\underset{\sim}{L}_1(\underset{\sim}{B};\lambda)$ as the C^* <u>cross-sectional algebra of</u> $\underset{\sim}{B}$, and denote it by $C^*(\underset{\sim}{B})$.

Thus the *-representations of $\underset{\sim}{B}$ are in natural one-to-one correspondence with those of $C^*(\underset{\sim}{B})$.

<u>Definition</u>. Given a C^*-algebra A, by a C^*-<u>algebraic bundle structure for</u> A <u>over</u> G we mean a C^*-algebraic bundle $\underset{\sim}{B}$ over G together with a *-isomorphism Φ of A onto $C^*(\underset{\sim}{B})$.

Two C^*-algebraic bundle structures $\underset{\sim}{B},\Phi$ and $\underset{\sim}{B}',\Phi'$ for A over G are <u>isomorphic</u> if there is an isometric isomorphism ψ of $\underset{\sim}{B}$ and $\underset{\sim}{B}'$ such that

$$\Psi\circ\Phi = \Phi' ,$$

where Ψ is the *-isomorphism of $C^*(\underset{\sim}{B})$ onto $C^*(\underset{\sim}{B}')$ induced by ψ.

Clearly two isomorphic C^*-algebraic bundle structures for A should be regarded as essentially the same.

<u>Remark 1</u>. Suppose that the group G is discrete. If $\underset{\sim}{B}$ is a C^*-algebraic bundle over G, it follows from the existence of a norm-preserving *-representation of $\underset{\sim}{B}$ (see the proof of Prop. 18.1) that

each fiber B_x can be identified with a closed linear subspace B'_x of $C^*(\underset{\sim}{B})$, and that the $\{B'_x\}$ $(x \in G)$ are linearly independent and have dense linear span in $C^*(\underset{\sim}{B})$. Now let $\underset{\sim}{B}, \Phi$ be a C^*-algebraic bundle structure over G for some C^*-algebra A. If $B''_x = \Phi^{-1}(B'_x)$, it follows from the last observation that the B''_x $(x \in G)$ are linearly independent closed linear subspaces of A whose linear span is dense in A. Furthermore, the knowledge of the B''_x $(x \in G)$ clearly determines $\underset{\sim}{B}, \Phi$ to within isomorphism.

Assume now that G is <u>finite</u>. Then the converse of the last paragraph holds. That is, suppose we are given a C^*-algebra A, and a collection $\{B''_x\}$ $(x \in G)$ of linearly independent closed subspaces of A whose linear span is A and which satisfy

$$B''_x B''_y \subset B''_{xy}, (B''_x)^* = B''_{x^{-1}} (x,y \in G) . \qquad \ldots(1)$$

Then there is a C^*-algebraic bundle structure $\underset{\sim}{B}, \Phi$ for A over G which gives rise to the B''_x as in the last paragraph. This is easy to verify.

However, if G is discrete but infinite, this is not necessarily the case. For example, let G be a discrete infinite non-amenable group, and let A be the C^*-algebra generated by the left regular representation R of G. For each x in G let B''_x be the one-dimensional subspace of A spanned by R_x. Then (1) holds; and the B''_x are linearly independent and span a dense subspace of A. However, the $\{B''_x\}$ do not determine a C^*-algebraic bundle structure $\underset{\sim}{B}, \Phi$ for A over G. For if they did, $\underset{\sim}{B}$ would be the group bundle of G; and since G is not amenable, there are *-representations of the group bundle (i.e., unitary representations of G) which do not extend to *-representations of A.

Remark 2. A C^*-algebraic bundle structure $\underset{\sim}{B}, \Phi$ for a C^*-algebra A is said to be saturated if $\underset{\sim}{B}$ is saturated. To specify a saturated C^*-algebraic bundle structure for a given C^*-algebra A is somewhat analogous to specifying a system of imprimitivity for a given unitary representation of a group.

Given a C^*-algebra A, it is a worthwhile endeavor to try to classify to within isomorphism all possible saturated C^*-algebraic bundle structures for A. If no such structures exist (except the trivial structure over the one-element group), A can be called primitive.

Remark 3. Perhaps the greatest advantage accruing from the knowledge of a non-trivial saturated C^*-algebraic bundle structure $\underset{\sim}{B}, \Phi$ for a C^*-algebra A is that we can then apply the "Mackey normal subgroup analysis" to the *-representation theory of A. Indeed, in a subsequent publication we shall show how, by an extension of the Mackey analysis, the irreducible *-representations of any saturated Banach *-algebraic bundle $\underset{\sim}{B}$ over G can be analyzed in terms of the irreducible *-representations of the unit fiber *-algebra B_e and the irreducible cocycle representations of subgroups of G. Now if $\underset{\sim}{B}, \Phi$ is a saturated C^*-algebraic bundle structure for A, the *-representations of A are in natural correspondence with the *-representations of $\underset{\sim}{B}$; and so the analysis just mentioned becomes an analysis of the *-representations of A.

Remark 4. Let A be a C^*-algebra and G a locally compact group; and suppose we are given a C^*-algebraic bundle structure $\underset{\sim}{B}, \Phi$ for A over G. Then, for any unitary representation U of G and any

*-representation T of A, the tensor product operation applied in $\underset{\sim}{B}$ enables us to construct a new *-representation S of A. To be precise, let T' be the *-representation of $\underset{\sim}{B}$ corresponding to the *-representation $T \circ \Phi^{-1}$ of $C^*(\underset{\sim}{B})$; form the tensor product *-representation $U \otimes T'$ of $\underset{\sim}{B}$ as in §14; let S' be the *-representation of $C^*(\underset{\sim}{B})$ corresponding to $U \otimes T'$; and, finally, set $S = S' \circ \Phi$. Let us denote S by $U \bar{\otimes} T$. The operation $\bar{\otimes}$ clearly depends only on the isomorphism class of $\underset{\sim}{B}, \Phi$. Question: Does $\bar{\otimes}$ determine the isomorphism class of $\underset{\sim}{B}, \Phi$? At present we do not know. An affirmative answer to this question would be a convincing demonstration of the importance of the tensor product operation of §14.

It would also be natural to ask for an axiomatic characterization of those "tensor product operations" $\bar{\otimes}$ which arise in this way from C^*-algebraic bundle structures. In the context of group bundles this is essentially the well known duality theory of (non-commutative) locally compact groups studied by Stinespring, Tatsuuma, Takesaki, Kac and others. For a bibliography of this theory the reader may consult Takesaki [2] (see also the later publication of Vainerman and Kac [1]). In the context of semidirect product bundles the problem has been largely solved by Landstad [1].

<u>A C^*-algebraic bundle structure for the CAR algebra</u>.

To conclude this section we present an example of a C^*-algebraic bundle structure for a particular class of C^*-algebras of importance in physics. These C^*-algebras will be direct limits of finite-dimensional matrix *-algebras.

Let $X_1, X_2, \cdots$ be a sequence of Hilbert spaces, X_n being of finite dimension $d_n > 1$; and let A_n be the C^*-algebra

$\underset{\sim}{O}(X_1)\otimes\cdots\otimes\underset{\sim}{O}(X_n) \cong \underset{\sim}{O}(X_1\otimes\cdots\otimes X_n)$. If $m < n$ we take $F_{nm}:A_m \to A_n$ to be the *-isomorphism $a \mapsto a \otimes 1_{nm}$ $(a \in A_m)$, where 1_{nm} stands for the identity operator on $X_{m+1}\otimes\cdots\otimes X_n$. Considering the A_n as pairwise disjoint, we introduce into $\cup_{n=1}^{\infty} A_n$ the smallest equivalence relation $\sim$ containing all pairs $(a, F_{nm}(a))$, where $m < n$ and $a \in A_m$. Let $\tilde{a}$ be the equivalence class containing a , and denote by E the set of all $\sim$-equivalence classes. Now let α and β be any two elements of E ; we can then choose a and b so that $\tilde{a} = \alpha$, $\tilde{b} = \beta$, and a and b belong to the same A_n . Since the F_{nm} are isometric *-isomorphisms, the equations

$$\begin{aligned}
\alpha + \beta &= (a+b)^{\sim} , \\
\lambda\alpha &= (\lambda a)^{\sim} \qquad (\lambda \in \mathbb{C}) , \\
\alpha\beta &= (ab)^{\sim} \\
\alpha^* &= (a^*)^{\sim} , \\
\|\alpha\| &= \|a\|
\end{aligned}$$

are meaningful, and define in E the structure of a pre-C^*-algebra. The completion A of E is thus a C^*-algebra; it is the direct limit of the directed system $\{A_n\}$, $\{F_{nm}\}$. We refer to A as the $(d_1,d_2,\cdots)$ Glimm algebra. It is a simple C^*-algebra with unit. In the special case that all $d_n = 2$, A is called the CAR algebra (CAR standing for 'canonical anticommutation relations', from its application to physics).

We shall now exhibit a (semidirect product) C^*-algebraic bundle structure for the $(d_1,d_2,\cdots)$ Glimm algebra A .

For each $n = 1,2,\cdots$ let F_n be a finite group of order d_n , with unit e_n ; and let F be the (compact) Cartesian product group $\Pi_{n=1}^{\infty} F_n$. Let G be the dense subgroup of F consisting of those x

such that $x_n = e_n$ for all but finitely many n ; G will be considered to carry the discrete topology. Let B be the commutative C^*-algebra $\underset{\sim}{C}(F)$; and form the ι-semidirect product bundle $\underset{\sim}{C} = B \times_\iota G$, where ι is the action of G on B by left translation:

$$(\iota_x \varphi)(y) = \varphi(x^{-1}y) \qquad (\varphi \in B;\ x \in G;\ y \in F) .$$

We shall show that the C^* cross-sectional algebra of $\underset{\sim}{C}$ is isomorphic to A .

For each positive integer p , let G_p be the finite subgroup of G consisting of those x such that $x_n = e_n$ for all $n > p$; thus $G_p \cong F_1 \times F_2 \times \cdots \times F_p$. Let $\pi_p : F \to G_p$ be the surjective homomorphism $x \mapsto (x_1, x_2, \cdots ; x_p, e_{p+1}, e_{p+2}, \cdots)$; and let B_p be the *-subalgebra of B consisting of those φ such that $\varphi(x)$ depends only on $\pi_p(x)$ $(x \in F)$; thus $B_p \cong \underset{\sim}{C}(G_p)$. Let $\underset{\sim}{L}^p(\underset{\sim}{C})$ be the space of those f in $\underset{\sim}{L}(\underset{\sim}{C})$ which vanish outside G_p and whose values all lie in B_p . Since $\iota_x(B_p) = B_p$ for $x \in G_p$, $\underset{\sim}{L}^p(\underset{\sim}{C})$ is a *-subalgebra of $\underset{\sim}{L}(\underset{\sim}{C})$. By Remark 4 of §13, $\underset{\sim}{L}^p(\underset{\sim}{C})$ is *-isomorphic with the $G_p \times G_p$ total matrix *-algebra, and so can (and will) be regarded as a C^*-algebra. If $i_{pq} : \underset{\sim}{L}^q(\underset{\sim}{C}) \to \underset{\sim}{L}^p(\underset{\sim}{C})$ is the identity injection (for $q < p$), one verifies without difficulty that the directed system $\{\underset{\sim}{L}^p(\underset{\sim}{C})\}$, $\{i_{pq}\}$ is isomorphic in an obvious sense with the directed system $\{A_n\},\{F_{nm}\}$ from which the Glimm algebra A was constructed. Consequently A is *-isomorphic with the direct limit of the system $\{\underset{\sim}{L}^p(\underset{\sim}{C})\},\{i_{pq}\}$, that is, with the completion D of the pre-C^*-algebra $D_0 = \bigcup_{p=1}^{\infty} \underset{\sim}{L}^p(\underset{\sim}{C})$.

Now $G = \bigcup_p G_p$, and by the Stone-Weierstrass Theorem $\bigcup_p B_p$ is dense in B . It follows from this that D_0 is dense in $\underset{\sim}{L}_1(\underset{\sim}{C})$ (the $\underset{\sim}{L}_1$ cross-sectional algebra of $\underset{\sim}{C}$). Now from the fact that *-representations of Banach *-algebras are norm-decreasing, we deduce that the

$L_1(C)$ norm majorizes the C^* norm on each $L^p(C)$, and hence on all of D_0 . It follows that the identity map $D_0 \to D_0$ extends to a continuous *-homomorphism $i: L_1(C) \to D$ whose range is dense in D .

I claim that D is *-isomorphic to the C^*-completion of $L_1(C)$. To prove this, it is enough by the last paragraph to take a *-representation T of $L_1(C)$, and to show that it factors through i , that is, can be written in the form $S \circ i$ where S is a *-representation of D . But for this we have only to observe that for each p $T|L^p(C)$ is a *-representation of $L^p(C)$ and hence norm-decreasing with respect to the C^* norm of $L^p(C)$, and hence that T is norm-decreasing with respect to the C^* norm on all of D_0 .

So $C^*(C)$ is *-isomorphic to D , and hence to A . Thus, if Φ is a *-isomorphism of A and $C^*(C)$, then C, Φ is a C^*-algebraic bundle structure for the Glimm algebra A over the discrete group G .

These Glimm algebras were first studied by Glimm [2]. The above C^*-algebraic bundle structure for the Glimm algebras is due to Takesaki [3].

§20. The regional topology of *-representations of bundles.

In this section we translate the regional topology, as defined in §6 for representations of *-algebras, into a regional topology for *-representations of Banach *-algebraic bundles.

Fix a Banach *-algebraic bundle $B = (B, \pi, \cdot, ^*)$ over a locally compact group G , with unit e and left Haar measure λ . Let T and T' be the set of all (concrete) *-representations of B and of the L_1 cross-sectional algebra $L_1(B;\lambda)$ respectively. We have seen in §15 that the map sending each T in T into its integrated form

T' is a bijection of $\underset{\sim}{T}$ onto $\underset{\sim}{T}'$. Since $\underset{\sim}{L}_1(\underset{\sim}{B};\lambda)$ is a *-algebra, $\underset{\sim}{T}'$ is equipped with the regional topology defined in §6.

Definition. By the regional topology of $\underset{\sim}{T}$ we mean the topology which makes the bijection $T \mapsto T'$ a homeomorphism with respect to the regional topology of $\underset{\sim}{T}'$.

Remark. This definition generalizes the usual treatment of the topology of the space of unitary representations of a group (see for example Dixmier [1], §18).

Since $\underset{\sim}{L}(\underset{\sim}{B})$ is a dense *-subalgebra of $\underset{\sim}{L}_1(\underset{\sim}{B};\lambda)$, by Prop. 6.4 we could equally well have defined the regional topology of $\underset{\sim}{T}$ as that which makes $T \mapsto T'|\underset{\sim}{L}(\underset{\sim}{B})$ a homeomorphism with respect to the regional topology of the space of *-representations of $\underset{\sim}{L}(\underset{\sim}{B})$. Lakewise, if $C^*(\underset{\sim}{B})$ is the C^*-completion of $\underset{\sim}{L}_1(\underset{\sim}{B};\lambda)$, and T^c is the *-representation of $C^*(\underset{\sim}{B})$ corresponding to the integrated form T' of any element T of $\underset{\sim}{T}$, then $T \mapsto T^c$ is a homeomorphism with respect to the regional topologies of $\underset{\sim}{T}$ and of the space of *-representations of $C^*(\underset{\sim}{B})$.

In this section we shall usually denote the above three representations T, T', T^c by the same letter T.

Functionals of positive type.

The role played by positive linear functionals on *-algebras is played in the theory of group representations by functions of positive type. These have an easy generalization to Banach *-algebraic bundles, which we now introduce.

By a linear functional on $\underset{\sim}{B}$ we shall mean a function $p: B \mapsto \mathbb{C}$ whose restriction to B_x is linear for each x in G. If in addition p is continuous on B, then by Prop. 10.2 it is norm-continuous on each fiber B_x ; and we can form the norms $\|p|B_x\|$. We shall say that the continuous linear functional p is <u>bounded</u> if

$$\sup\{\|p|B_x\|: x \in G\} < \infty ; \qquad \ldots(1)$$

and in that case the supremum in (1) is denoted by $\|p\|$.

<u>Definition</u>. By a <u>functional of positive type on</u> $\underset{\sim}{B}$ we mean a continuous linear functional p on $\underset{\sim}{B}$ satisfying the inequality

$$\sum_{i,j=1}^{n} p(b_j^* b_i) \geq 0 \qquad \ldots(2)$$

for any finite sequence $b_1,\cdots,b_n$ of elements of B .

<u>Remark</u>. In particular, $p|B_e$ is a continuous positive linear functional on B_e .

<u>Remark</u>. If $G = \{e\}$, so that $B = B_e$, then the left side of (2) is $p((\Sigma_{i=1}^n b_i)^*(\Sigma_{i=1}^n b_i))$; and a functional of positive type becomes just a continuous positive linear functional on B_e .

<u>Remark</u>. If $\underset{\sim}{B}$ is the group bundle of G and p is a functional of positive type on $\underset{\sim}{B}$, the equation

$$q(x) = p(1,x) \qquad (x \in G) \qquad \ldots(3)$$

defines a continuous complex function q on G satisfying

$$\sum_{i,j=1}^{n} \lambda_i \overline{\lambda_j}\, q(x_j^{-1}x_i) \geq 0$$

for all $x_1,\cdots,x_n$ in G and $\lambda_1,\cdots,\lambda_n$ in $\mathbb{C}$ -- that is, a function of positive type on G in the usual sense. It is easy to see that the relation (3) identifies functions of positive type on G with functionals of positive type on the group bundle of G .

Our main goal in this section is to express the regional topology of the space of *-representations of $\underset{\sim}{B}$ in terms of the uniform-on-compacta convergence of their associated functionals of positive type. Before doing this, however, we shall mention without proof some facts about functionals of positive type which generalize well-known phenomena in the group situation, but which will not be needed in what follows.

In general (for example, if $\underset{\sim}{B}$ has trivial multiplication) a functional of positive type on $\underset{\sim}{B}$ need not be bounded. However, if $\underset{\sim}{B}$ has an approximate unit, every functional p of positive type on $\underset{\sim}{B}$ is bounded, and in fact satisfies $p(b^*) = \overline{p(b)}$ $(b \in B)$.

The most important functionals of positive type are those which arise from *-representations. If T is a *-representation of $\underset{\sim}{B}$ and $\xi \in X(T)$, the map $p_{T,\xi}: B \longrightarrow \mathbb{C}$ given by

$$p_{T,\xi}(b) = \langle T_b \xi, \xi \rangle \qquad (b \in B) \qquad \ldots (4)$$

is evidently a bounded functional of positive type. Conversely, it can be shown that, if $\underset{\sim}{B}$ has an approximate unit, every functional p of positive type on $\underset{\sim}{B}$ is of the form (4) for some cyclic *-representation T of $\underset{\sim}{B}$ and some cyclic vector ξ for T . As in the case of *-algebras, the functional $p_{T,\xi}$ uniquely determines the "cyclic pair" (T,ξ) up to unitary equivalence.

For brevity's sake we omit the proof of these important facts, since they are not needed for the main purpose of this section.

<u>Definition</u>. Let us denote $\{p_{T,\xi}: T$ is a non-degenerate *-representation of $\underset{\sim}{B}$, $\xi \in X(T)$, $\|\xi\| = 1\}$ by $\underset{\sim}{P}(\underset{\sim}{B})$. We equip $\underset{\sim}{P}(\underset{\sim}{B})$ with the topology of uniform convergence on compact subsets of B .

By §14(1)

$$\|p\| \leq 1 \quad \text{for all} \quad p \quad \text{in} \quad \underset{\sim}{P}(\underset{\sim}{B}) \ . \qquad \ldots(5)$$

Each $p = p_{T,\xi}$ in $\underset{\sim}{P}(\underset{\sim}{B})$ gives rise to a positive linear functional α_p on $\underset{\sim}{L}_1(\underset{\sim}{B};\lambda)$:

$$\begin{aligned} \alpha_p(f) &= \int_G p(f(x))d\lambda x \\ &= \int \langle T_{f(x)}\xi, \xi\rangle \, d\lambda x \\ &= \langle T_f\xi, \xi\rangle \qquad (f \in \underset{\sim}{L}_1(\underset{\sim}{B};\lambda)) \ . \qquad \ldots(6) \end{aligned}$$

Since $\|\xi\| = 1$, we have $\|\alpha_p\| \leq 1$. Let $\gamma\colon \underset{\sim}{L}_1(\underset{\sim}{B};\lambda) \to C^*(\underset{\sim}{B})$ be the natural *-homomorphism of $\underset{\sim}{L}_1(\underset{\sim}{B};\lambda)$ into $C^*(\underset{\sim}{B})$. Then p also gives rise to a continuous positive linear functional α_p^c on $C^*(\underset{\sim}{B})$ determined by the relation

$$\alpha_p = \alpha_p^c \circ \gamma \ , \qquad \ldots(7)$$

that is,

$$\alpha_p^c(f) = \langle T_f\xi, \xi\rangle \qquad (f \in C^*(\underset{\sim}{B})) \ . \qquad \ldots(8)$$

Since $C^*(\underset{\sim}{B})$ is a C^*-algebra and $\|\xi\| = 1$, it follows from (8) that

$$\|\alpha_p^c\| = 1 \quad \text{for} \quad p \in \underset{\sim}{P}(\underset{\sim}{B}) \ . \qquad \ldots(9)$$

I claim that a net $\{p_i\}$ approaches p in $\underset{\sim}{P}(\underset{\sim}{B})$ if and only if

$$p_i(f(x)) \to p(f(x)) \quad \text{uniformly on compact subsets of} \quad G \qquad \ldots(10)$$

for every f in $\underset{\sim}{C}(\underset{\sim}{B})$.

Indeed: If C is a compact subset of G and $f \in \underset{\sim}{C}(\underset{\sim}{B})$, $f(C)$ is compact in B , so (10) is certainly necessary. Conversely, assume

(10); and let D be any compact subset of B. Thus $C = \pi(D)$ is compact in G. Given $\varepsilon > 0$, the compactness of D permits us to choose finitely many continuous cross-sections $f_1, \cdots, f_n$ such that every element b of D satisfies

$$\|b - f_j(\pi(b))\| < \varepsilon \qquad \ldots(11)$$

for some $j = 1, \cdots, n$. Now by (10) we have for all sufficiently large i

$$|p_i(f_j(x)) - p(f_j(x))| < \varepsilon \text{ for all } x \text{ in } C \text{ and all } j = 1, \cdots, n . \qquad \ldots(12)$$

Combining (11) and (12) with the fact that $\|p_i\| \le 1$ and $\|p\| \le 1$ we get for all sufficiently large i

$$|p_i(b) - p(b)| < 3\varepsilon \text{ for all } b \text{ in } D .$$

This and the arbitrariness of D and ε show that $p_i \to p$ in $\underset{\sim}{P}(\underset{\sim}{B})$; and the claim is proved.

Before proceeding to the next proposition we mention two lemmas.

Lemma 20.1. *If X is a Hilbert space, $\varepsilon > 0$, ξ is a unit vector in X, $a \in \underset{\sim}{O}(X)$, $\|a\| \le 1$, and $|\langle a\xi, \xi\rangle - 1| < \varepsilon$, then*

$$\|a\xi - \xi\|^2 < \varepsilon^2 + 2\varepsilon .$$

Proof. We can write $a\xi$ in the form $\lambda\xi + \eta$, where $\lambda \in \mathbb{C}$ and $\eta \perp \xi$. Then $\lambda = \langle a\xi, \xi\rangle$ and $|\lambda - 1| < \varepsilon$. Since $|\lambda|^2 + \|\eta\|^2 = \|a\xi\|^2 \le 1$, we have

$$\|\eta\|^2 \le 1 - |\lambda|^2 = (1 - |\lambda|)(1 + |\lambda|) < 2\varepsilon .$$

So $\|a\xi - \xi\|^2 = |\lambda - 1|^2 + \|\eta\|^2 < \varepsilon^2 + 2\varepsilon$. ∎

Lemma 20.2 (Gelfand). Let X be a Banach space, and $\{\alpha_i\}$ a norm-bounded net of elements of X^* such that $\alpha_i \to 0$ pointwise on X. Then $\alpha_i(\xi) \to 0$ uniformly for ξ in any norm-compact subset of X.

We omit the proof of this well-known fact.

Proposition 20.3. The map $p \mapsto \alpha_p$ is a homeomorphism on $\underset{\sim}{P}(\underset{\sim}{B})$ with respect to the topology of pointwise convergence of linear functionals on $\underset{\sim}{L}_1(\underset{\sim}{B};\lambda)$ (or, equivalently, on $\underset{\sim}{L}(\underset{\sim}{B})$). Likewise, the map $p \mapsto \alpha_p^c$ is a homeomorphism on $\underset{\sim}{P}(\underset{\sim}{B})$ with respect to the topology of pointwise convergence of linear functionals on $C^*(\underset{\sim}{B})$.

Proof. It is enough to prove the continuity of three maps: (I) the map $p \mapsto \alpha_p$; (II) the map $\alpha_p \mapsto \alpha_p^c$; (III) the map $\alpha_p^c \mapsto p$. (In all of these, p runs over $\underset{\sim}{P}(\underset{\sim}{B})$.)

(I) Let $p_i \to p$ in $\underset{\sim}{P}(\underset{\sim}{B})$; then evidently $\alpha_{p_i}(f) \to \alpha_p(f)$ for every f in $\underset{\sim}{L}(\underset{\sim}{B})$. Since $\{\|\alpha_{p_i}\|\}$ is bounded and $\underset{\sim}{L}(\underset{\sim}{B})$ is dense in $\underset{\sim}{L}_1(\underset{\sim}{B};\lambda)$, it follows that $\alpha_{p_i} \to \alpha_p$ pointwise on $\underset{\sim}{L}_1(\underset{\sim}{B};\lambda)$. So map (I) is continuous on $\underset{\sim}{P}(\underset{\sim}{B})$.

(II) Let $\alpha_{p_i} \to \alpha_p$ pointwise on $\underset{\sim}{L}_1(\underset{\sim}{B};\lambda)$ (where $p_i, p \in \underset{\sim}{P}(\underset{\sim}{B})$). By (7) $\alpha_{p_i}^c \to \alpha_p^c$ pointwise on range(γ). Since range(γ) is dense in $C^*(\underset{\sim}{B})$, it follows from (9) that $\alpha_{p_i}^c \to \alpha_p^c$ pointwise on $C^*(\underset{\sim}{B})$. So map (II) is continuous.

(III) To prove the continuity of (III), we have to observe that each b in B acts as a natural multiplier on $C^*(\underset{\sim}{B})$. Indeed: b acts as a multiplier m_b on $\underset{\sim}{L}_1(\underset{\sim}{B};\lambda)$ by §16(1),(2); and it is easy to deduce from Prop. 14.3 (applied with $G = \{e\}$) that the equations

$$m_b' \ \gamma(f) = \gamma(m_b f) \ , \quad \gamma(f) m_b' = \gamma(f m_b) \qquad \dots (13)$$

$(f \in \underset{\sim}{L}_1(\underset{\sim}{B};\lambda))$ define a multiplier m_b' of $C^*(\underset{\sim}{B})$ with

$$\|m_b'\| \leq \|b\| . \qquad \ldots (14)$$

Furthermore, the map $b \mapsto m_b'$ is continuous on B with respect to the strong multiplier topology. Indeed: If $\varphi \in \underset{\sim}{L}(\underset{\sim}{B})$, then by Prop. 14.4 the maps $b \mapsto m_b\varphi$ and $b \mapsto \varphi m_b$ are continuous with respect to the $\underset{\sim}{L}_1(\underset{\sim}{B};\lambda)$ norm and hence also with respect to the $C^*(\underset{\sim}{B})$ norm. The strong continuity of $b \mapsto m_b'$ now follows from this fact together with (14) and the denseness of $\gamma(\underset{\sim}{L}(\underset{\sim}{B}))$ in $C^*(\underset{\sim}{B})$.

Now let $\{p_i\}$ and p be elements of $\underset{\sim}{P}(\underset{\sim}{B})$ given by

$$p(b) = \langle T_b\xi, \xi\rangle , \quad p_i(b) = \langle T_b^i\xi^i, \xi^i\rangle \quad (b \in B) \qquad \ldots (15)$$

where $\{T^i\}$, T are non-degenerate *-representations of $\underset{\sim}{B}$ and ξ^i and ξ are unit vectors in $X(T^i)$ and $X(T)$ respectively. We assume that $\alpha_{p_i}^c \to \alpha_p^c$ pointwise on $C^*(\underset{\sim}{B})$. To prove that $p_i \to p$ in $\underset{\sim}{P}(\underset{\sim}{B})$, we shall take a compact subset D of B and show that

$$p_i(b) \to p(b) \text{ uniformly for } b \in D . \qquad \ldots (16)$$

Clearly we may as well assume $\|b\| \leq 1$ for $b \in D$.

Now by Dixmier [1], 1.7.2, $C^*(\underset{\sim}{B})$ has an approximate unit consisting of self-adjoint elements of norm no greater than 1. Hence, given $\epsilon > 0$, and picking $\delta > 0$ so that $\delta^2 + 2\delta < \epsilon^2$, we can find an element φ of $C^*(\underset{\sim}{B})$ satisfying

$$\|\varphi\|_c \leq 1, \ \varphi^* = \varphi , \qquad \ldots (17)$$

$$|\langle T_\varphi\xi, \xi\rangle - 1| < \delta . \qquad \ldots (18)$$

From (15), (18), and the assumption that $\alpha_{p_i}^c \to \alpha_p^c$ pointwise, we conclude that

$$|\langle T^i_\varphi \xi^i, \xi^i\rangle - 1| < \delta \quad \text{for all large enough} \quad i. \qquad \ldots(19)$$

Since $\|\xi\| = \|\xi^i\| = 1$, $\|T_\varphi\| \leq 1$, and $\|T^i_\varphi\| \leq 1$ (by (17)), it follows from (18), (19), Lemma 20.1, and the definition of δ, that

$$\|T_\varphi \xi - \xi\| < \varepsilon, \|T^i_\varphi \xi^i - \xi^i\| < \varepsilon \quad \text{for all large} \quad i. \qquad \ldots(20)$$

Let us write φb instead of $\varphi m'_b$. By the continuity of $b \mapsto \varphi b$ mentioned above, the set $\{\varphi b : b \in D\}$ is norm-compact in $C^*(\underline{B})$. From this and Lemma 20.2 applied to the net $\{\alpha^c_{p_i} - \alpha^c_p\}$, we find:

$$\langle T^i_{\varphi b} \xi^i, \xi^i\rangle \to \langle T_{\varphi b}\xi, \xi\rangle \quad \text{uniformly for} \quad b \in D.$$

Since $\varphi^* = \varphi$ this implies that for all large enough i

$$|\langle T^i_b \xi^i, T^i_\varphi \xi^i\rangle - \langle T_b \xi, T_\varphi \xi\rangle| < \varepsilon \quad \text{for all} \quad b \quad \text{in} \quad D. \qquad \ldots(21)$$

Combining (15), (20), and (21) (and recalling that $\|b\| \leq 1$ for $b \in D$) we find that for all large i

$$|p_i(b) - p(b)| < 3\varepsilon \quad \text{for all} \quad b \quad \text{in} \quad D.$$

Thus (16) has been proved. Therefore $p_i \to p$; and the mapping (III) is continuous.

The continuity of the maps (I), (II), (III) establishes the proposition. □

__Remark__. The above proof closely imitates that given in Dixmier [1], 13.5.2 for the group case.

By means of the above proposition, many of the results of §6, when transferred to the context of Banach *-algebraic bundles, can be stated in terms of uniform-on-compacta convergence of functionals of positive type.

To begin with, we shall reformulate the original definition of the regional topology of the space of all *-representations of $\underset{\sim}{B}$ in the light of Prop. 20.3.

Let $\underset{\sim}{T}^0$ be the family of all non-degenerate *-representations of $\underset{\sim}{B}$. Let T be a fixed *-representation in $\underset{\sim}{T}^0$, and let Z be a fixed family of vectors in $X(T)$ which generates $X(T)$ (i.e., the smallest closed T-stable subspace containing Z is $X(T)$ itself). If $\varepsilon > 0$, D is a compact subset of B, and $\xi_1,\cdots,\xi_n$ is any finite sequence of vectors in $X(T)$, let $U''(T;\varepsilon;\{\xi_i\};D)$ be the set of all T' in $\underset{\sim}{T}^0$ such that there exist vectors $\xi'_1,\cdots,\xi'_n$ in $X(T')$ satisfying (i) $|\langle\xi'_i,\xi'_j\rangle - \langle\xi_i,\xi_j\rangle| < \varepsilon$ for all $i,j = 1,\cdots,n$, and (ii) $|\langle T'_b\xi'_i,\xi'_j\rangle - \langle T_b\xi_i,\xi_j\rangle| < \varepsilon$ for all $i,j = 1,\cdots,n$ and all b in D.

<u>Theorem 20.4</u>. <u>The set</u> $\underset{\sim}{U}''$ <u>of all</u> $U''(T;\varepsilon;\{\xi_i\};D)$, <u>where</u> ε <u>runs over all positive numbers,</u> D <u>over all compact subsets of</u> B, <u>and</u> $\{\xi_i\}$ <u>over all finite sequences of vectors in</u> Z, <u>is a basis of neighborhoods of</u> T <u>in the regional topology of</u> $\underset{\sim}{T}^0$.

<u>Proof</u>. For each S in $\underset{\sim}{T}^0$ let $\tilde{S}$ be the integrated form of S on $\underset{\sim}{L}(\underset{\sim}{B})$. It is clear from Prop. 6.1 that every regional neighborhood of $\tilde{S}$ in $\tilde{\underset{\sim}{T}}^0$ contains the image under $S \mapsto \tilde{S}$ of some set in $\underset{\sim}{U}''$. So it is enough to show that every set in $\underset{\sim}{U}''$ is in fact a regional neighborhood of T. That is, we fix $U'' = U''(T;\varepsilon;\{\xi_i\};D) \in \underset{\sim}{U}''$, and also a net $\{T^\nu\}$ converging to T in the regional topology of $\underset{\sim}{T}^0$; and we shall show that some subnet of $\{T^\nu\}$ lies entirely in U''.

By §6(6),(7) (applied to the integrated forms) we can replace

$\{T^\nu\}$ by a subnet and find vectors $\{\xi_i^\nu\}$ $(i = 1,\cdots,n)$ in $X(T^\nu)$ such that: (i) $\langle\xi_i^\nu,\xi_j^\nu\rangle \underset{\nu}{\longrightarrow} \langle\xi_i,\xi_j\rangle$ for all $i,j = 1,\cdots,n$, and (ii) $\langle\tilde{T}_f^\nu\xi_i^\nu,\xi_j^\nu\rangle \underset{\nu}{\longrightarrow} \langle\tilde{T}_f\xi_i,\xi_j\rangle$ for all $i,j = 1,\cdots,n$ and f in $L_1(B;\lambda)$. It follows from (i) and (ii) that if $c_1,\cdots,c_n$ are any fixed complex numbers, and if we put $\eta^\nu = \Sigma_{i=1}^n c_i\xi_i^\nu$, $\eta = \Sigma_{i=1}^n c_i\xi_i$, then $\|\eta^\nu\| \to \|\eta\|$ and $\langle\tilde{T}_f^\nu\eta^\nu,\eta^\nu\rangle \to \langle\tilde{T}_f\eta,\eta\rangle$ for all f in $L_1(B;\lambda)$. From this and Prop. 20.3 (trivially extended to cover the case that $\|\eta^\nu\| \to \|\eta\|$ instead of $\|\eta^\nu\| = \|\eta\| = 1$) we deduce that

$$\langle T_b^\nu\eta^\nu,\eta^\nu\rangle \to \langle T_b\eta,\eta\rangle \qquad \ldots(22)$$

uniformly in b on compact subsets of B. In particular, for fixed i,j, (22) holds when $\eta = \xi_i \pm \xi_j$ and $\eta^\nu = \xi_i^\nu \pm \xi_j^\nu$, and also when $\eta = \xi_i \pm i\xi_j$ and $\eta^\nu = \xi_i^\nu \pm i\xi_j^\nu$. Hence the polarization identity applied to (22) shows that for each $i,j = 1,\cdots,n$

$$\langle T_b^\nu\xi_i^\nu,\xi_j^\nu\rangle \to \langle T_b\xi_i,\xi_j\rangle \text{ uniformly in } b \text{ on compact sets.} \qquad \ldots(23)$$

Now from (i) and (23) it follows that T^ν lies in U'' for all large ν. This completes the proof. □

A functional p of positive type on B will be said to be <u>associated with</u> a *-representation T of B if $p = p_{T,\xi}$ (see (4)) for some ξ in $X(T)$.

If T is cyclic, Theorem 20.4 takes the following form:

<u>Corollary 20.5</u>. <u>Let T be a cyclic *-representation of B, with unit cyclic vector ξ; and let V be a family of non-degenerate *-representations of B. Then T belongs to the regional closure of V if and only if there is a net $\{q_i\}$ of elements of $P(B)$, each associated with some *-representation in V, such that</u>

$$q_i(b) \longrightarrow p_{T,\xi}(b)$$

<u>uniformly in</u> b <u>on each compact subset of</u> B .

A one-dimensional *-representation of $\underset{\sim}{B}$ will be regarded as a continuous map $\varphi: B \to \mathbb{C}$ which is linear on each fiber and satisfies $\varphi(bc) = \varphi(b)\,\varphi(c)$, $\varphi(b^*) = \overline{\varphi(b)}$ $(b, c \in B)$. Thus φ itself is the unique element of $\underset{\sim}{P}(\underset{\sim}{B})$ associated with φ . Therefore we have by Cor. 20.5:

<u>Corollary 20.6</u>. <u>The regional topology, relativized to the space of all non-zero one-dimensional *-representations of</u> $\underset{\sim}{B}$, <u>coincides with the topology of uniform convergence on compact subsets of</u> B .

<u>Remark</u>. It would be useful to characterize the elements $p_{T,\xi}$ of the set $\underset{\sim}{P}(\underset{\sim}{B})$ in terms of their behavior on B , rather than in terms of T and ξ . Let us assume that $\underset{\sim}{B}$ has an approximate unit consisting of elements $\{u_i\}$ such that $\|u_i\| \le 1$ (for all i). In that case it can be shown that $\underset{\sim}{P}(\underset{\sim}{B})$ <u>consists of all functionals</u> p <u>of positive type on</u> $\underset{\sim}{B}$ <u>such that</u> $\|p\| = 1$ (<u>or, equivalently,</u> $\|p|B_e\| = 1$). In particular, if $\underset{\sim}{B}$ is the group bundle of G , $\underset{\sim}{P}(\underset{\sim}{B})$ consists of all functions q of positive type on G such that $q(e) = 1$. We shall omit the proof of this fact, since it will not be needed for what follows.

<u>The continuity of restriction and the tensor product</u>.

We conclude this section with some observations on the continuity of operations on *-representations of Banach *-algebraic bundles.

<u>Proposition 20.7</u>. <u>Let</u> H <u>be a closed subgroup of</u> G , <u>and</u> $\underset{\sim}{B}_H$ <u>the reduction of</u> $\underset{\sim}{B}$ <u>to</u> H . <u>For any non-degenerate *-representation</u> T

of $\underset{\sim}{B}$, the restriction $T|\underset{\sim}{B}_H$ of T to $\underset{\sim}{B}_H$ is a non-degenerate *-representation of $\underset{\sim}{B}_H$; and the map $T \mapsto T|\underset{\sim}{B}_H$ is continuous with respect to the regional topologies of the spaces of *-representations of $\underset{\sim}{B}$ and of $\underset{\sim}{B}_H$.

Proof. The non-degeneracy of $T|\underset{\sim}{B}_H$ follows from Prop. 14.1. The continuity of the map $T \mapsto T|\underset{\sim}{B}_H$ is an immediate consequence of Theorem 20.4. □

We shall now establish the continuity of the inner tensor product operation $\otimes$ studied in §14.

Let $\underset{\sim}{U}$ be the space of all unitary representations of G; and as before let $\underset{\sim}{T}^0$ be the space of all non-degenerate *-representations of $\underset{\sim}{B}$.

Proposition 20.8. The inner tensor product operation $\gamma: (U,T) \mapsto U \otimes T$ is continuous on $\underset{\sim}{U} \times \underset{\sim}{T}^0$ to $\underset{\sim}{T}^0$ with respect to the regional topologies of $\underset{\sim}{U}$ and $\underset{\sim}{T}^0$.

Proof. Suppose that $U^\nu \to U$ in $\underset{\sim}{U}$ and $T^\nu \to T$ in $\underset{\sim}{T}^0$, and that W is an element of some basis of neighborhoods of $U \otimes T$. It is enough to show that some subnet of $\{U^\nu \otimes T^\nu\}$ lies entirely in W. Thus, by Theorem 20.4 it is enough to take n-termed sequences $\xi_1,\cdots,\xi_n$ and $\eta_1,\cdots\eta_n$ of elements of $X(U)$ and $X(T)$ respectively, and show that, on passing to a subnet of $\{U^\nu \otimes T^\nu\}$, we can find elements $\xi_1^\nu,\cdots,\xi_n^\nu$ of $X(U^\nu)$ and $\eta_1^\nu,\cdots,\eta_n^\nu$ of $X(T^\nu)$ such that for all i,j

$$\langle \xi_i^\nu \otimes \eta_i^\nu, \xi_j^\nu \otimes \eta_j^\nu \rangle \underset{\nu}{\longrightarrow} \langle \xi_i \otimes \eta_i, \xi_j \otimes \eta_j \rangle \qquad \ldots(24)$$

and

$$\langle (U^\nu \otimes T^\nu)_b(\xi_i^\nu \otimes \eta_i^\nu), \xi_j^\nu \otimes \eta_j^\nu \rangle \underset{\nu}{\longrightarrow} \langle (U \otimes T)_b(\xi_i \otimes \eta_i), \xi_j \otimes \eta_j \rangle \ldots(25)$$

uniformly in b on compact subsets of B .

To obtain such vectors ξ_i^ν and η_i^ν , we apply Theorem 20.4 to U and T separately. Thus we can pass twice to a subnet and find elements $\xi_1^\nu, \cdots, \xi_n^\nu$ of $X(U^\nu)$ and $\eta_1^\nu, \cdots, \eta_n^\nu$ of $X(T^\nu)$ such that for all i,j

$$\langle \xi_i^\nu, \xi_j^\nu \rangle \to \langle \xi_i, \xi_j \rangle , \qquad \ldots(26)$$

$$\langle U_x^\nu \xi_i^\nu, \xi_j^\nu \rangle \to \langle U_x \xi_i, \xi_j \rangle \text{ uniformly on compact sets} , \qquad \ldots(27)$$

$$\langle \eta_i^\nu, \eta_j^\nu \rangle \to \langle \eta_i, \eta_j \rangle , \qquad \ldots(28)$$

$$\langle T_b^\nu \eta_i^\nu, \eta_j^\nu \rangle \to \langle T_b \eta_i, \eta_j \rangle \text{ uniformly on compact sets} . \qquad \ldots(29)$$

Now, multiplying (26) and (28), we get (24). Similarly, multiplying (27) and (29), we get (25). □

CHAPTER III

INDUCED REPRESENTATIONS AND IMPRIMITIVITY FOR BANACH *-ALGEBRAIC BUNDLES

§21. The framework of the induction process in the bundle context. Positivity with respect to a Banach *-algebraic bundle.

The well-known classical definition of induced representations of a finite group G has been set down in the Introduction and in §4(7), (8), (9). This definition was generalized by Mackey [4] to arbitrary separable locally compact groups (the restriction of separability being removed by Blattner [1]). Generalizing the observation concerning finite groups made in §4, Rieffel [1] showed that the Mackey-Blattner definition of induced representations of locally compact groups is a special case of the abstract inducing process described in §4. The first part of the present chapter will generalize the Mackey-Blattner construction to develop a theory of induced representations of arbitrary Banach *-algebraic bundles over locally compact groups, exhibiting them as special cases of Rieffel's abstract inducing process of §4.

Throughout this section G is a fixed locally compact group with unit element e, left Haar measure λ, and modular function Δ; and H is a closed subgroup of G, with left Haar measure ν and modular function δ.

Measures on the coset space.

To begin with, we shall remind the reader of certain facts about the relationship between functions on G and measures on G/H.

Definition. An H-rho-function (or simply rho-function if no ambiguity can arise) on G is a complex-valued function ρ on G satisfying

$$\rho(xh) = \Delta(h)^{-1}\ \delta(h)\ \rho(x) \qquad \ldots(1)$$

for all x in G and h in H.

Notice that continuous rho-functions exist in abundance. Indeed, for any f in $\underset{\sim}{L}(G)$, the equation

$$\rho_f(x) = \int_H \Delta(h)\ \delta(h)^{-1}\ f(xh)\ d\nu h \quad (x \in G) \qquad \ldots(2)$$

defines a continuous rho-function ρ.

In fact we have:

Proposition 21.1. *There exists a continuous everywhere positive rho-function on* G.

For the proof see Loomis [1]; also Bourbaki [4], Chap. 7, §2, n° 5, Théorème 2.

Now every non-negative continuous rho-function gives rise in a canonical manner to a regular Borel measure on the coset space G/H.

Proposition 21.2. *Let* ρ *be any non-negative continuous rho-function on* G. *Then there is a unique regular Borel measure on* G/H, *which we shall call* $\rho^{\#}$, *such that the double integration formula*

$$\int_G \rho(x)\ f(x)d\lambda x = \int_{G/H} d\rho^{\#}(xH) \int_H f(xh)d\nu h \qquad \ldots(3)$$

holds for all f *in* $\underset{\sim}{L}(G)$.

For the proof see Mackey [4], Lemma 1.5.

If $0 \leq f \in \underset{\sim}{L}(G)$ and ρ_f is the rho-function given by (2), one verifies that

$$\int_{G/H} g(\alpha)\ d(\rho_f)^{\#}\alpha = \int_G f(x)\ g(xH)d\lambda x \qquad \ldots(4)$$

for all g in $\underset{\sim}{L}(G/H)$.

The next two propositions are vital for the manipulation of measures on coset spaces. Their simple proof is left to the reader (see Loomis [1]).

Proposition 21.3. Let ρ be a non-negative continuous rho-function on G, x an element of G, and $x\rho: y \mapsto \rho(x^{-1}y)$ the left x-translate of ρ. Then $x\rho$ is a rho-function; and $(x\rho)^{\#} = x(\rho^{\#})$ (the image of $\rho^{\#}$ under the action of x on G/H).

Proposition 21.4. Let ρ be a non-negative continuous rho-function on G, and φ a non-negative continuous function on G/H. Then $\sigma: x \mapsto \varphi(xH)\,\rho(x)$ is a rho-function; and

$$d\sigma^{\#}\alpha = \varphi(\alpha)\, d\rho^{\#}\alpha .$$

From Propositions 21.3 and 21.4 we easily obtain:

Proposition 21.5. Suppose that ρ is a continuous everywhere positive rho-function on G. Then $\rho^{\#}$ is G-quasi-invariant; that is, for all x in G, $\rho^{\#}$ and $x(\rho^{\#})$ have the same null sets.

The basic conditional expectation.

Now let $\underset{\sim}{B} = (B, \pi, \cdot, {}^*)$ be a Banach *-algebraic bundle over G. By Theorem 10.5 $\underset{\sim}{B}$ automatically has enough continuous cross-sections. As usual, $\underset{\sim}{L}(\underset{\sim}{B})$ denotes the *-algebra of all continuous cross-sections of $\underset{\sim}{B}$ with compact support, with the operations $*$ and * defined in §13(1),(2). Every *-representation T of $\underset{\sim}{B}$ gives rise to a *-representation T' of $\underset{\sim}{L}(\underset{\sim}{B})$, its integrated form (see §15(1)). We shall usually write T_f instead of T'_f for $f \in \underset{\sim}{L}(\underset{\sim}{B})$. Similarly $\underset{\sim}{B}_H$, the reduction of $\underset{\sim}{B}$ to the closed subgroup H, is a Banach *-algebraic bundle over H ; and every *-representation of $\underset{\sim}{B}_H$ has an integrated form which is a *-representation of $\underset{\sim}{L}(\underset{\sim}{B}_H)$.

We shall often want to specialize our results to the group context treated by Mackey, Blattner, and others. For brevity the phrase 'the group case' will be used to refer to the situation when $\underset{\sim}{B}$ is the group bundle. In that case $\underset{\sim}{L}(\underset{\sim}{B}) = \underset{\sim}{L}(G)$; and non-degenerate *-representations of $\underset{\sim}{B}$ are essentially just unitary representations of G .

In the context of the Banach *-algebraic bundle $\underset{\sim}{B}$, the inducing process will consist in taking a (non-degenerate) *-representation S of $\underset{\sim}{B}_H$ and constructing from it a (non-degenerate) *-representation T of $\underset{\sim}{B}$. To fit this into the pattern of the abstract inducing process of §4, it will be necessary to set up an $\underset{\sim}{L}(\underset{\sim}{B}_H)$-rigged $\underset{\sim}{L}(\underset{\sim}{B})$-module $\underset{\sim}{M}$. Then T will be defined as that *-representation of $\underset{\sim}{B}$ (if such exists) whose integrated form is induced via $\underset{\sim}{M}$ from the integrated form of S . The example of finite groups in §4 suggests that $\underset{\sim}{M}$ is going to come from an $\underset{\sim}{L}(\underset{\sim}{B})$, $\underset{\sim}{L}(\underset{\sim}{B}_H)$ conditional expectation p ; and that p will essentially be the operation of restricting a function on G to H . With this in mind, we make the following definition:

<u>Definition</u>. Let $p\colon \underset{\sim}{L}(\underset{\sim}{B}) \to \underset{\sim}{L}(\underset{\sim}{B}_H)$ be the linear map given by

$$p(f)(h) = f(h)(\Delta(h))^{\frac{1}{2}}(\delta(h))^{-\frac{1}{2}} (f \in \underset{\sim}{L}(\underset{\sim}{B});\ h \in H) . \qquad \ldots(5)$$

Except for the factor $\Delta(h)^{\frac{1}{2}}\ \delta(h)^{-\frac{1}{2}}$, p is just the operation of restriction to H . The insertion of this factor makes p self-adjoint; that is, we have

$$p(f^*) = (p(f))^* \qquad (f \in \underset{\sim}{L}(\underset{\sim}{B})) . \qquad \ldots(6)$$

Indeed: If $h \in H$ and $f \in \underset{\sim}{L}(\underset{\sim}{B})$,

$$p(f^*)(h) = f(h^{-1})^* \; \Delta(h^{-1}) \; \Delta(h)^{\frac{1}{2}} \; \delta(h)^{-\frac{1}{2}}$$
$$= f(h^{-1})^* \; \Delta(h^{-1})^{\frac{1}{2}} \; \delta(h^{-1})^{-\frac{1}{2}} \; \delta(h)^{-1}$$
$$= (p(f)(h^{-1}))^* \; \delta(h^{-1})$$
$$= (p(f))^*(h) \; .$$

So (6) holds.

In order to speak of an $\underset{\sim}{L}(\underset{\sim}{B})$, $\underset{\sim}{L}(\underset{\sim}{B}_H)$ conditional expectation, we must make $\underset{\sim}{L}(\underset{\sim}{B}_H)$ act to the right on $\underset{\sim}{L}(\underset{\sim}{B})$. This we do as follows:

$$(f\varphi)(x) = \int_H f(xh^{-1}) \; \varphi(h) \, (\delta(h)\Delta(h))^{-\frac{1}{2}} d\nu h \qquad \ldots(7)$$

$(f \in \underset{\sim}{L}(\underset{\sim}{B});\ \varphi \in \underset{\sim}{L}(\underset{\sim}{B}_H);\ x \in G)$. The integrand on the right of (7) is continuous on H to B_x with compact support; so the right side exists as a B_x-valued Bochner integral. The cross-section $f\varphi$ of $\underset{\sim}{B}$ defined by (7) has compact support, and is continuous by Lemma 10.14. So $f\varphi \in \underset{\sim}{L}(\underset{\sim}{B})$. Obviously $f\varphi$ is linear in f and in φ. I claim that (7) makes the linear space underlying $\underset{\sim}{L}(\underset{\sim}{B})$ into a right $\underset{\sim}{L}(\underset{\sim}{B}_H)$-module, that is,

$$(f\varphi_1)\varphi_2 = f(\varphi_1 * \varphi_2) \qquad (f \in \underset{\sim}{L}(\underset{\sim}{B});\ \varphi_1,\varphi_2 \in \underset{\sim}{L}(\underset{\sim}{B}_H)) \; . \qquad \ldots(8)$$

Indeed: By (7) and Prop. 10.13,

$$((f\varphi_1)\varphi_2)(x) = \int_H \int_H f(xh^{-1}k^{-1}) \varphi_1(k) \varphi_2(h) (\delta(kh)\Delta(kh))^{-\frac{1}{2}} d\nu k \; d\nu h \; .$$

Making the substitutions $h \mapsto h^{-1}$ and $k \mapsto kh$, and using Fubini's Theorem 10.15, we get

$$((f\varphi_1)\varphi_2)(x) = \int_H \int_H f(xk^{-1}) \; \varphi_1(kh) \; \varphi_2(h^{-1}) (\delta(k)\Delta(k))^{-\frac{1}{2}} \; d\nu h \; d\nu k$$
$$= (f(\varphi_1 * \varphi_2))(x) \; ;$$

and (8) is proved.

We are now in a position to assert the crucial relationship:

$$p(f\varphi) = p(f) * \varphi \qquad (f \in \underset{\sim}{L}(\underset{\sim}{B});\ \varphi \in \underset{\sim}{L}(\underset{\sim}{B}_H)) \ . \qquad \ldots(9)$$

This is easily verified. Relations (6) and (9) now show that p is an $\underset{\sim}{L}(\underset{\sim}{B})$, $\underset{\sim}{L}(\underset{\sim}{B}_H)$ conditional expectation in the sense of §4.

Definition. We call p the basic $\underset{\sim}{L}(\underset{\sim}{B})$, $\underset{\sim}{L}(\underset{\sim}{B}_H)$ conditional expectation.

A calculation similar to that which led to (8) shows that the right action (7) of $\underset{\sim}{L}(\underset{\sim}{B}_H)$ commutes with left multiplication by elements of $\underset{\sim}{L}(\underset{\sim}{B})$:

$$f * (g\varphi) = (f * g)\varphi \qquad (f,g \in \underset{\sim}{L}(\underset{\sim}{B});\ \varphi \in \underset{\sim}{L}(\underset{\sim}{B}_H)) \qquad \ldots(10)$$

We also notice for later use that $f\varphi$ is separately continuous in f and φ in the inductive limit topologies of $\underset{\sim}{L}(\underset{\sim}{B})$ and $\underset{\sim}{L}(\underset{\sim}{B}_H)$.

Remark 0. In the group case, when $\underset{\sim}{L}(\underset{\sim}{B}) = \underset{\sim}{L}(G)$ and $\underset{\sim}{L}(\underset{\sim}{B}_H) = \underset{\sim}{L}(H)$, we can form the measure convolution $f*\varphi$ whenever $f \in \underset{\sim}{L}(G)$ and $\varphi \in \underset{\sim}{L}(H)$; and (7) becomes, not $f\varphi = f*\varphi$ as one might have expected, but

$$f\varphi = f * \alpha(\varphi) \ , \qquad \ldots(11)$$

where α is the automorphism (not in general a *-automorphism) of $\underset{\sim}{L}(H)$ given by

$$\alpha(\varphi)(h) = \varphi(h)\ \Delta(h)^{\frac{1}{2}}\ \delta(h)^{-\frac{1}{2}} \qquad \ldots(12)$$

The presence of the α in (11) is necessary in order to ensure (9).

Positivity with respect to $\underset{\sim}{B}$.

We now make a very important definition:

Definition. A *-representation S of $\underset{\sim}{B}_H$ is said to be positive with respect to $\underset{\sim}{B}$ (or $\underset{\sim}{B}$-positive) if the integrated form of S is positive with respect to the basic conditional expectation p , that is, if

$$S_{p(f^{*}*f)} \geq 0 \quad \text{for all } f \text{ in } L(\underset{\sim}{B}) .$$

Remark 1. Any *-representation of $\underset{\sim}{B}$ itself is obviously $\underset{\sim}{B}$-positive.

Remark 2. In general S need not be positive with respect to $\underset{\sim}{B}$. However there are important classes of Banach *-algebraic bundles (including the group bundles) for which $\underset{\sim}{B}$-positivity always holds. See for example Cor. 22.4 and Thm. 27.8.

Remark 3. Suppose that S is $\underset{\sim}{B}$-positive. Let $\underset{\sim}{X}$ be the Hilbert space deduced from the operator inner product $V: (f,g) \mapsto S_{p(g^{*}*f)}$ on $L(\underset{\sim}{B})$; and let $f \tilde{\otimes} \xi$ be the image of $f \otimes \xi$ in $\underset{\sim}{X}$. It was observed in §1 that $\xi \mapsto f \tilde{\otimes} \xi$ is continuous for each fixed f in $L(\underset{\sim}{B})$. I claim that for each fixed ξ in $X(S)$, $f \mapsto f \tilde{\otimes} \xi$ is continuous on $L(\underset{\sim}{B})$ to $\underset{\sim}{X}$ with respect to the inductive limit topology of $L(\underset{\sim}{B})$. Indeed: Since $f \mapsto f \tilde{\otimes} \xi$ is linear, it is enough to take a net $\{f_i\}$ of elements of $L(\underset{\sim}{B})$ converging to 0 uniformly on G and all vanishing outside the same compact set, and show that $\|f_i \tilde{\otimes} \xi\| \to 0$. But our assumption implies that the $\{f_i^{*} * f_i\}$ converge to 0 uniformly on G and all vanish outside the same compact set; and this in turn implies that

$$\|f_i \tilde{\otimes} \xi\|^2 = \langle S_{p(f^{*}*f)} \xi, \xi \rangle \to 0 .$$

Now our next goal is as follows: Starting from an arbitrary $\underset{\sim}{B}$-positive *-representation S of $\underset{\sim}{B}_H$, we shall show (in §25) that the integrated form of S is inducible via p to a *-representation T of

$\underset{\sim}{L}(\underset{\sim}{B})$, and that T is the integrated form of a (unique) *-representation, also called T , of $\underset{\sim}{B}$. This T will then be called the *-representation of $\underset{\sim}{B}$ induced from S .

§22. Equivalent conditions for positivity.

We keep the notation of §21.

Let S be a $\underset{\sim}{B}$-positive *-representation of $\underset{\sim}{B}_H$. It turns out that the Hilbert space in which the *-representation T of $\underset{\sim}{B}$ induced by S will act has a natural presentation as the $\underset{\sim}{L}_2$ space of a Hilbert bundle over the coset space G/H . From this presentation we shall be able to deduce (in §26) that, in the group case, T coincides with the induced representation of G as constructed by Mackey and Blattner. To obtain the required Hilbert bundle over G/H one needs a different formulation of the definition of $\underset{\sim}{B}$-positivity. Most of this section is devoted to obtaining this equivalent formulation.

Recall that, for each coset α in G/H , $\underset{\sim}{B}_\alpha$ denotes the (Banach bundle) reduction of $\underset{\sim}{B}$ to the closed subset α of G . The underlying set of $\underset{\sim}{B}_\alpha$ is $B_\alpha = \cup_{x\in\alpha} B_x$; and $\underset{\sim}{L}(\underset{\sim}{B}_\alpha)$ is of course the linear space of all continuous cross-sections of $\underset{\sim}{B}_\alpha$ with compact support.

Let us fix a coset α in G/H . It is very easy to give to $\underset{\sim}{L}(\underset{\sim}{B}_\alpha)$ the structure of an $\underset{\sim}{L}(\underset{\sim}{B}_H)$-rigged space. Indeed, one defines a right $\underset{\sim}{L}(\underset{\sim}{B}_H)$-module structure for $\underset{\sim}{L}(\underset{\sim}{B}_\alpha)$ as follows:

$$(f\varphi)(x) = \int_H f(xh)\ \varphi(h^{-1})d\nu h \qquad \ldots(1)$$

$(f \in \underset{\sim}{L}(\underset{\sim}{B}_\alpha);\ \varphi \in \underset{\sim}{L}(\underset{\sim}{B}_H);\ x \in \alpha)$, and an $\underset{\sim}{L}(\underset{\sim}{B}_H)$-valued rigging $[\ ,\ ]_\alpha$ as follows:

$$[f,g]_\alpha(h) = \int_H (f(xk))^*\ g(xkh)d\nu k \qquad \ldots(2)$$

$(f,g \in \underset{\sim}{L}(\underset{\sim}{B}_\alpha);\ h \in H)$, where x is a fixed element of α . The left-invariance of ν shows that the right side of (2) is actually independent of the choice of x . One verifies immediately that $(\underset{\sim}{L}(\underset{\sim}{B}_\alpha),$ $[\ ,\ ]_\alpha)$ satisfies the postulates for a $\underset{\sim}{L}(\underset{\sim}{B}_H)$-rigged space.

Proposition 22.1. Let S be a *-representation of $\underset{\sim}{B}_H$. The following three conditions are equivalent:

(I) S is $\underset{\sim}{B}$-positive.

(II) For every coset α in G/H , S is positive with respect to $(\underset{\sim}{L}(\underset{\sim}{B}_\alpha), [\ ,\ ]_\alpha)$; that is,

$$S_{[f,f]_\alpha} \geq 0 \quad \text{for all } \alpha \text{ in } G/H \text{ and } f \text{ in } \underset{\sim}{L}(\underset{\sim}{B}_\alpha) \ . \qquad \ldots(3)$$

(III) For every coset α in G/H , every positive integer n , all $b_1,\cdots,b_n$ in B_α , and all $\xi_1,\cdots,\xi_n$ in $X(S)$,

$$\sum_{i,j=1}^{n} \langle S_{b_j^* b_i} \xi_i, \xi_j\rangle \geq 0 \ . \qquad \ldots(4)$$

Note. (4) makes sense since $b_j^* b_i \in B_H$ whenever b_i and b_j both belong to the same B_α .

Proof. The first step is to express $\langle S_{p(g^* * f)} \xi, \eta\rangle$ in a new form. Choose a continuous everywhere positive H-rho function ρ on G (see Prop. 21.1); and let $\rho^\#$ be the corresponding regular Borel measure on G/H (Prop. 21.2). I claim that for all f,g in $\underset{\sim}{L}(\underset{\sim}{B})$ and ξ,η in $X(S)$

$$\langle S_{p(g^* * f)} \xi, \eta\rangle =$$

$$\int_{G/H} d\rho^\#(xH) \int_H \int_H (\rho(xh)\rho(xk))^{-\frac{1}{2}} \langle S_{(g(xh))^* f(xk)} \xi, \eta\rangle \, d\nu h \, d\nu k \ . \qquad \ldots(5)$$

Indeed: If $h \in H$ we have $(g^* * f)(h) = \int_G g(x)^* f(xh) d\lambda x$ $(B_h$-

valued Bochner integral); so by Prop. 10.13

$$\langle S_{(g^* * f)(h)}\xi,\eta\rangle = \int_G \langle S_{g(x)^* f(xh)}\xi,\eta\rangle\, d\lambda x \quad .$$

Thus

$$\begin{aligned}\langle S_{p(g^* * f)}\xi,\eta\rangle &= \int_H \Delta(h)^{\frac{1}{2}}\,\delta(h)^{-\frac{1}{2}}\,\langle S_{(g^* * f)(h)}\xi,\eta\rangle\, d\nu h\\ &= \int_H\int_G \Delta(h)^{\frac{1}{2}}\,\delta(h)^{-\frac{1}{2}}\,\langle S_{g(x)^* f(xh)}\xi,\eta\rangle\, d\lambda x\, d\nu h\\ &= \int_H\int_{G/H} I(xH,h)\, d\rho^{\#}(xH)\, d\nu h \quad (\text{by } \S 21(3)) ,\end{aligned}$$

where $I(xH,h) = \Delta(h)^{\frac{1}{2}}\,\delta(h)^{-\frac{1}{2}}\int_H \rho(xk)^{-1}\langle S_{g(xk)^* f(xkh)}\xi,\eta\rangle\, d\nu k$.

Now I is continuous on $(G/H)\times H$ with compact support. So, applying Fubini's Theorem, we get

$$\langle S_{p(g^* * f)}\xi,\eta\rangle =$$

$$= \int_{G/H} d\rho^{\#}(xH)\int_H\int_H \rho(xk)^{-1}\,\Delta(h)^{\frac{1}{2}}\,\delta(h)^{-\frac{1}{2}}\,\langle S_{g(xk)^* f(xkh)}\xi,\eta\rangle\, d\nu k\, d\nu h$$

$$= \int_{G/H} d\rho^{\#}(xH)\int_H\int_H \rho(xk)^{-1}\,\Delta(k^{-1}h)^{\frac{1}{2}}\,\delta(k^{-1}h)^{-\frac{1}{2}}\,\langle S_{g(xk)^* f(xh)}\xi,\eta\rangle\, d\nu h\, d\nu k$$

$$= \int_{G/H} d\rho^{\#}(xH)\int_H\int_H (\rho(xk)\rho(xh))^{-\frac{1}{2}}\,\langle S_{g(xk)^* f(xh)}\xi,\eta\rangle\, d\nu h\, d\nu k$$

(using the rho-function identity §21(1)). But this is (5), and the claim is proved.

Now assume (II); and let $\xi \in X(S)$, $f \in \underset{\sim}{L}(\underset{\sim}{B})$. If $x \in \alpha \in G/H$, we obtain from (2) and (3) (on replacing the f in (3) by $(\rho^{-\frac{1}{2}}f)|\alpha)$:

$$\int_H\int_H (\rho(xh)\rho(xk))^{-\frac{1}{2}}\,\langle S_{f(xh)^* f(xk)}\xi,\xi\rangle\, d\nu h\, d\nu k \geq 0 \ .$$

This and (5) together imply that

$$\langle S_{p(f^* * f)}\xi, \xi\rangle \geq 0$$

for all ξ in $X(S)$ and f in $\underset{\sim}{L}(\underset{\sim}{B})$. So (II) => (I).

Conversely, we shall show that (I) => (II). Let g be in $\underset{\sim}{L}(\underset{\sim}{B})$ and ξ in $X(S)$; and let σ be any continuous complex function on G which is constant on each coset xH. Assuming (I), we have by (5):

$$0 \leq \langle S_{p((\sigma g)^* * (\sigma g))}\xi, \xi\rangle$$

$$= \int_{G/H} d\rho^{\#}(xH)|\sigma(x)|^2 \int_H \int_H (\rho(xh)\rho(xk))^{-\frac{1}{2}} \langle S_{g(xh)^* g(xk)}\xi, \xi\rangle \, d\nu h \, d\nu k . \quad \ldots(6)$$

Since g is continuous with compact support, the inner double integral in (6) is continuous as a function on G/H. So by (6) and the arbitrariness of σ one deduces that

$$\int_H \int_H (\rho(xh)\rho(xk))^{-\frac{1}{2}} \langle S_{g(xh)^* g(xk)}\xi, \xi\rangle \, d\nu h \, d\nu k \geq 0$$

for all x in G. Replacing g by $\rho^{\frac{1}{2}} g$ in this inequality, we obtain by (2)

$$S_{[f,f]_\alpha} \geq 0 \quad \ldots(7)$$

for all α in G/H and all f in $\underset{\sim}{L}(\underset{\sim}{B}_\alpha)$ which are of the form $g|\alpha$ for some g in $\underset{\sim}{L}(\underset{\sim}{B})$. But by Theorem 10.7 every f in $\underset{\sim}{L}(\underset{\sim}{B}_\alpha)$ is of this form. So (7) holds for all f in $\underset{\sim}{L}(\underset{\sim}{B}_\alpha)$; and we have shown that (I) => (II).

We shall now show that (III) => (II). Assume (III); and let $x \in \alpha \in G/H$; $f \in \underset{\sim}{L}(\underset{\sim}{B}_\alpha)$; $\xi \in X(S)$. Then

$$\langle S_{[f,f]_\alpha}\xi, \xi\rangle = \int_H \int_H \langle S_{f(xk)^* f(xh)}\xi, \xi\rangle \, d\nu h \, d\nu k . \quad \ldots(8)$$

Now the right side of (8) can be approximated by finite sums of the form

$$\sum_{r,s=1}^{n} \nu(E_r)\nu(E_s) \langle S_{f(xh_s)^* f(xh_r)} \xi, \xi\rangle \qquad \ldots(9)$$

where the $E_1,\cdots,E_n$ are Borel subsets of H (with compact closure), and $h_r \in E_r$. By (III) the summation (9) is non-negative. So the right side of (8) is non-negative, and (II) holds.

Finally we must verify that (II) => (III). Fix $x \in \alpha \in G/H$; $\xi_1,\cdots,\xi_n \in X(S)$; $b_1,\cdots,b_n \in B_\alpha$. We can then choose $f_1',\cdots,f_n'$ in $\underset{\sim}{L}(\underset{\sim}{B}_\alpha)$ and $h_1,\cdots,h_n$ in H so that $f_i'(xh_i) = b_i$. Now let $f_i = \iota_i f_i'$, where ι_i is a non-negative element of $\underset{\sim}{L}(\alpha)$ which vanishes outside a small neighborhood U_i of xh_i and for which $\int_H \iota_i(xh)d\nu h = 1$. Assuming (II), we know from Prop. 4.1 that $(f,g) \mapsto S_{[g,f]_\alpha}$ is an operator inner product on $\underset{\sim}{L}(\underset{\sim}{B}_\alpha)$, and hence that

$$\begin{aligned} 0 &\le \sum_{i,j=1}^{n} \langle S_{[f_j,f_i]_\alpha} \xi_i, \xi_j\rangle \\ &= \sum_{i,j=1}^{n} \int_H\int_H \langle S_{f_j(xh)^* f_i(xk)} \xi_i, \xi_j\rangle \, d\nu h\, d\nu k \\ &= \sum_{i,j=1}^{n} \int_H \int_H \iota_j(xh)\iota_i(xk) \langle S_{f_j'(xh)^* f_i'(xk)} \xi_i, \xi_j\rangle \, d\nu h\, d\nu k . \end{aligned}$$

As the U_i shrink down to xh_i, the last expression approaches $\Sigma_{i,j=1}^{n} \langle S_{f_j'(xh_j)^* f_i'(xh_i)} \xi_i, \xi_j\rangle = \Sigma_{i,j=1}^{n} \langle S_{b_j^* b_i} \xi_i, \xi_j\rangle$. So the latter is non-negative, and (III) is proved.

We have now proved the equivalence of (I), (II), and (III). □

<u>Remark 1</u>. Let G_d be the same group as G only with the discrete topology, and $\underset{\sim}{B}^d$ the Banach *-algebraic bundle over G_d coinciding with $\underset{\sim}{B}$ except for its topology. In view of Prop. 4.1, condition (III)

of Prop. 22.1 asserts that for each α in G/H , S is positive with respect to the $\mathcal{L}((\mathcal{B}^d)_H)$-rigged space $(\mathcal{L}((\mathcal{B}^d)_\alpha), [\ ,\ ]^d_\alpha)$, constructed like $\mathcal{L}(\mathcal{B}_\alpha), [\ ,\ ]_\alpha$ except that we start from $\mathcal{B}^d$ instead of $\mathcal{B}$.

Thus the concept of $\mathcal{B}$-positivity of a *-representation of $\mathcal{B}_H$ is not altered when the topology of G is replaced by the discrete topology.

Corollary 22.2. *For any *-representation T of $\mathcal{B}$, the restriction of T to $\mathcal{B}_H$ is positive with respect to $\mathcal{B}$. More generally, if M is another closed subgroup of G with $H \subset M$, and if T is a *-representation of $\mathcal{B}_M$ which is positive with respect to $\mathcal{B}$, then $T|\mathcal{B}_H$ is positive with respect to $\mathcal{B}$.*

The proof is an obvious application of Prop. 22.1(III).

Corollary 22.3. *Let M be a closed subgroup of G with $H \subset M$; and let S be a *-representation of $\mathcal{B}_H$ which is positive with respect to $\mathcal{B}$. Then S is positive with respect to $\mathcal{B}_M$.*

This follows immediately from Prop. 22.1(II) or (III).

Corollary 22.4. *If $\mathcal{B}$ has enough unitary multipliers, then every *-representation of $\mathcal{B}_H$ is positive with respect to $\mathcal{B}$. This is the case if $\mathcal{B}$ is a semidirect product bundle (see §12(I)), in particular if $\mathcal{B}$ is the group bundle of G .*

Proof. Let S be a *-representation of $\mathcal{B}_H$. Take an element x of G , elements $b_1,\cdots,b_n$ of B_{xH} , and vectors $\xi_1,\cdots,\xi_n$ in $X(S)$. By hypothesis there is a unitary multiplier u of $\mathcal{B}$ of order x . Setting $c_i = u^*b_i$, we have $c_i \in B_H$ and $b_i = uc_i$. By the associative laws for multipliers, $b_j^*b_i = (c_j^*u^*)(uc_i) = c_j^*(u^*u)c_i = c_j^*c_i$. Thus

$$\sum_{i,j} \langle S_{b_j^* b_i} \xi_i, \xi_j \rangle = \sum_{i,j} \langle S_{c_j^* c_i} \xi_i, \xi_j \rangle$$

$$= \sum_{i,j} \langle S_{c_i} \xi_i, S_{c_j} \xi_j \rangle \geq 0 .$$

So by Prop. 22.1(III) S is $\underset{\sim}{B}$-positive. □

Remark 2. If $\underset{\sim}{B}$ is merely saturated, the conclusion of Cor. 22.4 need no longer hold. For example let $\underset{\sim}{B}$ be the saturated Banach *-algebraic bundle of §12(V) over the two-element group $\{1,-1\}$. The one-dimensional representation $\chi: (r,r) \rightarrow r$ of B_1 is not $\underset{\sim}{B}$-positive, since for $b = (1,-1) \in B_{-1}$ we have $\chi(b^*b) = -1$.

Remark 3. In Theorem 27.8 we shall see that if $\underset{\sim}{B}$ is a saturated C^*-algebraic bundle, every *-representation of $\underset{\sim}{B}_H$ is $\underset{\sim}{B}$-positive. (The example of §12(VI) shows that a saturated C^*-algebraic bundle need not have enough unitary multipliers.)

It follows easily from Prop. 1.3 that a Hilbert direct sum of $\underset{\sim}{B}$-positive *-representations of $\underset{\sim}{B}_H$ is again $\underset{\sim}{B}$-positive.

$\underset{\sim}{B}$-positivity and the regional topology.

We shall now discuss the relation between $\underset{\sim}{B}$-positivity and the regional topology (see §6).

Proposition 22.5. The set $\underset{\sim}{S}^+$ of all $\underset{\sim}{B}$-positive *-representations of $\underset{\sim}{B}_H$ is closed in the regional topology.

Proof. Let W belong to the regional closure of $\underset{\sim}{S}^+$; and let $f \in \underset{\sim}{L}(\underset{\sim}{B})$, $\xi \in X(W)$. By the definition of the regional topology, given $\epsilon > 0$ we can find a *-representation S in $\underset{\sim}{S}^+$ and a vector ξ' in $X(S)$ such that

$$|\langle W_{p(f^* * f)}\xi, \xi\rangle - \langle S_{p(f^* * f)}\xi', \xi'\rangle| < \varepsilon . \qquad \ldots (10)$$

Since $S \in \underset{\sim}{S}^+$, $\langle S_{p(f^* * f)}\xi', \xi'\rangle \geq 0$; and so by (10) $\langle W_{p(f^* * f)}\xi, \xi\rangle > -\varepsilon$. This is true for any $\varepsilon > 0$; hence $\langle W_{p(f^* * f)}\xi, \xi\rangle \geq 0$ for every f in $\underset{\sim}{L}(\underset{\sim}{B})$ and ξ in $X(W)$. Consequently $W \in \underset{\sim}{S}^+$. □

From Prop. 22.5 and the remark on direct sums preceding it we get:

Corollary 22.6. *If W is a *-representation of $\underset{\sim}{B}_H$ which is weakly contained in the set of all $\underset{\sim}{B}$-positive *-representations of $\underset{\sim}{B}_H$, then W is $\underset{\sim}{B}$-positive.*

Proposition 22.7. *Let S and W be two *-representations of $\underset{\sim}{B}_H$ such that (i) W is $\underset{\sim}{B}$-positive, and (ii) $\|S_f\| \leq \|W_f\|$ for all f in $\underset{\sim}{L}(\underset{\sim}{B}_H)$. Then S is $\underset{\sim}{B}$-positive.*

Proof. We may as well assume S and W to be non-degenerate. Taking the A of Prop. 6.8 to be the $\underset{\sim}{L}_1$ cross-sectional algebra of $\underset{\sim}{B}_H$ and the A_0 to be $\underset{\sim}{L}(\underset{\sim}{B}_H)$, we conclude from (ii) and Prop. 6.8 that S is weakly contained in W. From this and (i) the conclusion follows by Cor. 22.6. □

§23. The induced Hilbert bundle.

We maintain the notation and conventions of §21. In addition we will choose once for all a continuous everywhere positive H-rho function ρ on G (see Prop. 21.1), and denote by $\rho^\#$ the regular Borel measure on G/H constructed from ρ (as in Prop. 21.2).

We now fix a non-degenerate $\underset{\sim}{B}$-positive *-representation S of $\underset{\sim}{B}_H$, and write X for $X(S)$. Our first goal is to construct from S

a Hilbert bundle $\underset{\sim}{Y}$ over G/H. To this end, the first step is to construct a Hilbert space Y_α for each coset α in G/H. Let $x \in \alpha \in G/H$. We form the algebraic tensor product $\underset{\sim}{L}(\underset{\sim}{B}_\alpha) \otimes X$, and introduce into it the conjugate-bilinear form $\langle\ ,\ \rangle_\alpha$ given by

$$\langle f \otimes \xi, g \otimes \eta\rangle_\alpha = \int_H \int_H (\rho(xh)\rho(xk))^{-\frac{1}{2}} \langle S_{g(xk)^* f(xh)} \xi, \eta\rangle \, d\nu h \, d\nu k \qquad \ldots(1)$$

($f,g \in \underset{\sim}{L}(\underset{\sim}{B}_\alpha)$; $\xi, \eta \in X$). Notice that the right side of (1) depends only on α (not on x). In fact, if $f \mapsto f'$ is the linear automorphism of $\underset{\sim}{L}(\underset{\sim}{B}_\alpha)$ given by $f'(y) = \rho(y)^{-\frac{1}{2}} f(y)$ $(y \in \alpha)$, (1) can be written in the form

$$\langle f \otimes \xi, g \otimes \eta\rangle_\alpha = \langle S_{[g',f']_\alpha} \xi, \eta\rangle \qquad \ldots(2)$$

(recall §22(2)). It therefore follows from Props. 22.1(II) and 4.1 that $\langle\ ,\ \rangle_\alpha$ is positive. Let Y_α be the Hilbert space completion of the pre-Hilbert space $(\underset{\sim}{L}(\underset{\sim}{B}_\alpha) \otimes X)/N_\alpha$ (N_α being the null space of $\langle\ ,\ \rangle_\alpha$). This Y_α is going to be the fiber of $\underset{\sim}{Y}$ over α. We write $\langle\ ,\ \rangle_\alpha$ also for the inner product in Y_α, and $\|\ \|_\alpha$ for the norm in Y_α; and we denote by κ_α the quotient map of $\underset{\sim}{L}(\underset{\sim}{B}_\alpha) \otimes X$ into Y_α.

One easily verifies:

Proposition 23.1. $\kappa_\alpha(f \otimes \xi)$ is separately continuous in f (with the inductive limit topology of $\underset{\sim}{L}(\underset{\sim}{B}_\alpha)$) and in ξ.

Let Y be the disjoint union of the Y_α $(\alpha \in G/H)$.

Proposition 23.2. There is a unique topology for Y making $\underset{\sim}{Y} = (Y, \{Y_\alpha\})$ a Hilbert bundle over G/H such that for each f in $\underset{\sim}{L}(\underset{\sim}{B})$ and each ξ in X the cross-section

$$\alpha \mapsto \kappa_\alpha((f|\alpha) \otimes \xi) \qquad \ldots(3)$$

of $\underset{\sim}{Y}$ is continuous.

Proof. In order to apply Prop. 10.4 to the linear span of the family of cross-sections (3) and so complete the proof, it is enough to verify the following two facts: (I) If $f,g \in \underset{\sim}{L}(\underset{\sim}{B})$ and $\xi, \eta \in X$, then $\alpha \mapsto \langle (f|\alpha) \otimes \xi, (g|\alpha) \otimes \eta\rangle_\alpha$ is continuous on G/H; and (II) for each α in G/H, $\{\kappa_\alpha((f|\alpha) \otimes \xi): f \in \underset{\sim}{L}(\underset{\sim}{B}), \xi \in X\}$ has dense linear span in Y_α. The first of these facts results from a simple uniform continuity argument based on (1). The second is an immediate consequence of Thm. 10.7. □

Definition. The Hilbert bundle $\underset{\sim}{Y}$ whose existence has just been established is called the Hilbert bundle over G/H induced by S.

Remark. Strictly speaking $\underset{\sim}{Y}$ depends on the particular choice of ρ made at the beginning of this section, and so should be referred to as $\underset{\sim}{Y}^\rho$. But the dependence of $\underset{\sim}{Y}^\rho$ on ρ is not very serious. Indeed, let ρ' be another everywhere positive continuous H-rho function. Then by §21(1) there is a continuous positive-valued function σ on G/H such that $\rho'(x) = \sigma(xH)\rho(x)$ $(x \in G)$. If $\langle\ ,\ \rangle'_\alpha$ is constructed as in (1) from ρ', we have

$$\langle \zeta, \eta\rangle'_\alpha = \sigma(\alpha)^{-1} \langle \zeta, \eta\rangle_\alpha \qquad (\zeta, \eta \in \underset{\sim}{L}(\underset{\sim}{B}_\alpha) \otimes X) . \qquad \ldots(4)$$

It follows that the completed spaces $Y_\alpha^{\rho'}$ and Y_α^{ρ} are the same except that their inner products differ by the positive multiplicative constant $\sigma(\alpha)^{-1}$. The topologies of $\underset{\sim}{Y}^{\rho'}$ and $\underset{\sim}{Y}^\rho$ derived from Proposition 23.2 are the same.

In future we shall usually write $\underset{\sim}{Y}$ rather than $\underset{\sim}{Y}^\rho$.

The fiber Y_H of $\underset{\sim}{Y}$ over the coset H is essentially the same as the space X of S. Indeed, if $\varphi, \psi \in \underset{\sim}{L}(\underset{\sim}{B}_H)$ and $\xi, \eta \in X$,

$$\langle \varphi \otimes \xi, \psi \otimes \eta \rangle_H = \int_H \int_H (\rho(h)\rho(k))^{-\frac{1}{2}} \langle S_{\varphi(h)}\xi, S_{\psi(k)}\eta \rangle \, d\nu h \, d\nu k$$

$$= \rho(e)^{-1} \langle S_{\varphi'}\xi, S_{\psi'}\eta \rangle , \qquad \ldots (5)$$

where we have set $\varphi'(h) = \Delta(h)^{\frac{1}{2}}\delta(h)^{-\frac{1}{2}}\varphi(h)$ (and similarly for ψ'). It follows that the equation

$$F(\kappa_H(\varphi \otimes \xi)) = S_{\varphi'}\xi \qquad (\varphi \in \underset{\sim}{L}(\underset{\sim}{B}_H);\ \xi \in X) \qquad \ldots (6)$$

defines a linear map $F: Y_H \to X$ such that $\rho(e)^{-\frac{1}{2}}F$ is an isometry. Since S is non-degenerate, its integrated form is also non-degenerate, and hence F is <u>onto</u> X.

Notice that F is independent of the choice of ρ.

<u>An alternative construction of</u> Y_α

One can also build Hilbert spaces Y'_α over each coset α by starting from Prop. 22.1(III) instead of Prop. 22.1(II). To be specific, take $\alpha \in G/H$; let Z_α be the algebraic direct sum $\Sigma^{\oplus}_{x\in\alpha}(B_x \otimes X)$; and introduce into Z_α the conjugate-bilinear form $\langle\ ,\ \rangle'_\alpha$ given by:

$$\langle b \otimes \xi, c \otimes \eta \rangle'_\alpha = (\rho(x)\rho(y))^{-\frac{1}{2}} \langle S_{c^*b}\xi, \eta \rangle \qquad \ldots (7)$$

$(x,y \in \alpha;\ b \in B_x;\ c \in B_y;\ \xi,\eta \in X)$. By Prop. 22.1(III) $\langle\ ,\ \rangle'_\alpha$ is positive. So one can form a Hilbert space Y'_α by factoring out from Z_α the null space of $\langle\ ,\ \rangle'_\alpha$ and completing. Let $\kappa'_\alpha: Z_\alpha \to Y'_\alpha$ be the quotient map.

I claim that Y_α and Y'_α are canonically isomorphic. Indeed: From the continuity of S on B_H it is easy to see that $\kappa'_\alpha(b \otimes \xi)$

is continuous in b on B_α . So, if $f \in \underset{\sim}{L}(\underset{\sim}{B}_\alpha)$ and $\xi \in X$, the right side of the definition

$$F_\alpha(f \otimes \xi) = \int_H \kappa'_\alpha(f(xh) \otimes \xi)d\nu h \qquad \ldots(8)$$

(where $x \in \alpha$) exists as a Y'_α-valued Bochner integral. For $f,g \in \underset{\sim}{L}(\underset{\sim}{B}_\alpha)$ and $\xi, \eta \in X$, it follows from (1), (7), and (8) that

$$\langle F_\alpha(f \otimes \xi), F_\alpha(g \otimes \eta)\rangle'_\alpha = \langle \kappa_\alpha(f \otimes \xi), \kappa_\alpha(g \otimes \eta)\rangle_\alpha .$$

So there is a linear isometry $\tilde{F}_\alpha : Y_\alpha \to Y'_\alpha$ satisfying

$$\tilde{F}_\alpha(\kappa_\alpha(f \otimes \xi)) = F_\alpha(f \otimes \xi) = \int_H \kappa'_\alpha(f(xh) \otimes \xi)d\nu h . \qquad \ldots(9)$$

I claim that $\tilde{F}_\alpha$ is <u>onto</u> Y'_α . To see this it is enough to show that $\kappa'_\alpha(b \otimes \xi)$ belongs to the closure of range$(\tilde{F}_\alpha)$ whenever $x \in \alpha$, $b \in B_x$, and $\xi \in X$. This is proved by a standard argument based on (9), using cross-sections f that "peak" around x at the value b .

The isometry $\tilde{F}_\alpha$ is clearly independent of ρ .

<u>In future we shall identify Y'_α and Y_α by means of the isometry $\tilde{F}_\alpha$, writing $\kappa_\alpha(b \otimes \xi)$ instead of $\tilde{F}_\alpha^{-1}(\kappa'_\alpha(b \otimes \xi))$ $(b \in B_\alpha;\ \xi \in X)$. Thus by (7) we have</u>

$$\langle \kappa_\alpha(b \otimes \xi), \kappa_\alpha(c \otimes \eta)\rangle_\alpha = (\rho(x)\rho(y))^{-\frac{1}{2}} \langle S_{c^*b}\xi, \eta\rangle \qquad \ldots(10)$$

$(x,y \in \alpha;\ b \in B_x;\ c \in B_y;\ \xi, \eta \in X)$.

The topology of Y has a simple description in terms of the spaces Y'_α , whose proof we leave to the reader. (See Cor. 24.7.)

<u>Proposition 23.3</u>. <u>The topology of Y defined earlier is the unique topology which (i) makes $\underset{\sim}{Y}$ a Banach bundle, and (ii) for each fixed</u>

ξ in X makes the map $b \mapsto \varkappa_{\pi(b)H}(b \otimes \xi)$ (of B into Y) continuous.

In the case of the coset H, composing the map F of (6) with the identification of Y_H and Y'_H given by (9), we obtain

$$F(\varkappa_H(b \otimes \xi)) = \Delta(h)^{\frac{1}{2}}\delta(h)^{-\frac{1}{2}}S_b\xi \quad (h \in H;\ b \in B_h;\ \xi \in X) \ . \quad \ldots(11)$$

The L_2 cross-sectional space of the induced Hilbert bundle.

Having constructed $\mathcal{Y}$, one can form the cross-sectional Hilbert space $\mathcal{L}_2(\mathcal{Y};\ \rho^{\#})$ (see §10).

It is worth noticing that the elements of $\mathcal{L}_2(\mathcal{Y};\rho^{\#})$ and their norms are quite independent of ρ. Indeed, let ρ' be another everywhere positive continuous rho-function, related to ρ as in the context of (4) by the function σ on G/H. Then $d\rho'^{\#}\alpha = \sigma(\alpha)d\rho^{\#}\alpha$ (see Prop. 21.4). So in view of (4) the change from $\rho^{\#}$ to $\rho'^{\#}$ exactly compensates the difference in the inner products of the fibers of $\mathcal{Y}^{\rho'}$ and $\mathcal{Y}^{\rho}$. The details of the argument are left to the reader.

Now, remembering the original definition in §21 of the $\mathcal{B}$-positivity of S (in terms of the basic conditional expectation p of §21(5)), let us denote by V the operator inner product on $\mathcal{L}(\mathcal{B})$ constructed from S and p:

$$V_{f,g} = S_{p(g^* * f)} \qquad (f,g \in \mathcal{L}(\mathcal{B})) \ . \qquad \ldots(12)$$

As in §1, let $\mathcal{X}(V)$ be the Hilbert space deduced from V. We are going to show that $\mathcal{X}(V)$ and $\mathcal{L}_2(\mathcal{Y};\rho^{\#})$ are essentially the same Hilbert space. Thus the *-representation of $\mathcal{B}$ induced from S, as tentatively defined at the end of §21, will (if it exists) act on $\mathcal{L}_2(\mathcal{Y};\rho^{\#})$.

To see the identity of $\underset{\sim}{X}(V)$ and $\underset{\sim}{L}_2(\underset{\sim}{Y};\rho^{\#})$ recall from the definition of the topology of $\underset{\sim}{Y}$ that, for each f in $\underset{\sim}{L}(\underset{\sim}{B})$ and each ξ in X, the cross-section (3) of $\underset{\sim}{Y}$ is continuous and has compact support; call this cross-section $\kappa(f \otimes \xi)$. Since $\kappa(f \otimes \xi)$ is linear in f and in ξ, we have defined a linear map $\kappa : \underset{\sim}{L}(\underset{\sim}{B}) \otimes X \to \underset{\sim}{L}(\underset{\sim}{Y}) \subset \underset{\sim}{L}_2(\underset{\sim}{Y};\rho^{\#})$. Let $<\,,\,>_0$ and $<\,,\,>_2$ be the inner products of $\underset{\sim}{X}(V)$ and $\underset{\sim}{L}_2(\underset{\sim}{Y};\rho^{\#})$ respectively; and let $f \tilde{\otimes} \xi$ be the image of $f \otimes \xi$ in $\underset{\sim}{X}(V)$. Then, for f,g in $\underset{\sim}{L}(\underset{\sim}{B})$ and ξ,η in X,

$$<f \tilde{\otimes} \xi, g \tilde{\otimes} \eta>_0 = <S_{p(g^* * f)}\xi,\eta>$$

$$= \int_{G/H} d\rho^{\#}(xH) \int_H \int_H (\rho(xh)\rho(xk))^{-\frac{1}{2}} <S_{g(xh)^* f(xk)}\xi,\eta> d\nu h\, d\nu k \quad \text{(by §22(5))}$$

$$= \int_{G/H} <(f|\alpha) \otimes \xi, (g|\alpha) \otimes \eta>_\alpha \, d\rho^{\#}\alpha \qquad \text{(by (1))}$$

$$= <\kappa(f \otimes \xi), \kappa(g \otimes \eta)>_2 . \qquad \ldots(13)$$

This equality says that the equation

$$E(f \tilde{\otimes} \xi) = \kappa(f \otimes \xi) \qquad (f \in \underset{\sim}{L}(\underset{\sim}{B});\ \xi \in X) \qquad \ldots(14)$$

defines a linear isometry E of $\underset{\sim}{X}(V)$ into $\underset{\sim}{L}_2(\underset{\sim}{Y};\rho^{\#})$.

<u>Proposition 23.4</u>. E <u>is onto</u> $\underset{\sim}{L}_2(\underset{\sim}{Y};\rho^{\#})$.

<u>Proof</u>. It must be shown that the cross-sections $\kappa(f \otimes \xi)$ span a dense subspace Γ of $\underset{\sim}{L}_2(\underset{\sim}{Y};\rho^{\#})$.

The set of all $\kappa(f \otimes \xi)$ $(f \in \underset{\sim}{L}(\underset{\sim}{B});\ \xi \in X)$ is clearly closed under multiplication by continuous complex functions on G/H. Also, by Theorem 10.7 for each fixed α in G/H the linear span of $\{(\kappa(f \otimes \xi))(\alpha) : f \in \underset{\sim}{L}(\underset{\sim}{B}), \xi \in X\}$ coincides with the linear span of

$\{\kappa_\alpha(g \otimes \xi) : g \in \underset{\sim}{L}(\underset{\sim}{B}_\alpha), \xi \in X\}$, and so is dense in Y_α . We have thus verified that Γ satisfies the hypotheses of Prop. 10.12. So by the latter proposition Γ is dense in $\underset{\sim}{L}_2(\underset{\sim}{Y};\rho^\#)$. □

Thus E maps $\underset{\sim}{X}(V)$ linearly and isometrically onto $\underset{\sim}{L}_2(\underset{\sim}{Y},\rho^\#)$. If we wish, E can be considered as identifying the two spaces $\underset{\sim}{X}(V)$ and $\underset{\sim}{L}_2(\underset{\sim}{Y};\rho^\#)$.

§24. The action of $\underset{\sim}{B}$ on the induced Hilbert bundle.

We keep the notation of §§21-23.

In order to show that S is inducible to $\underset{\sim}{L}(\underset{\sim}{B})$ via p , we shall eventually (in §25) construct an explicit *-representation T of $\underset{\sim}{B}$ on $\underset{\sim}{L}_2(\underset{\sim}{Y};\rho^\#)$. The *-representation T will be derived from an action ι of $\underset{\sim}{B}$ on the Hilbert bundle $\underset{\sim}{Y}$, whose construction is our next project.

Lemma 24.1. *Let* α *be a coset in* G/H , *and* c *any element of* B . *Let* $b_1,\cdots,b_n$ *be elements of* B_α *and let* $\xi_1,\cdots,\xi_n$ *be vectors in* X . *Then*

$$\sum_{i,j=1}^{n} \langle S_{b_j^* c^* c b_i} \xi_i, \xi_j\rangle \leq \|c\|^2 \sum_{i,j=1}^{n} \langle S_{b_j^* b_i} \xi_i, \xi_j\rangle . \qquad \ldots(1)$$

Proof. Write S as a Hilbert direct sum $\Sigma_i^\oplus S^i$ of cyclic *-representations. Since S is $\underset{\sim}{B}$-positive so is each S^i ; and it is easy to see that, if each S^i satisfies (1), so does S . Therefore it is enough to prove (1) for each S^i , that is, to assume from the beginning that S is cyclic, with cyclic vector η .

Also it is enough by Remark 1 of §22 to assume that G carries the discrete topology. Let $\underset{\sim}{L}_1(\underset{\sim}{B})$ and $\underset{\sim}{L}_1(\underset{\sim}{B}_H)$ be the $\underset{\sim}{L}_1$ cross-sectional algebras of $\underset{\sim}{B}$ and $\underset{\sim}{B}_H$ (with respect to counting measure on G and H).

Since G is discrete, the restriction map $p: \underset{\sim}{L}_1(\underset{\sim}{B}) \to \underset{\sim}{L}_1(\underset{\sim}{B}_H)$ (sending f into $f|H$) is continuous; and so the functional $w: f \mapsto \langle S_{p(f)}\eta, \eta\rangle$ on $\underset{\sim}{L}_1(\underset{\sim}{B})$ is continuous on $\underset{\sim}{L}_1(\underset{\sim}{B})$. From this and the original definition (§21) of $\underset{\sim}{B}$-positivity it follows that w is a positive linear functional on the Banach *-algebra $\underset{\sim}{L}_1(\underset{\sim}{B})$. Thus (see Rickart [1], Theorem 4.5.2) w is admissible and so generates a *-representation U of $\underset{\sim}{L}_1(\underset{\sim}{B})$. Let $f \mapsto \tilde{f}$ be the canonical quotient map of $\underset{\sim}{L}_1(\underset{\sim}{B})$ into $X(U)$, so that

$$\langle \tilde{f}, \tilde{g}\rangle_{X(U)} = w(g^* * f) . \qquad \ldots(2)$$

By the continuity of the operators S_b, it is sufficient to prove (1) when the ξ_i range only over a dense subspace of X. We can therefore assume that $\xi_i = S_{\varphi_i}\eta$, where $\varphi_i \in \underset{\sim}{L}_1(\underset{\sim}{B}_H)$. Recalling that B and $\underset{\sim}{L}_1(\underset{\sim}{B}_H)$ are both inside $\underset{\sim}{L}_1(\underset{\sim}{B})$, we then have (omitting the convolution symbol $*$):

$$\begin{aligned}
\sum_{i,j=1}^{n} \langle S_{b_j^* c^* c b_i}\xi_i, \xi_j\rangle &= \sum_{i,j} \langle S_{\varphi_j^* b_j^* c^* c b_i \varphi_i}\eta, \eta\rangle \\
&= \sum_{i,j} w(\varphi_j^* b_j^* c^* c b_i \varphi_i) \\
&= \sum_{i,j} \langle (c b_i \varphi_i)^\sim, (c b_j \varphi_j)^\sim\rangle_{X(U)} \quad \text{(by (2))} \\
&= \sum_{i,j} \langle U_{c b_i}\tilde{\varphi}_i, U_{c b_j}\tilde{\varphi}_j\rangle_{X(U)} \\
&= \Big\| \sum_i U_{c b_i}\tilde{\varphi}_i \Big\|^2 \\
&= \Big\| U_c \sum_i U_{b_i}\tilde{\varphi}_i \Big\|^2
\end{aligned}$$

$$\leq \|U_c\|^2 \|\sum_i U_{b_i}\tilde{\varphi}_i\|^2$$

$$\leq \|c\|^2 \sum_{i,j} \langle U_{b_i}\tilde{\varphi}_i, U_{b_j}\tilde{\varphi}_j\rangle_{X(U)}$$

$$= \|c\|^2 \sum_{i,j} \langle S_{b_j^* b_i}\xi_i, \xi_j\rangle$$

(reversing the first four steps of this calculation). □

<u>Proposition 24.2</u>. <u>For each</u> x <u>in</u> G, c <u>in</u> B_x, <u>and</u> α <u>in</u> G/H, <u>there is a unique continuous linear map</u> $\iota_c^{(\alpha)}: Y_\alpha \to Y_{x\alpha}$ <u>satisfying</u>

$$\iota_c^{(\alpha)}(\kappa_\alpha(b \otimes \xi)) = \kappa_{x\alpha}(cb \otimes \xi) \qquad \ldots(3)$$

<u>for all</u> b <u>in</u> B_α <u>and</u> ξ <u>in</u> X.

<u>Proof</u>. The uniqueness follows from the denseness of the linear span of $\{\kappa_\alpha(b \otimes \xi)\}$. To prove the existence of $\iota_c^{(\alpha)}$, we take $\xi_1,\cdots,\xi_n$ in X, $y_1,\cdots,y_n$ in α, and $b_i \in B_{y_i}$, and argue as follows:

$$\|\sum_{i=1}^{n} \kappa_{x\alpha}(cb_i \otimes \xi_i)\|_\alpha^2$$

$$= \sum_{i,j} (\rho(xy_i)\rho(xy_j))^{-\frac{1}{2}} \langle S_{b_j^* c^* cb_i}\xi_i, \xi_j\rangle \qquad \text{(by §23(10))}$$

$$\leq \|c\|^2 \sum_{ij} (\rho(xy_i)\rho(xy_j))^{-\frac{1}{2}} \langle S_{b_j^* b_i}\xi_i, \xi_j\rangle \qquad \ldots(4)$$

(by (1), replacing ξ_i by $\rho(xy_i)^{-\frac{1}{2}}\xi_i$). Let y be a fixed element of α. By §21(1) $\rho(y_i)^{-1}\rho(xy_i) = \rho(y)^{-1}\rho(xy)$. So $(\rho(xy_i)\rho(xy_j))^{-\frac{1}{2}} = (\rho(y_i)\rho(y_j))^{-\frac{1}{2}}\rho(xy)^{-1}\rho(y)$. Substituting this into (4) and again using §23(10) we get

$$\|\sum_{i=1}^{n} \kappa_{x\alpha}(cb_i \otimes \xi_i)\|_\alpha^2 \leq \|c\|^2 \rho(xy)^{-1}\rho(y) \|\sum_{i=1}^{n} \kappa_\alpha(b_i \otimes \xi)\|_\alpha^2 . \qquad \ldots(5)$$

From this follows the existence of the continuous map $\iota_c^{(\alpha)}$ satisfying (3). □

In view of §21(1) the quantity $\rho(xy)^{-1}\rho(y)$ occurring in (5) depends only on x and yH ; call it $\sigma(x,yH)$. Thus σ is a positive-valued continuous function on $G \times G/H$; and by (5)

$$\|\iota_c^{(\alpha)}\|^2 \leq \|c\|^2 \sigma(x,\alpha) . \qquad \ldots(6)$$

Notice that $\iota_c^{(\alpha)}$ has been defined in terms of the alternative description of Y_α in §23. In terms of the original description of Y_α we have

$$\iota_c^{(\alpha)}(\kappa_\alpha(f \otimes \xi)) = \kappa_{x\alpha}(cf \otimes \xi) \qquad \ldots(7)$$

($x \in G$; $c \in B_x$; $\alpha \in G/H$; $f \in \underline{L}(\underline{B}_\alpha)$; $\xi \in X$). Here cf is the element of $\underline{L}(\underline{B}_{x\alpha})$ given by $(cf)(y) = cf(x^{-1}y)$ $(y \in x\alpha)$.

To prove (7), recall that, in view of the identification $Y'_\alpha \cong Y_\alpha$ equation §23(9) can be written

$$\kappa_\alpha(f \otimes \xi) = \int_H \kappa_\alpha(f(yh) \otimes \xi)d\nu h \qquad \ldots(8)$$

(where $y \in \alpha$) . Applying $\iota_c^{(\alpha)}$ to both sides of this, we have by Prop. 10.13

$$\begin{aligned}\iota_c^{(\alpha)}(\kappa_\alpha(f \otimes \xi)) &= \int \iota_c^{(\alpha)}(\kappa_\alpha(f(yh) \otimes \xi))d\nu h \\ &= \int \kappa_{x\alpha}(cf(yh) \otimes \xi)d\nu h \qquad \text{(by (3))} \\ &= \kappa_{x\alpha}(cf \otimes \xi) \qquad \text{(by (8))} .\end{aligned}$$

So (7) is proved.

__Definition.__ If $c \in B$, let ι_c be the union of the maps $\iota_c^{(\alpha)}$ ($\alpha \in$

G/H). That is, $\iota_c\colon Y \to Y$ is the map whose restriction to each fiber Y_α is $\iota_c^{(\alpha)}$. Thus, if $c \in B_x$,

$$\iota_c(Y_\alpha) \subset Y_{x\alpha} . \qquad \dots(9)$$

Notice that the definition of ι is independent of ρ.

We know that ι_c is linear on each fiber of $\underset{\sim}{Y}$, and is linear in c on each fiber of $\underset{\sim}{B}$. Evidently

$$\iota_b \iota_c = \iota_{bc} \qquad (b,c \in B) . \qquad \dots(10)$$

As regards involution, I claim that for each $x \in G$, $b \in B_x$, $\alpha \in G/H$, $\zeta \in Y_\alpha$, and $\zeta' \in Y_{x\alpha}$,

$$\langle \iota_b \zeta, \zeta' \rangle_{x\alpha} = \sigma(x,\alpha) \langle \zeta, \iota_{b^*} \zeta' \rangle_\alpha . \qquad \dots(11)$$

Indeed: By continuity and linearity it is enough to prove (11) assuming that $\zeta = \kappa_\alpha(c \otimes \xi)$, $\zeta' = \kappa_{x\alpha}(d \otimes \eta)$ $(\xi, \eta \in X;\ c \in B_\alpha;\ d \in B_{x\alpha})$. Now $\pi(d) = x\pi(c)h$ for some h in H; and so

$$\rho(x\pi(c))\rho(\pi(d))$$

$$= \rho(\pi(c))\rho(x^{-1}\pi(d))[(\rho(\pi(c)))^{-1}\rho(x\pi(c))][(\rho(\pi(c)h))^{-1}\rho(x\pi(c)h)]$$

$$= \rho(\pi(c))\rho(x^{-1}\pi(d))\sigma(x,\alpha)^{-2} .$$

Therefore by §23(10)

$$\langle \iota_b \zeta, \zeta' \rangle_{x\alpha} = (\rho(x\pi(c))\rho(\pi(d)))^{-\frac{1}{2}} \langle S_{d^*bc}\xi, \eta \rangle$$

$$= \sigma(x,\alpha)(\rho(\pi(c))\rho(x^{-1}\pi(d)))^{-\frac{1}{2}} \langle S_{d^*bc}\xi, \eta \rangle$$

$$= \sigma(x,\alpha) \langle \zeta, \iota_{b^*}\zeta' \rangle_\alpha ;$$

and (11) is proved.

Recall from §23(6) the definition of the bijection $F: Y_H \to X$.

<u>Proposition 24.3</u>. *If* $\xi \in X$, $\alpha \in G/H$, *and* $b \in B_\alpha$,

$$\iota_b(F^{-1}\xi) = \varkappa_\alpha(b \otimes \xi) . \qquad \ldots(12)$$

<u>Proof</u>. Suppose that $h \in H$, $c \in B_h$, $d \in B_\alpha$, and $\eta \in X$; then by §23(10) and §21(1)

$$\langle bc \otimes \xi, d \otimes \eta \rangle_\alpha = \Delta(h)^{\frac{1}{2}}\delta(h)^{-\frac{1}{2}} \langle b \otimes S_c\xi, d \otimes \eta \rangle_\alpha .$$

By the arbitrariness of d and η this implies

$$\varkappa_\alpha(bc \otimes \xi) = \Delta(h)^{\frac{1}{2}}\delta(h)^{-\frac{1}{2}}\varkappa_\alpha(b \otimes S_c\xi) . \qquad \ldots(13)$$

Now, if b is fixed, both sides of (12) are continuous in ξ. So it is enough to prove (12) assuming that $\xi = F(\varkappa_H(c \otimes \eta))$, where $\eta \in X$, $h \in H$, $c \in B_H$. But then

$$\begin{aligned}
\varkappa_\alpha(b \otimes \xi) &= \Delta(h)^{\frac{1}{2}}\delta(h)^{-\frac{1}{2}}\varkappa_\alpha(b \otimes S_c\eta) && \text{(by §23(11))} \\
&= \varkappa_\alpha(bc \otimes \eta) && \text{(by (13))} \\
&= \iota_b\varkappa_H(c \otimes \eta) = \iota_b(F^{-1}\xi) ;
\end{aligned}$$

and (12) is proved. □

<u>Remark</u>. If in the above proposition $b \in B_h$ $(h \in H)$, then by (12) and §23(11)

$$\iota_b(F^{-1}\xi) = \Delta(h)^{\frac{1}{2}}\delta(h)^{-\frac{1}{2}} F^{-1}(S_b\xi) . \qquad \ldots(14)$$

<u>Corollary 24.4</u>. *For each* α *in* G/H, *the linear span of* $\{\iota_b\zeta : b \in B_\alpha, \zeta \in Y_H\}$ *is dense in* Y_α.

Proof. This follows from (12) and the definition of Y_α . □

In this connection we notice:

Proposition 24.5. *If* $\underset{\sim}{B}$ *is saturated, and if* $\alpha \in G/H$ *and* $x \in G$, *then the linear span* L *of* $\{\iota_b\zeta : b \in B_x, \zeta \in Y_\alpha\}$ *is dense in* $Y_{x\alpha}$.

Proof. We have observed in §23 that $c \mapsto \kappa_{x\alpha}(c \otimes \xi)$ is continuous on $B_{x\alpha}$ for each ξ . So it is enough to fix ξ and show that L contains $\{\kappa_{x\alpha}(c \otimes \xi) : c \in M\}$ for some dense subset M of $B_{x\alpha}$. But this follows from (3) and the saturation of $\underset{\sim}{B}$, when we take $M = \{bd: b \in B_x, d \in B_\alpha\}$. □

We come now to the very important continuity property of ι :

Proposition 24.6. *The map* $(b,\zeta) \mapsto \iota_b\zeta$ *is continuous on* $B \times Y$ *to* Y .

Proof. We shall first show that, if $f \in \underset{\sim}{L}(\underset{\sim}{B})$ and $c_i \to c$ in B , then

$$c_i f \to cf \quad \text{uniformly on } G . \qquad \ldots(15)$$

(Here, as usual, $(cf)(y) = cf(\pi(c)^{-1}y)$, $(c_i f)(y) = c_i f(\pi(c_i)^{-1}y)$.)

Indeed: $\|(cf)(y) - (df)(y)\| = \|cf(\pi(c)^{-1}y) - df(\pi(d)^{-1}y)\|$ is continuous in c,d and y , and vanishes when $c = d$. So given $\varepsilon > 0$ and $c \in B$, there is a neighborhood U of c such that $\|(cf)(y) - (df)(y)\| < \varepsilon$ for all d in U and all y in any pre-assigned compact set $K \subset G$. Taking K to be so large that cf and $c_i f$ vanish outside K for large enough i , we obtain (15).

The next step is to show that, for fixed ξ in X and f in $\underset{\sim}{L}(\underset{\sim}{B})$, the function

$$(c,\beta) \mapsto \varkappa_\beta((cf|\beta) \otimes \xi) \qquad \ldots(16)$$

is continuous on $B \times G/H$ to Y. This follows from two facts: First, by the definition of the topology of Y, (16) is continuous in β when c is fixed. Secondly, by (15), as $c_i \to c$ in B, $\varkappa_\beta((c_i f|\beta) \otimes \xi) \to \varkappa_\beta((cf|\beta) \otimes \xi)$ uniformly for β in G/H.

We are now ready to prove the continuity of ι. Let $b_r \to b$ in B and $\zeta_r \to \zeta$ in Y. It must be shown that

$$\iota_{b_r}\zeta_r \to \iota_b\zeta. \qquad \ldots(17)$$

Put $x = \pi(b)$, $x_r = \pi(b_r)$; and suppose $\zeta \in Y_\alpha$, $\zeta_r \in Y_{\alpha_r}$.

Let $\varepsilon > 0$. By assertion (II) in the proof of Prop. 23.2, there are elements $f_1,\cdots,f_n$ in $\underset{\sim}{L}(\underset{\sim}{B})$ and $\xi_1,\cdots,\xi_n$ in X such that

$$\|\zeta - \sum_{i=1}^{n} \varkappa_\alpha((f_i|\alpha) \otimes \xi_i)\|_\alpha < \varepsilon. \qquad \ldots(18)$$

By continuity this implies that

$$\|\zeta_r - \sum_{i=1}^{n} \varkappa_{\alpha_r}((f_i|\alpha_r) \otimes \xi_i)\|_{\alpha_r} < \varepsilon \qquad \ldots(19)$$

for all large enough r. If k is a positive constant majorizing $\|b\|\,\sigma(x,\alpha)^{\frac{1}{2}}$ and all the $\|b_r\|\,\sigma(x_r,\alpha_r)^{\frac{1}{2}}$, (18) and (19) imply by (6) and (7) that

$$\|\iota_b\zeta - \sum_{i=1}^{n} \varkappa_{x\alpha}((bf_i|x\alpha) \otimes \xi_i)\|_{x\alpha} < k\varepsilon, \qquad \ldots(20)$$

$$\|\iota_{b_r}\zeta_r - \sum_{i=1}^{n} \varkappa_{x_r\alpha_r}((b_r f_i|x_r\alpha_r) \otimes \xi_i)\|_{x_r\alpha_r} < k\varepsilon. \qquad \ldots(21)$$

Now (17) follows from (20), (21), and the continuity of (16). □

Corollary 24.7. *The map* $(b,\xi) \mapsto \kappa_{\pi(b)H}(b \otimes \xi)$ *is continuous on* $B \times X$ *to* Y .

Proof. Combine Props. 23.3 and 23.6. □

§25. The definition of the induced representation.

We shall now use the action ι of $\underset{\sim}{B}$ on $\underset{\sim}{Y}$, defined in §24, to obtain a *-representation of $\underset{\sim}{B}$ acting on $\underset{\sim}{L}_2(\underset{\sim}{Y};\rho^{\#})$. This will be essentially the induced representation of $\underset{\sim}{B}$.

Fix $b \in B_x$ $(x \in G)$. If f is any cross-section (continuous or not) of $\underset{\sim}{Y}$, then

$$T_b'f : \alpha \mapsto \iota_b f(x^{-1}\alpha) \qquad (\alpha \in G/H) \qquad \ldots(1)$$

is a cross-section of $\underset{\sim}{Y}$ in virtue of §24(9).

Lemma 25.1. *If the cross-section* f *of* $\underset{\sim}{Y}$ *is continuous, then so is* $T_b'f$. *If* f *is locally* $\rho^{\#}$*-measurable, then so is* $T_b'f$. *If* $f \in \underset{\sim}{L}_2(\underset{\sim}{Y};\rho^{\#})$, *then so is* $T_b'f$, *and*

$$\|T_b'f\|_2 \le \|b\| \; \|f\|_2 \qquad \ldots(2)$$

(*where* $\| \; \|_2$ *is the norm in* $\underset{\sim}{L}_2(\underset{\sim}{Y};\rho^{\#})$).

Proof. The first statement follows from Prop. 24.6.

To prove the second we recall from §10 the definition of local $\rho^{\#}$-measurability. Let K be a compact subset of G/H , and assume that f is locally $\rho^{\#}$-measurable. Thus there is a sequence $\{g_n\}$ of continuous cross-sections of $\underset{\sim}{Y}$ such that $g_n(\alpha) \to f(\alpha)$ $\rho^{\#}$-almost everywhere on $x^{-1}K$ (where $x = \pi(b)$). Recalling from Prop. 21.5 that $\rho^{\#}$ is quasi-invariant, we see from (1) that $(T_b'g_n)(\alpha) \to$

$(T'_b f)(\alpha)$ $\rho^{\#}$-almost everywhere on K. By the first statement of the lemma the $T'_b g_n$ are continuous. So $T'_b f$ is locally $\rho^{\#}$-measurable.

To prove the last statement it is enough, in view of the preceding paragraph, to verify the norm inequality (2). To begin with, we observe from Props. 21.3 and 21.4 that

$$d\rho^{\#}(x^{-1}\alpha) = \sigma(x,x^{-1}\alpha)d\rho^{\#}\alpha . \qquad \dots(3)$$

Therefore, if $b \in B_x$ and $f \in \underset{\sim}{L}_2(\underset{\sim}{Y};\rho^{\#})$,

$$\int_{G/H} \|(T'_b f)(\alpha)\|_\alpha^2 \, d\rho^{\#}\alpha = \int \|\iota_b f(x^{-1}\alpha)\|_\alpha^2 \, d\rho^{\#}\alpha$$

$$\leq \|b\|^2 \int \sigma(x,x^{-1}\alpha) \|f(x^{-1}\alpha)\|_{x^{-1}\alpha}^2 \, d\rho^{\#}\alpha \qquad \text{(by §24(6))}$$

$$= \|b\|^2 \int \|f(\alpha)\|_\alpha^2 \, d\rho^{\#}\alpha \qquad \text{(by (3))} .$$

But this is (2). □

<u>Definition</u>. If $b \in B$, we shall define T_b as the restriction to $\underset{\sim}{L}_2(\underset{\sim}{Y};\rho^{\#})$ of the map T'_b of (1). By Lemma 25.1, T_b is a bounded linear operator on $\underset{\sim}{L}_2(\underset{\sim}{Y};\rho^{\#})$, with

$$\|T_b\| \leq \|b\| . \qquad \dots(4)$$

<u>Proposition 25.2</u>. <u>The map</u> $T: b \mapsto T_b$ <u>is a non-degenerate</u> *-<u>representation of</u> $\underset{\sim}{B}$ <u>on</u> $\underset{\sim}{L}_2(\underset{\sim}{Y};\rho^{\#})$.

<u>Proof</u>. T is obviously linear on each fiber. That $T_b T_c = T_{bc}$ follows immediately from §24(10). If $b \in B_x$ and $f,g \in \underset{\sim}{L}_2(\underset{\sim}{Y};\rho^{\#})$,

$$\langle T_{b^*}f,g\rangle_2 = \int \langle \iota_{b^*}f(x\alpha),g(\alpha)\rangle_\alpha \, d\rho^{\#}\alpha$$

$$= \int \sigma(x^{-1},x\alpha) \, \langle f(x\alpha), \iota_b g(\alpha)\rangle_{x\alpha} \, d\rho^{\#}\alpha \quad \text{(by §24(11))}$$

$$= \int \langle f(x\alpha), \iota_b g(\alpha)\rangle_{x\alpha} \, d\rho^{\#}(x\alpha) \text{ (by (3))}$$

$$= \langle f, T_b g\rangle_2 .$$

Therefore $(T_b)^* = T_{b^*}$.

Thus T will be a *-representation of $\underset{\sim}{B}$ if we can show that $b \mapsto T_b f$ is continuous on B for each f in $\underset{\sim}{L}_2(\underset{\sim}{Y}; \rho^{\#})$. By (4) it is enough to prove this when f belongs to the dense subspace $\underset{\sim}{L}(\underset{\sim}{Y})$ of $\underset{\sim}{L}_2(\underset{\sim}{Y}; \rho^{\#})$. But if $b_i \to b$ in B and $f \in \underset{\sim}{L}(\underset{\sim}{Y})$, it follows by the same argument that was used to prove §24(15) that $T_{b_i} f \to T_b f$ uniformly on G/H with uniform compact support, and hence that $T_{b_i} f \to T_b f$ in $\underset{\sim}{L}_2(\underset{\sim}{Y}; \rho^{\#})$. Therefore T is a *-representation.

To see that T is non-degenerate, let Γ be the linear span (in $\underset{\sim}{L}(\underset{\sim}{Y})$) of $\{T_b f : f \in \underset{\sim}{L}(\underset{\sim}{Y}), b \in B\}$. If $\varphi \in \underset{\sim}{C}(G/H)$, $b \in B_x$, and $f \in \underset{\sim}{L}(\underset{\sim}{Y})$, we verify that $\varphi(T_b f) = T_b g$, where $g(\alpha) = \varphi(x\alpha) f(\alpha)$. Therefore Γ is closed under multiplication by continuous complex functions on G/H. Also, it follows from Cor. 24.4 that $\{g(\alpha): g \in \Gamma\}$ is dense in Y_α for each fixed α. Consequently Γ is dense in $\underset{\sim}{L}_2(\underset{\sim}{Y}; \rho^{\#})$ by Prop. 10.12, and T is non-degenerate. □

<u>Definition</u>. The T of Prop. 25.2 is called <u>the *-representation of $\underset{\sim}{B}$ concretely induced by</u> the $\underset{\sim}{B}$-positive non-degenerate *-representation S of $\underset{\sim}{B}_H$, and will be denoted by $\mathrm{Ind}_{\underset{\sim}{B}_H \uparrow \underset{\sim}{B}}(S)$.

We use the term 'concretely' here to distinguish T from the formally different abstract inducing process of §4. To fulfill the promise made at the end of §21, we shall now show that the integrated form of the *-representation $b \mapsto E^{-1} T_b E$ of $\underset{\sim}{B}$ on $\underset{\sim}{X}(V)$ (E being as in §23(14)) is just the result of abstractly inducing S to $\underset{\sim}{L}(\underset{\sim}{B})$ via p. That is, keeping the notation of the paragraph preceding

Prop. 23.4, we must prove:

Proposition 25.3. *Let* T^0 *denote the integrated form of* $T = \text{Ind}_{\underset{\sim}{B}_H \uparrow \underset{\sim}{B}}(S)$. *Then, if* $f, g \in \underset{\sim}{L}(\underset{\sim}{B})$ *and* $\xi \in X$, *we have*

$$E((f*g) \tilde{\otimes} \xi) = T_f^0(E(g \tilde{\otimes} \xi)) . \qquad \ldots(5)$$

Proof. As usual let $(bg)(y) = bg(x^{-1}y)$ $(x, y \in G;\ b \in B_x)$. From §24 (7) and the definition of $\varkappa(g \otimes \xi)$ we check that

$$T_b(\varkappa(g \otimes \xi)) = \varkappa(bg \otimes \xi) . \qquad \ldots(6)$$

Now by §16(3)

$$f*g = \int_G (f(x)g)d\lambda x , \qquad \ldots(7)$$

the right side being an $\underset{\sim}{L}(\underset{\sim}{B})$-valued integral with respect to the inductive limit topology. Also, for fixed ξ , the linear map

$$\varphi \mapsto \varkappa(\varphi \otimes \xi) \qquad (\varphi \in \underset{\sim}{L}(\underset{\sim}{B})) \qquad \ldots(8)$$

is continuous from $\underset{\sim}{L}(\underset{\sim}{B})$ to $\underset{\sim}{L}_2(\underset{\sim}{Y};\rho^{\#})$ in the inductive limit topology of the former. So, applying (8) to both sides of (7) (and invoking Prop. 10.16), we get

$$\begin{aligned} \varkappa((f*g) \otimes \xi) &= \int_G \varkappa(f(x)g \otimes \xi)d\lambda x \\ &= \int T_{f(x)}\varkappa(g \otimes \xi)d\lambda x \quad \text{(by (6))} \\ &= T_f^0\, \varkappa(g \otimes \xi) . \end{aligned}$$

This combined with §23(14) gives (5). □

We have now finally proved the following result.

Theorem 25.4. Let S be a $\mathcal{B}$-positive non-degenerate *-representation of $\mathcal{B}_H$. Then: (I) The integrated form of S is (abstractly) inducible to $\mathcal{L}(\mathcal{B})$ via the basic conditional expectation p of §21(5). (II) $\mathrm{Ind}^p_{\mathcal{L}(\mathcal{B}_H)\uparrow\mathcal{L}(\mathcal{B})}(S)$ is the integrated form of a (unique) non-degenerate *-representation $\tilde{T}$ of $\mathcal{B}$. (III). $\tilde{T}$ is unitarily equivalent with the concretely induced *-representation $T = \mathrm{Ind}_{\mathcal{B}_H\uparrow\mathcal{B}}(S)$.

Remark. The hypothesis that S is non-degenerate is of no importance. If S is degenerate, with non-degenerate part S^0, both the abstract and the concrete inducing processes give the same result when applied to S^0 as when applied to S.

In view of Thm. 25.4(III) we shall usually omit the word 'concretely' in the above definition of $\mathrm{Ind}_{\mathcal{B}_H\uparrow\mathcal{B}}(S)$. The latter will be called the bundle formulation of the inducing construction, as contrasted with the "abstract" or "global" formulation of §21.

If $H = G$ we have the following simple result.

Proposition 25.5. If S is a non-degenerate *-representation of $\mathcal{B}$, then

$$\mathrm{Ind}_{\mathcal{B}\uparrow\mathcal{B}}(S) \cong S .$$

Proof. In this case $\mathcal{L}_2(\mathcal{Y};\rho^\#) = Y_G = Y_H$ and $T_b = \iota_b$; and the result follows from §24(14) (since $\Delta(h)^{\frac{1}{2}}\,\delta(h)^{-\frac{1}{2}}$ is now 1). □

Proposition 25.6. If the bundle space B of $\mathcal{B}$ is second-countable and X $(= X(S))$ is separable, then $\mathcal{L}_2(\mathcal{Y};\rho^\#)$ (the space of $\mathrm{Ind}_{\mathcal{B}_H\uparrow\mathcal{B}}(S)$) is separable.

Proof. By Prop. 10.10 there is a countable subset Γ of $\mathcal{L}(\mathcal{B})$ which is dense in $\mathcal{L}(\mathcal{B})$ in the inductive limit topology. Let X_c be a countable dense subset of X . Keeping the notation of the paragraph preceding Prop. 23.4, let Γ be the countable subfamily of $\underset{\sim}{X}(V)$ consisting of all finite sums of elements of the form $f \tilde{\otimes} \xi$, where $f \in \Gamma$ and $\xi \in X_c$. Since by Remark 3 of §21 $f \tilde{\otimes} \xi$ is separately continuous in f (with the inductive limit topology) and ξ , the $\underset{\sim}{X}(V)$-closure of $\tilde{\Gamma}$ contains all $g \tilde{\otimes} \eta$ $(g \in \underset{\sim}{L}(\underset{\sim}{B});\ \eta \in X)$ and hence is equal to $\underset{\sim}{X}(V)$. So $\underset{\sim}{X}(V)$ is separable. Now apply Prop. 23.4. □

§26. The Mackey and Blattner descriptions of induced representations of groups.

In this section we shall see how the very general description of induced representations given in §25 reduces in the group case to the more familiar constructions of Mackey and Blattner. First we shall deduce from it the Blattner construction, and next the Mackey construction. We shall conclude the section with an analogous "Blattner description" of induced representations of semidirect product bundles.

Suppose in §§21-25 that we are in the group case, that is, $\underset{\sim}{B}$ is the group bundle of G . Then $\underset{\sim}{B}_H$ is the group bundle of H ; and the non-degenerate *-representation S of $\underset{\sim}{B}_H$ with which we started §21 is just a unitary representation of H . By Cor. 22.4 S is automatically $\underset{\sim}{B}$-positive. So we can form $T = \mathrm{Ind}_{\underset{\sim}{B}_H \uparrow \underset{\sim}{B}}(S)$; and this is a non-degenerate *-representation of $\underset{\sim}{B}$, that is, a unitary representation of G . Thus, in the group case, the inducing operation carries an arbitrary unitary representation S of the closed subgroup H of G into a unitary representation T of the whole group G . <u>In this case we shall write</u> $\mathrm{Ind}_{H\uparrow G}(S)$ <u>for</u> T , instead of the more cumbersome $\mathrm{Ind}_{\underset{\sim}{B}_H \uparrow \underset{\sim}{B}}(S)$.

We shall now see how the general formulation of the inducing operation given in §25 reduces in the group case to the more classical formulation of Mackey and Blattner.

Since $\underset{\sim}{B}$ is now the group bundle $\mathbb{C} \times G$ of the locally compact group G, an element of B_x is of the form (t,x), where $t \in \mathbb{C}$. We recall that, if T is a non-degenerate *-representation of $\underset{\sim}{B}$, the unitary representation of G with which T is identified is just $x \mapsto T_{(1,x)}$.

Fix a unitary representation S of the closed subgroup H of G, acting in the Hilbert space X. We shall continue to use the notation of §§21-25 without explicit reference.

Our basic tool will be the map $D: G \times X \to Y$ given by

$$D(x,\xi) = \kappa_{xH}((1,x) \otimes \xi) \quad (x \in G;\ \xi \in X). \qquad \ldots(1)$$

By Prop. 24.3 this can be written in the form

$$D(x,\xi) = \iota_{(1,x)}(F^{-1}\xi). \qquad \ldots(2)$$

Denoting by D_x the map $\xi \mapsto D(x,\xi)$, we notice that

$$D_x(X) \subset Y_{xH}. \qquad \ldots(3)$$

I claim that $D_x: X \to Y_{xH}$ is a bijection and that $\rho(x)^{\frac{1}{2}}D_x$ is an isometry. To see this, one checks from §23(10) that

$$\langle D_x(\xi), D_x(\eta)\rangle_{xH} = \rho(x)^{-1}\langle\xi,\eta\rangle \quad (\xi,\eta \in X;\ x \in G), \qquad \ldots(4)$$

whence $\rho(x)^{\frac{1}{2}}D_x$ is an isometry. By §24(13)

$$D_{xh} = \Delta(h)^{\frac{1}{2}}\,\delta(h)^{-\frac{1}{2}}\,D_x S_h \quad (x \in G;\ h \in H). \qquad \ldots(5)$$

By the definition of Y_α, the latter is spanned by $\{D_x(\xi): x \in \alpha, \xi \in X\}$. So (5) implies that, for each fixed x in α, $D_x(X)$ is dense in Y_{xH}. Since $\rho(x)^{\frac{1}{2}} D_x$ is an isometry, $D_x(X)$ is closed and hence must be equal to Y_{xH}. So the claim is proved.

Let f be any cross-section of $\underset{\sim}{Y}$. Since each $D_x: X \to Y_{xH}$ is a bijection, we can construct from f a function $\varphi: G \to X$ as follows:

$$\varphi(x) = D_x^{-1}(f(xH)) \qquad (x \in G). \qquad \ldots(6)$$

If $x \in G$ and $h \in H$, then by (5) $\varphi(xh) = D_{xh}^{-1} f(xH) = \delta(h)^{\frac{1}{2}} \Delta(h)^{-\frac{1}{2}} S_{h^{-1}} D_x^{-1} f(xH)$, or

$$\varphi(xh) = \delta(h)^{\frac{1}{2}} \Delta(h)^{-\frac{1}{2}} S_{h^{-1}}(\varphi(x)) \qquad (x \in G;\ h \in H). \qquad \ldots(7)$$

Conversely, let $\varphi: G \to X$ be any map satisfying (7). Combining (5) and (7) we find that $D_{xh}\varphi(xh) = D_x\varphi(x)$ for all x in G and $h \in H$. Hence φ generates a cross-section f of $\underset{\sim}{Y}$:

$$f(xH) = D_x(\varphi(x)) \qquad (x \in G). \qquad \ldots(8)$$

Obviously the constructions $f \mapsto \varphi$ and $\varphi \mapsto f$ described in (6) and (8) are each other's inverses. So (6) and (8) set up a one-to-one correspondence $f \leftrightarrow \varphi$ between cross-sections f of $\underset{\sim}{Y}$ and functions $\varphi: G \to X$ satisfying (7).

Notice how the action of G behaves under this correspondence. Let $y \in G$; and let T'_y stand for the map $T'_{(1,y)}$ of §25(1), sending the cross-section f of $\underset{\sim}{Y}$ into the cross-section $\alpha \mapsto \iota_{(1,y)} f(y^{-1}\alpha)$. If $f \leftrightarrow \varphi$, $f' = T'_y f$, and $f' \leftrightarrow \varphi'$, then for all x in G

$$D_x\varphi'(x) = f'(xH) \qquad \text{(by (8))}$$

$$= \iota_{(1,y)} f(y^{-1}xH)$$

$$= \iota_{(1,y)} D_{y^{-1}x} \varphi(y^{-1}x) \qquad \text{(by (8)}$$

$$= \iota_{(1,y)} \varkappa_{y^{-1}xH}((1,y^{-1}x) \otimes \varphi(y^{-1}x)) \qquad \text{(by (1))}$$

$$= \varkappa_{xH}((1,x) \otimes \varphi(y^{-1}x))$$

$$= D_x(\varphi(y^{-1}x)) \; . \qquad \text{(by (1))} \; .$$

Since D_x is one-to-one, this implies that

$$\varphi'(x) = \varphi(y^{-1}x) \qquad (x \in G) \; . \qquad \ldots(9)$$

Thus, under the correspondence (6), (8), the action of T'_y goes into simple y-translation of the functions φ .

Does the correspondence $f \longleftrightarrow \varphi$ of (6), (8) preserve interesting properties of functions? In particular, what is the image under $f \mapsto \varphi$ of the important space $\mathcal{L}_2(\mathcal{Y};\rho^{\#})$?

Proposition 26.1. Let $f \longleftrightarrow \varphi$ as in (6), (8). Then: (I) f is a continuous cross-section of $\mathcal{Y}$ if and only if φ is continuous on G ; (II) f vanishes locally $\rho^{\#}$-almost everywhere if and only if φ vanishes locally λ-almost everywhere; (III) f is locally $\rho^{\#}$-measurable if and only if φ is locally λ-measurable.

Statement (I) follows from Prop. 24.6. As to the meaning of (II) we recall that, if μ is any regular measure on a locally compact Hausdorff space Z , a statement is said to hold locally μ-almost everywhere on Z if for each compact subset K of Z it holds μ-almost everywhere on K . For brevity we shall omit the proofs of

(II) and (III). They involve lengthy measure-theoretic considerations. Besides, the above proposition is not required for the development of the main theme of this work.

Now let f, φ be as in (6), (8). By (4)

$$\|f(xH)\|_{xH}^2 = \rho(x)^{-1}\|\varphi(x)\|^2 \quad (x \in G) . \qquad \ldots(10)$$

In particular the right side of (10) depends only on the coset xH. If φ is locally λ-measurable, then by (10) and Prop. 26.1 $\rho(x)^{-1}\|\varphi(x)\|^2$ is locally $\rho^{\#}$-measurable as a function of xH, and will be $\rho^{\#}$-summable if and only if $f \in \underset{\sim}{L}_2(\underset{\sim}{Y};\rho^{\#})$. Combining this fact with Prop. 26.1 and (9), we obtain the following theorem, which is essentially the Blattner formulation (Blattner [1]) of Mackey's construction of induced representations of groups.

Theorem 26.2. Let S be a unitary representation of H, acting on the Hilbert space X. Define $\underset{\sim}{Z}$ to be the linear space of all those functions $\varphi: G \to X$ such that: (i) φ is locally λ-measurable, (ii) the relation (7) holds for all x in G and h in H, and (iii) the numerical function $xH \mapsto \rho(x)^{-1}\|\varphi(x)\|^2$ on G/H is $\rho^{\#}$-summable. Then, if we identify functions in $\underset{\sim}{Z}$ which differ only on a locally λ-null set, $\underset{\sim}{Z}$ becomes a Hilbert space under the inner product

$$\langle\varphi,\psi\rangle_{\underset{\sim}{Z}} = \int_{G/H} \rho(x)^{-1}\langle\varphi(x),\psi(x)\rangle_X \, d\rho^{\#}(xH) . \qquad \ldots(11)$$

(By (10) the integrand in (11) depends only on xH.) For each y in G the left translation operator T_y:

$$(T_y\varphi)(x) = \varphi(y^{-1}x) \qquad (\varphi \in \underset{\sim}{Z};\ x \in G) \qquad \ldots(12)$$

is a unitary operator on $\underset{\sim}{Z}$; and $y \mapsto T_y$ $(y \in G)$ is a unitary representation of G on $\underset{\sim}{Z}$, unitarily equivalent (under the correspondence $f \longleftrightarrow \varphi$ of (6), (8)) with $\mathrm{Ind}_{H\uparrow G}(S)$.

Remark. By an observation in §23, the definition of $\underset{\sim}{Z}$ and of its inner product (11) is quite independent of the particular choice of the rho-function ρ .

To obtain the Mackey description of induced representations, we begin by setting

$$\sigma'(y,xH) = \rho(x)^{-1}\rho(y^{-1}x) \qquad (x,y \in G) .$$

Thus σ' is a continuous function on $G \times (G/H)$. By Props. 21.3 and 21.4, $\sigma'(y,\cdot)$ is the Radon-Nikodym derivative of $y\rho^{\#}$ (the y-translate $W \mapsto \rho^{\#}(y^{-1}W)$ of $\rho^{\#}$) with respect to $\rho^{\#}$; that is, if $y \in G$,

$$d(y\rho^{\#})m = \sigma'(y,m)d\rho^{\#}m .$$

Now, to obtain the Mackey formulation from the Blattner formulation, we have only to apply the transformation which sends the function $\varphi\colon G \to X$ into the function $x \mapsto \rho(x)^{-\frac{1}{2}}\varphi(x)$. Under this transformation Theorem 26.2 becomes:

Theorem 26.3. Let S be a unitary representation of H acting on the Hilbert space X . Define $\underset{\sim}{W}$ to be the linear space of all those functions $\varphi\colon G \to X$ such that: (i) φ is locally λ-measurable, (ii) for all x in G and h in H

$$\varphi(xh) = S_{h^{-1}}(\varphi(x)) , \qquad \ldots(13)$$

and (iii) the numerical function $xH \mapsto \|\varphi(x)\|^2$ (which is well defined

<u>since by</u> (13) $\|\varphi(x)\|^2$ <u>depends only on</u> xH) <u>is</u> $\rho^{\#}$-<u>summable</u>. <u>Then, if functions in</u> $\underset{\sim}{W}$ <u>which differ only on a locally</u> λ-<u>null set are identified,</u> $\underset{\sim}{W}$ <u>becomes a Hilbert space under the inner product</u>

$$\langle\varphi,\psi\rangle_{\underset{\sim}{W}} = \int_{G/H} \langle\varphi(x),\psi(x)\rangle_X \, d\rho^{\#}(xH) \ . \qquad \ldots(14)$$

<u>For each</u> y <u>in</u> G <u>the equation</u>

$$(T_y\varphi)(x) = (\sigma'(y,xH))^{\frac{1}{2}} \varphi(y^{-1}x) \qquad (\varphi \in \underset{\sim}{W};\ x \in G) \qquad \ldots(15)$$

<u>defines a unitary operator on</u> $\underset{\sim}{W}$; <u>and</u> $y \mapsto T_y$ <u>is a unitary representation of</u> G <u>on</u> $\underset{\sim}{W}$, <u>unitarily equivalent to</u> $\mathrm{Ind}_{H\uparrow G}(S)$.

The easy derivation of this from Theorem 26.2 is left to the reader. We shall refer to Theorem 26.3 as the <u>Mackey formulation</u> of the inducing construction, as contrasted with the <u>Blattner formulation</u> in Theorem 26.2, and the (more general) <u>bundle formulation</u> of §25.

The Mackey formulation has the advantage that it contains explicit reference only to the quasi-invariant measure $\rho^{\#}$ on G/H , not to the rho-function ρ itself. Its disadvantage lies in the Radon-Nikodym factor $\sigma'(y,xH)^{\frac{1}{2}}$ occurring in (15), and also in the fact that different choices of ρ lead to different Hilbert spaces $\underset{\sim}{W}$ in Theorem 26.3.

<u>Remark</u>. For convenience we have always assumed that ρ is continuous, and have therefore obtained the Mackey formulation, strictly speaking, only for those quasi-invariant measures $\rho^{\#}$ which come from a continuous ρ . As Mackey has shown, this is quite inessential (at least for second-countable groups): Formulae (13), (14), and (15) define the induced representation $\mathrm{Ind}_{H\uparrow G}(S)$ no matter what quasi-invariant measure $\rho^{\#}$ we start with.

Remark. Suppose that there exists a non-zero G-invariant regular Borel measure μ on G/H. As is well known (see Weil [1], p. 45), this amounts to saying that $\Delta(h) \equiv \delta(h)$ $(h \in H)$, that is, that we can take $\rho(x) \equiv 1$. In this case $\rho^{\#} = \mu$ (if μ is properly normalized); and there is no difference between the Mackey and the Blattner descriptions of $T = \mathrm{Ind}_{H\uparrow G}(S)$. According to both, the Hilbert space $X(T)$ consists of all functions $\varphi: G \to X$ such that (i) φ is locally λ-measurable, (ii) $\varphi(xh) = S_{h^{-1}}(\varphi(x))$ $(x \in G;\ h \in H)$, and (iii) $\|\varphi\|^2 = \int_{G/H} \|\varphi(x)\|^2 d\mu(xH) < \infty$. The operators of T are those of left translation: $(T_y\varphi)(x) = \varphi(y^{-1}x)$ $(\varphi \in X(T);\ x,y \in G)$.

As one might expect, there are Blattner and Mackey descriptions not merely of induced representations of groups, but also of induced representations of more general classes of Banach *-algebraic bundles. Let us see how the Blattner description works for semidirect product bundles.

Let A be a Banach *-algebra, and $\underset{\sim}{B}$ the ι-semidirect product $A \underset{\iota}{\times} G$ as in §12(I). Thus $\underset{\sim}{B}_H$ is the $(\iota|H)$-semidirect product of A and H. We shall take a non-degenerate *-representation S of $\underset{\sim}{B}_H$, and recall from Cor. 22.4 that S is automatically $\underset{\sim}{B}$-positive. By Prop. 14.6 S gives rise to a unitary representation W of H and a non-degenerate *-representation R of A (both acting on $X(S)$) such that

$$W_h R_a W_h^{-1} = R_{\iota_h(a)},$$

$$S_{(a,h)} = R_a W_h$$

$(a \in A;\ h \in H)$.

We now form the induced unitary representation $V = \mathrm{Ind}_{H\uparrow G}(W)$, using the Blattner description of V; thus V acts by left translation on the space $\underset{\sim}{Z}$ defined in Theorem 26.2. Furthermore, for each a in A, we define an operator Q_a on $\underset{\sim}{Z}$ as follows:

$$(Q_a\varphi)(x) = R_{\iota_{x^{-1}}(a)}(\varphi(x)) \qquad (\varphi \in \underset{\sim}{Z};\ x \in G) . \qquad \ldots(16)$$

The reader can now verify the following result, which constitutes the Blattner description of $\mathrm{Ind}_{\underset{\sim}{B}_H \uparrow \underset{\sim}{B}}(S)$:

Proposition 26.4. Equation (16) defines a non-degenerate *-representation $Q: a \mapsto Q_a$ of A on $\underset{\sim}{Z}$ satisfying

$$V_x Q_a V_x^{-1} = Q_{\iota_x(a)} \qquad (x \in G;\ a \in A) .$$

Thus (see Prop. 14.6) the equation

$$T_{(a,x)} = Q_a V_x \qquad (x \in G;\ a \in A)$$

defines a non-degenerate *-representation T of $\underset{\sim}{B}$ acting on $\underset{\sim}{Z}$. This T is unitarily equivalent with $\mathrm{Ind}_{\underset{\sim}{B}_H \uparrow \underset{\sim}{B}}(S)$.

Thus the operators of T are given by

$$(T_{(a,x)}\varphi)(y) = R_{\iota_{y^{-1}}(a)}\varphi(x^{-1}y)$$

$$= S_{(\iota_{y^{-1}}(a),e)}\varphi(x^{-1}y) \qquad \ldots(17)$$

$(\varphi \in \underset{\sim}{Z};\ a \in A;\ x,y \in G)$.

Remark. For a "Blattner description" of induced representations of the more general so-called homogeneous Banach *-algebraic bundles, see Fell [6], §11.

§27. $\underset{\sim}{B}$-positive *-representations and the bundle C^*-completion.

As a first application of the inducing construction, we propose to establish an important connection (Prop. 27.3) between $\underset{\sim}{B}$-positive *-representations and the C^*-completion of a Banach *-algebraic bundle $\underset{\sim}{B}$. This will lead to Theorem 27.9 on $\underset{\sim}{B}$-positive *-representations in saturated bundles.

Throughout this section we fix a Banach *-algebraic bundle $\underset{\sim}{B} = (B,\pi,\cdot,{}^*)$ over a locally compact group G with unit e and left Haar measure λ. The unit fiber *-algebra B_e will be denoted by A.

Let $\underset{\sim}{C} = (C,\pi',\cdot,{}^*)$ be the bundle C^*-completion of $\underset{\sim}{B}$ (with norm $\| \|_C$), and $\rho: B \longrightarrow C$ the canonical quotient map (see §18). We remarked in §18 that C_e is not in general the C^*-completion of A; in other words, not every *-representation S of A is obtained by composing ρ with a *-representation of C_e. The S which we obtain in this way are, as we shall soon show, just the $\underset{\sim}{B}$-positive ones.

Lemma 27.1. *Let H be a closed subgroup of G and S a $\underset{\sim}{B}$-positive non-degenerate *-representation of $\underset{\sim}{B}_H$; and put $T = \operatorname{Ind}_{\underset{\sim}{B}_H \uparrow \underset{\sim}{B}}(S)$. Then*

$$\|S_a\| \leq \|T_a\| \quad \text{for all } a \text{ in } A. \qquad \ldots(1)$$

Proof. We adopt the notation of §§23, 24 for the Hilbert bundle $\underset{\sim}{Y}$ over G/H induced by S and the action ι of $\underset{\sim}{B}$ on $\underset{\sim}{Y}$. For $a \in A$ we have

$$(T_a f)(\alpha) = \iota_a(f(\alpha)) \qquad (f \in \underset{\sim}{L}_2(\underset{\sim}{Y};\rho^{\#});\ \alpha \in G/H). \qquad \ldots(2)$$

Identifying Y_{eH} with $X(S)$ by means of the mapping F of §23(6), we have by §24(14)

$$\iota_a \xi = S_a \xi \qquad (a \in A;\ \xi \in X(S)). \qquad \ldots(3)$$

Take a non-zero vector ξ in $X(S)$; and choose a continuous cross-section φ of $\underset{\sim}{Y}$ with $\varphi(eH) = \xi$. Let $\{u_i\}$ be an "approximate unit" on G/H, that is, a net of elements of $\underset{\sim}{L}_+(G/H)$ such that (i) $\int u_i \, d\rho^\# = 1$ for all i, and (ii) for any (G/H)-neighborhood W of the coset eH, u_i vanishes outside W for all large enough i. We now set $\varphi_i(\alpha) = u_i(\alpha)^{\frac{1}{2}}\varphi(\alpha)$ $(\alpha \in G/H)$. Thus $\varphi_i \in \underset{\sim}{L}_2(\underset{\sim}{Y};\rho^\#)$; and one verifies that

$$\|\varphi_i\|_2 \xrightarrow[i]{} \|\varphi(eH)\| = \|\xi\| . \quad \ldots(4)$$

Let a be an element of A. Then $\alpha \mapsto \iota_a\varphi(\alpha)$ is continuous; and by (2) and (3), the same argument that led to (4) gives:

$$\|T_a\varphi_i\|_2 \to \|S_a\xi\| . \quad \ldots(5)$$

Combining (4) and (5) we find that $\|T_a\| \geq \|\xi\|^{-1}\|S_a\xi\|$. Since ξ was an arbitrary non-zero vector in $X(S)$, this implies (1). □

Combining the last lemma with Prop. 6.8 we obtain the following useful fact:

Proposition 27.2. *Let H be a closed subgroup of G and S a non-degenerate $\underset{\sim}{B}$-positive *-representation of $\underset{\sim}{B}_H$; and put $T = \mathrm{Ind}_{\underset{\sim}{B}_H\uparrow\underset{\sim}{B}}(S)$. Then $S|A$ is weakly contained in $T|A$.*

Proposition 27.3. *For any *-representation S of A, the following two conditions are equivalent:* (I) *S is $\underset{\sim}{B}$-positive.* (II) *There is a (unique) *-representation S' of C_e such that $S_a = S'_{\rho(a)}$ for all a in A.*

Proof. Clearly (II) amounts to asserting that

$$\|S_a\| \leq \|\rho(a)\|_c \quad \text{for all } a \text{ in } A. \qquad \ldots(6)$$

We may as well suppose that S is non-degenerate.

Assume (I). We can then form the induced *-representation $T = \mathrm{Ind}_{A\uparrow B}(S)$ of $\underset{\sim}{B}$. By the definition of $\underset{\sim}{C}$, $\|T_b\| \leq \|\rho(b)\|_c$ for all b in B. On the other hand, by Lemma 27.1 $\|S_a\| \leq \|T_a\|$ for all a in A. The last two facts imply (6), and hence (II).

Conversely, assume (II). Thus (6) holds. By the definition of $\underset{\sim}{C}$ there is a *-representation W of $\underset{\sim}{B}$ such that $\|W_b\| = \|\rho(b)\|_c$ $(b \in B)$. Combining this with (6) we find:

$$\|S_a\| \leq \|W_a\| \qquad (a \in A). \qquad \ldots(7)$$

Now by Cor. 22.2 $W|A$ is $\underset{\sim}{B}$-positive. Hence by (7) and Prop. 22.7 S is $\underset{\sim}{B}$-positive; and we have shown that (II) => (I). □

<u>Corollary 27.4</u>. <u>The map</u> $H: A_c \to C_e$ <u>of Remark 2 of §18 is one-to-one (hence an isometric</u> *<u>-isomorphism) if and only if every</u> *<u>-representation of</u> A <u>is</u> $\underset{\sim}{B}$<u>-positive</u>.

By Cor. 22.4 (I) and (II) hold if $\underset{\sim}{B}$ has enough unitary multipliers.

Proposition 27.3 has the following generalization:

<u>Proposition 27.5</u>. <u>Let</u> H <u>be any closed subgroup of</u> G, <u>and</u> S <u>a</u> *<u>-representation of</u> $\underset{\sim}{B}_H$. <u>Then the following two conditions are equivalent:</u> (I) $S|A$ <u>is</u> $\underset{\sim}{B}$<u>-positive</u>. (II) <u>There is a (unique)</u> *<u>-representation</u> S' <u>of</u> $\underset{\sim}{C}_H$ <u>such that</u> $S_b = S'_{\rho(b)}$ <u>for all</u> b <u>in</u> $\underset{\sim}{B}_H$. <u>If</u> (I) <u>and</u> (II) <u>hold, then</u> S <u>is</u> $\underset{\sim}{B}$<u>-positive if and only if</u> S' <u>is</u> $\underset{\sim}{C}$<u>-positive</u>.

<u>Proof</u>. (II) => (I) by Prop. 27.3.

Let us assume (I). Then by Prop. 27.3

$$\|S_a\| \leq \|\rho(a)\|_c \quad \text{for all } a \text{ in } A . \qquad \ldots(8)$$

If $b \in B_H$ then $b^*b \in A$, and so by (8)

$$\|S_b\|^2 = \|S_{b^*b}\| \leq \|\rho(b^*b)\|_c = \|\rho(b)\|_c^2 . \qquad \ldots(9)$$

From this it follows that S gives rise to a map $S' : C_H \longrightarrow \underset{\sim}{O}(X(S))$ which is linear on each fiber and satisfies

$$S'_{\rho(b)} = S_b \qquad (b \in B_H) , \qquad \ldots(10)$$

$$\|S'_d\| \leq \|d\|_c \qquad (d \in C_H) . \qquad \ldots(11)$$

I claim that S' is a *-representation of $\underset{\sim}{C}_H$. Indeed: An easy continuity argument shows that S' preserves multiplication and *. It remains only to prove the strong continuity of S'. Let $d_i \longrightarrow d$ in C_H; and put $x_i = \pi'(d_i)$, $x = \pi'(d)$. Given $\varepsilon > 0$, choose a continuous cross-section φ of $\underset{\sim}{B}$ such that

$$\|\rho(\varphi(x)) - d\|_c < \varepsilon . \qquad \ldots(12)$$

Since $\rho: B \longrightarrow C$ is continuous, the map $y \longmapsto \rho(\varphi(y))$ is continuous, and so by (12)

$$\|\rho(\varphi(x_i)) - d_i\|_c < \varepsilon \quad \text{for all large enough } i . \qquad \ldots(13)$$

Since S is strongly continuous on B_H,

$$S_{\varphi(y)} \longrightarrow S_{\varphi(x)} \quad \text{strongly as } y \longrightarrow x \text{ in } H . \qquad \ldots(14)$$

Combining (10), (11), (12), (13) and (14), we obtain for any unit vector ξ in $X(S)$,

$$\|S'_{d_i}\xi - S'_d\xi\| \le \|d_i - \rho(\varphi(x_i))\|_C + \|d - \rho(\varphi(x))\|_C$$
$$+ \|S_{\varphi(x_i)}\xi - S_{\varphi(x)}\xi\|$$
$$< 3\varepsilon \quad \text{for all large enough } i .$$

So $S'_{d_i} \to S'_d$ strongly; and we have shown that S' is a *-representation of $\underset{\sim}{C}_H$.

Thus by (10) (II) holds; and we have shown that (I) => (II). Consequently (I) <=> (II).

If we assume (I) and (II), the last statement of the proposition follows from Prop. 22.1(III) by an easy density argument. □

The most interesting result of this section concerns saturated bundles. To obtain it we need the following lemma on C^*-algebras, which we prove here though it is well known to experts in the field (see for example Prosser [1]).

Lemma 27.6. *Let D be a C^*-algebra, D_+ the set of all positive elements of D, and P a norm-closed subset of D_+ such that* (i) *P is a cone (i.e., P is closed under addition and multiplication by non-negative reals), and* (ii) *if $a \in D$ and $b \in P$ then $a^*ba \in P$. Then:* (I) *The linear span I of P in D is a closed *-ideal of D, and* (II) $I \cap D_+ = P$.

Proof. We begin by proving the following technical fact: If $\varepsilon > 0$ and $0 \le a \le b$ in D, then

$$\|a^{\frac{1}{2}}(b + \varepsilon 1)^{-\frac{1}{2}}\| \le 1 . \qquad \ldots(15)$$

(Here $\mathbb{1}$ is either the unit of D or the unit adjoined to D.) Indeed: Since $a + \varepsilon\mathbb{1} \leq b + \varepsilon\mathbb{1}$, we have $(b + \varepsilon\mathbb{1})^{-1} \leq (a + \varepsilon\mathbb{1})^{-1}$ (Dixmier [1], 1.6.8). So

$$a^{\frac{1}{2}}(b + \varepsilon\mathbb{1})^{-1}a^{\frac{1}{2}} \leq a^{\frac{1}{2}}(a + \varepsilon\mathbb{1})^{-1}a^{\frac{1}{2}} = a(a + \varepsilon\mathbb{1})^{-1} . \qquad \ldots(16)$$

Since $t(t + \varepsilon)^{-1} \leq 1$ whenever $0 \leq t \in \mathbb{R}$, the functional representation of a in D shows that $\|a(a + \varepsilon\mathbb{1})^{-1}\| \leq 1$. This and (16) imply (see Dixmier [1], 1.6.9) that

$$\|a^{\frac{1}{2}}(b + \varepsilon\mathbb{1})^{-1}a^{\frac{1}{2}}\| \leq 1 .$$

This says that $\|cc^*\| \leq 1$, or $\|c\| \leq 1$, where $c = a^{\frac{1}{2}}(b + \varepsilon\mathbb{1})^{-\frac{1}{2}}$. So (15) is proved.

Next I assert that

$$b \in D_+,\ b \leq a \in P \Longrightarrow b \in P . \qquad \ldots(17)$$

Indeed: Denote $a + \varepsilon\mathbb{1}$ by a_ε $(0 < \varepsilon \in \mathbb{R})$. Since $b^{\frac{1}{2}}a_\varepsilon^{-\frac{1}{2}} \in D$, hypothesis (ii) says that

$$c_\varepsilon = b^{\frac{1}{2}}a_\varepsilon^{-\frac{1}{2}}a\, a_\varepsilon^{-\frac{1}{2}}b^{\frac{1}{2}} \in P . \qquad \ldots(18)$$

Now

$$\|c_\varepsilon - b\| = \|b^{\frac{1}{2}}a_\varepsilon^{-\frac{1}{2}}(a - a_\varepsilon)a_\varepsilon^{-\frac{1}{2}}b^{\frac{1}{2}}\|$$

$$\leq \|b^{\frac{1}{2}}a_\varepsilon^{-\frac{1}{2}}\|^2\ \|a - a_\varepsilon\| . \qquad \ldots(19)$$

Since $\|a - a_\varepsilon\| \to 0$ as $\varepsilon \to 0$, it follows from (15) and (19) that $c_\varepsilon \to b$ as $\varepsilon \to 0$. This, (18), and the closedness of P give (17).

Now put $J = \{a \in D: a^*a \in P\}$. I claim that

$$J \text{ is a right ideal of } D . \qquad \ldots(20)$$

Indeed: Let $a,b \in J$. Since

$$(a + b)^*(a + b) \leq 2a^*a + 2b^*b ,$$

hypothesis (i) and (17) show that $a+b \in J$. Hypothesis (i) also implies that J is closed under scalar multiplication. So J is a linear subspace of D. Hypothesis (ii) evidently implies that J is a right ideal of D. So (20) holds.

Let K be the linear span of $\{b^*a: a,b \in J\}$. Since $a \in P \Rightarrow a^{\frac{1}{2}} \in J \Rightarrow a \in K$, we have $I \subset K$. Conversely, let $b,a \in J$. By the polarization identity b^*a is a linear combination of elements of the form c^*c, where c is a linear combination of b and a. Since J is linear, such c are in J, so $c^*c \in P$, whence $b^*a \in I$. We have shown that

$$K = I . \qquad \ldots(21)$$

I is obviously self-adjoint. By (20) and (21) I is a (two-sided) ideal of D. Thus the norm-closure E of I is a closed *-ideal of D.

We shall now prove that

$$I \cap D_+ = P . \qquad \ldots(22)$$

Take $a \in I \cap D_+$. Since $a \in I$ and P is a cone we can write

$$a = a_1 - a_2 + ia_3 - ia_4 ,$$

where the a_i are in P. Since $a^* = a$, we have $ia_3 - ia_4 = 0$. Thus $0 \leq a \leq a_1 \in P$, whence $a \in P$ by (17). We have shown that $I \cap D_+ \subset P$. The reverse inclusion being trivial, (22) holds.

By Dixmier [1], 1.7.2, E has an approximate unit $\{u_i\}$ consisting of elements of $I \cap D_+$. By (22)

$$u_i \in P . \qquad \ldots(23)$$

Now let a be any element of $E \cap D_+$. Since $a^{\frac{1}{2}} \in E$ we have $a^{\frac{1}{2}}u_i \to a^{\frac{1}{2}}$, hence

$$a^{\frac{1}{2}}u_i^2 a^{\frac{1}{2}} = (a^{\frac{1}{2}}u_i)(a^{\frac{1}{2}}u_i)^* \to a . \qquad \ldots(24)$$

By (23) and hypothesis (ii) $a^{\frac{1}{2}}u_i^2 a^{\frac{1}{2}} = (a^{\frac{1}{2}}u_i^{\frac{1}{2}})u_i(a^{\frac{1}{2}}u_i^{\frac{1}{2}})^* \in P$. Combining this with (24) and the closedness of P we get $a \in P$. Thus $E \cap D_+ \subset P$. Since $P \subset E \cap D_+$ trivially, we have proved

$$E \cap D_+ = P . \qquad \ldots(25)$$

Since any C^*-algebra is the linear span of its positive elements, (25) implies that E is the linear span of P, that is, $I = E$. So I was closed all the time, and conclusion (I) is proved. Equality (25) now gives conclusion (II). □

We now return to the bundle context.

<u>Lemma 27.7</u>. <u>Suppose that $\mathcal{B}$ is a saturated C^*-algebraic bundle over G; and let x be a fixed element of G. There exists an approximate unit $\{u_r\}$ of $\mathcal{B}$ such that each u_r is of the form</u>

$$\sum_{i=1}^{n} b_i^* b_i , \qquad \ldots(26)$$

<u>where</u> $n = 1,2,\cdots$ <u>and the</u> $b_1,\cdots,b_n$ <u>are in</u> B_x.

<u>Proof</u>. Let A as usual be the unit fiber C^*-algebra B_e; and let Q be the subset of A consisting of all elements of the form (26) (for fixed x). Thus Q is a cone, and by the definition of a C^*-algebraic

bundle the elements of Q are positive in A. Replacing the b_j in (26) by $b_j c$, we see that

$$a \in Q,\ c \in A \Longrightarrow c^*ac \in Q\ . \qquad \ldots(27)$$

It follows that the norm-closure $\bar{Q}$ of Q also has property (27). Also, by the polarization identity, the linear span of Q in A is the same as the linear span of $B_{x^{-1}}B_x$; and by saturation this is dense in A. Hence by Lemma 27.6 (applied to $\bar{Q}$), $\bar{Q}$ contains all positive elements of A. From this and the existence of positive approximate units in A (Dixmier [1], 1.7.2) we conclude that A has an approximate unit $\{u_i\}$ such that $u_i \in Q$ for all i. By the proof of Prop. 11.6 this $\{u_i\}$ is an approximate unit of $\underset{\sim}{B}$. □

<u>Theorem 27.8</u>. <u>Let $\underset{\sim}{B}$ be a saturated C^*-algebraic bundle over the locally compact group G; and let H be any closed subgroup of G. Then every *-representation of $\underset{\sim}{B}_H$ is positive with respect to $\underset{\sim}{B}$.</u>

<u>Proof</u>. Let S be a *-representation of $\underset{\sim}{B}_H$. Let $x \in \alpha \in G/H$; and take elements $c_1, \cdots c_n$ of B_α and vectors $\xi_1, \cdots, \xi_n$ in $X(S)$. By Lemma 27.7 (applied to x^{-1} instead of x) there is an approximate unit $\{u_r\}$ of $\underset{\sim}{B}$ such that for each r

$$u_r = \sum_{t=1}^{m_r} (b_t^r)^* b_t^r\ , \qquad \ldots(28)$$

where each b_t^r belongs to $B_{x^{-1}}$.

Now for each i,j we have $c_j^* c_i = \lim_r c_j^* u_r c_i$. So by (28)

$$\sum_{i,j=1}^{n} \langle S_{c_j^* c_i} \xi_i, \xi_j \rangle$$

$$= \lim_r \sum_{t=1}^{m_r} \sum_{i,j=1}^{n} \langle S_{c_j^*(b_t^r)^* b_t^r c_i} \xi_i, \xi_j \rangle \ . \qquad \ldots(29)$$

Since $b_t^r c_i \in B_{x^{-1}} B_\alpha \subset B_H$, the last summation in (29) becomes

$$\sum_{i,j=1}^{n} \langle S_{b_t^r c_i} \xi_i, S_{b_t^r c_j} \xi_j \rangle$$

$$= \| \sum_i S_{b_t^r c_i} \xi_i \|^2 \geq 0 \ .$$

Therefore by (29) $\Sigma_{i,j} \langle S_{c_j^* c_i} \xi_i, \xi_j \rangle \geq 0$; and so by Prop. 22.1(III) S is $\underset{\sim}{B}$-positive. □

Remark. Does the above theorem remain true when the saturation hypothesis is omitted? We do not know. If it does remain true, then the hypothesis of saturation can also be omitted in the following theorem.

Theorem 27.9. *Let $\underset{\sim}{B}$ be any saturated Banach *-algebraic bundle over the locally compact group G ; and let H be any closed subgroup of G. For any *-representation S of $\underset{\sim}{B}_H$, the following two conditions are equivalent: (I) S is $\underset{\sim}{B}$-positive; (II) $S|B_e$ is $\underset{\sim}{B}$-positive.*

Proof. (I) => (II) by Cor. 22.2.

Assume (II). Form the bundle C^*-completion $\underset{\sim}{C}$ of $\underset{\sim}{B}$ as in §18; and let $\rho: B \to C$ be the canonical quotient map. By (II) and Prop. 27.5 there is a *-representation S' of $\underset{\sim}{C}_H$ such that $S_b = S'_{\rho(b)}$ $(b \in \underset{\sim}{B}_H)$. As we observed in §18, $\underset{\sim}{C}$ is saturated. Hence by Thm. 27.8 S' is $\underset{\sim}{C}$-positive. Therefore by the last statement of Prop. 27.5 S is $\underset{\sim}{B}$-positive. □

Remark. Keeping the assumptions of this section we might make the following conjecture: Let H be any closed subgroup of G and S a *-representation of $\underset{\sim}{B}_H$. Then S is $\underset{\sim}{B}$-positive if and only if S is weakly contained in $\{T|B_H : T$ is a *-representation of $\underset{\sim}{B}\}$.

It follows from Prop. 27.2 and Cor. 22.6 that the conjecture is true if $H = \{e\}$. More generally it is true if H is normal in G. We do not know whether it holds in general.

§28. Elementary properties of induced representations of Banach *-algebraic bundles.

Again throughout this section $\underset{\sim}{B} = (B, \pi, \cdot, ^*)$ is a Banach *-algebraic bundle over a locally compact group G (with unit e, left Haar measure λ, and modular function Δ).

Direct sums.

Proposition 28.1. *Let H be a closed subgroup of G; and let $\{S^i\}$ $(i \in I)$ be an indexed collection of *-representations of $\underset{\sim}{B}_H$, with Hilbert direct sum $S = \Sigma^{\oplus}_{i \in I} S^i$. Then S is $\underset{\sim}{B}$-positive if and only if each S^i is $\underset{\sim}{B}$-positive; and in that case*

$$\mathrm{Ind}_{\underset{\sim}{B}_H \uparrow \underset{\sim}{B}}(S) \cong \sum_{i \in I}^{\oplus} \mathrm{Ind}_{\underset{\sim}{B}_H \uparrow \underset{\sim}{B}}(S^i) . \qquad \ldots(1)$$

Proof. By Prop. 5.2 (see Theorem 25.4). □

Remark. One can be more explicit about the equivalence (1). For each i let $\underset{\sim}{Y}^i$ be the Hilbert bundle over G/H induced by S^i. Then the Hilbert bundle $\underset{\sim}{Y}$ induced by S can be identified with a naturally constructed "Hilbert bundle direct sum" of the $\underset{\sim}{Y}^i$; and the corresponding identification of $\underset{\sim}{L}_2(\underset{\sim}{Y}; \rho^\#)$ with $\Sigma^{\oplus}_{i \in I} \underset{\sim}{L}_2(\underset{\sim}{Y}^i; \rho^\#)$ is the unitary equivalence which realizes (1). The details are left to the reader.

Remark. It follows from Prop. 28.1 that for an induced *-representation $\mathrm{Ind}_{\underset{\sim}{B}_H \uparrow \underset{\sim}{B}}(S)$ of $\underset{\sim}{B}$ to be irreducible, it is necessary that S be irreducible. However the irreducibility of S is by no means sufficient for that of the induced representation.

Regional continuity of the inducing operation.

Proposition 28.2. *Let H be a closed subgroup of G; and let $\underset{\sim}{S}$ be the space of all non-degenerate $\underset{\sim}{B}$-positive *-representations of $\underset{\sim}{B}_H$. Then the inducing map $S \mapsto \mathrm{Ind}_{\underset{\sim}{B}_H \uparrow \underset{\sim}{B}}(S)$ is continuous on $\underset{\sim}{S}$ with respect to the regional topologies of $\underset{\sim}{S}$ and of the space of *-representations of $\underset{\sim}{B}$.*

Note. The regional topologies for the spaces of *-representations of $\underset{\sim}{B}$ and $\underset{\sim}{B}_H$ were defined at the beginning of §20.

Proof. This follows from Theorem 6.9 and the description (Theorem 25.4) of $\mathrm{Ind}_{\underset{\sim}{B}_H \uparrow \underset{\sim}{B}}(S)$ in terms of the abstract inducing process. □

Remark. This proposition is closely related to Theorem 4.1 of Fell [3] and Theorem 4.2 of Fell [4], though it is not strictly a generalization of either of them (since, for one thing, as we saw in §6, the regional and inner hull-kernel topologies are different).

Induced representations and dense embedding.

Suppose that $\underset{\sim}{C} = (C, \pi', \cdot, ^*)$ is another Banach *-algebraic bundle over the same group G, and that $\Phi: B \rightarrow C$ satisfies the following conditions: (i) for each x in G, $\Phi | B_x$ is a linear map of B_x into C_x; (ii) $\Phi(b_1 b_2) = \Phi(b_1)\Phi(b_2)$ $(b_1, b_2 \in B)$; (iii) $\Phi(b^*) =$

$(\Phi(b))^*$ $(b \in B)$; (iv) $\Phi: B \to C$ is continuous; (v) $\Phi(B_x)$ is dense in C_x for each x in G .

Proposition 28.3. Let H be a closed subgroup of G and S a non-degenerate *-representation of $\underset{\sim}{C}_H$. Then $S': b \mapsto S_{\Phi(b)}$ $(b \in B_H)$ is a non-degenerate *-representation of $\underset{\sim}{B}_H$, and is $\underset{\sim}{B}$-positive if and only if S is $\underset{\sim}{C}$-positive. If S is $\underset{\sim}{C}$-positive, then

$$\mathrm{Ind}_{\underset{\sim}{B}_H \uparrow \underset{\sim}{B}}(S') \cong (\mathrm{Ind}_{\underset{\sim}{C}_H \uparrow \underset{\sim}{C}}(S)) \circ \Phi . \qquad \ldots (2)$$

Proof. Conditions (i)-(iv) show that S' is a *-representation, and condition (v) implies that S' is non-degenerate. The equivalence of the positivity of S and of S' follows from (v) and Prop. 22.1(III).

Assume now that S and S' are positive. To prove (2) we shall use the "abstract" description of the inducing process. Let $\Psi: \underset{\sim}{L}(\underset{\sim}{B}) \to \underset{\sim}{L}(\underset{\sim}{C})$ be the linear map given by $\Psi(f)(x) = \Phi(f(x))$ $(f \in \underset{\sim}{L}(\underset{\sim}{B});\ x \in G)$, and $\Psi_H: \underset{\sim}{L}(\underset{\sim}{B}_H) \to \underset{\sim}{L}(\underset{\sim}{C}_H)$ the corresponding map for H: $\Psi_H(\varphi)(h) = \Phi(\varphi(h))$ $(\varphi \in \underset{\sim}{L}(\underset{\sim}{B}_H);\ h \in H)$. Conditions (i)-(iv) imply that Ψ and Ψ_H are *-homomorphisms of the cross-sectional algebras. Notice that

$$S'_\varphi = S_{\Psi_H(\varphi)} \qquad (\varphi \in \underset{\sim}{L}(\underset{\sim}{B}_H)) . \qquad \ldots (3)$$

Let p denote the conditional expectation of §21(5) both on $\underset{\sim}{L}(\underset{\sim}{B})$ and on $\underset{\sim}{L}(\underset{\sim}{C})$. We shall write X for $X(S)$, and T and T' for $\mathrm{Ind}_{\underset{\sim}{C}_H \uparrow \underset{\sim}{C}}(S)$ and $\mathrm{Ind}_{\underset{\sim}{B}_H \uparrow \underset{\sim}{B}}(S')$ respectively.

Now $X(T')$ is the completion of $\underset{\sim}{L}(\underset{\sim}{B}) \otimes X$ with respect to the positive form $\langle\ ,\ \rangle_0$ given by:

$$\langle f' \otimes \xi, g' \otimes \eta \rangle_0 = \langle S'_{p(g'^* * f')} \xi, \eta \rangle \qquad (f', g' \in \underset{\sim}{L}(\underset{\sim}{B});\ \xi, \eta \in X) ; \ldots (4)$$

and $X(T)$ is the completion of $\underset{\sim}{L}(\underset{\sim}{C}) \otimes X$ with respect to the positive form $\langle\ ,\ \rangle_{00}$ given by:

$$\langle f \otimes \xi,\ g \otimes \eta\rangle_{00} = \langle S_{p(g^{*}*f)}\xi, \eta\rangle \quad (f,g \in \underset{\sim}{L}(\underset{\sim}{C});\ \xi,\eta \in X)\ . \qquad \ldots(5)$$

Observe that, if $\xi,\eta \in X$, $f',g' \in \underset{\sim}{L}(\underset{\sim}{B})$, $f = \Psi(f')$, and $g = \Psi(g')$, then $p(g^{*}*f) = p(\Psi(g'^{*}*f')) = \Psi_H(p(g'^{*}*f'))$, and so by (3)

$$\langle S_{p(g^{*}*f)}\xi, \eta\rangle = \langle S'_{p(g'^{*}*f')}\xi, \eta\rangle\ .$$

From this and (4), (5) we conclude that the map F sending the image $f' \tilde{\otimes} \xi$ of $f' \otimes \xi$ in $X(T')$ into the image $\Psi(f') \tilde{\otimes} \xi$ of $\Psi(f') \otimes \xi$ in $X(T)$ is a linear isometry of $X(T')$ into $X(T)$. Now by (v) and Prop. 10.9 $\Psi(\underset{\sim}{L}(\underset{\sim}{B}))$ is dense in $\underset{\sim}{L}(\underset{\sim}{C})$ in the inductive limit topology. Also we have seen in Remark 3 of §21 that the linear map $f \mapsto f \tilde{\otimes} \xi$ of $\underset{\sim}{L}(\underset{\sim}{C})$ into $X(T)$ is continuous in the inductive limit topology. From the last two facts it follows that the range of F is dense in $X(T)$ and hence equal to $X(T)$.

Thus $F: X(T') \to X(T)$ is a linear isometric surjection. If $f',g' \in \underset{\sim}{L}(\underset{\sim}{B})$ and $\xi \in X$ we have

$$\begin{aligned} T_{\Psi(f')}F(g' \tilde{\otimes} \xi) &= T_{\Psi(f')}(\Psi(g') \tilde{\otimes} \xi) \\ &= (\Psi(f') * \Psi(g')) \tilde{\otimes} \xi \\ &= \Psi(f' * g') \tilde{\otimes} \xi \\ &= F((f' * g') \tilde{\otimes} \xi) \\ &= FT'_{f'}(g' \tilde{\otimes} \xi)\ . \end{aligned}$$

This shows that F intertwines T' and $T \circ \Psi$ (as *-representations of $\underset{\sim}{L}(\underset{\sim}{B})$). Since $T \circ \Psi$ is clearly the integrated form of $T \circ \Phi$, it follows that $T' \cong T \circ \Phi$. □

Remark. Thus, in the context of this proposition, it makes no difference whether we induce S within $\underset{\sim}{B}$ or within $\underset{\sim}{C}$.

Remark. The hypotheses of this proposition hold when $\underset{\sim}{C}$ is the bundle C^*-completion of $\underset{\sim}{B}$ and Φ is the canonical quotient map ρ of §18.

Positive functionals and induced representations.

For the next theorem we will suppose that $\underset{\sim}{B}$ has an approximate unit.

Let H be a closed subgroup of G, with left Haar measure ν and modular function δ. Let $p: \mathcal{L}(\underset{\sim}{B}) \to \mathcal{L}(\underset{\sim}{B}_H)$ be the basic conditional expectation defined in §21(5).

Suppose we are given a positive linear functional q on the *-algebra $\mathcal{L}(\underset{\sim}{B}_H)$ which is continuous in the inductive limit topology. By Theorem 16.1 q generates a *-representation S of $\underset{\sim}{B}_H$; and S is non-degenerate by Prop. 16.2 (since $\underset{\sim}{B}$ and hence $\underset{\sim}{B}_H$ has an approximate unit). In this situation the following theorem is of considerable interest.

Theorem 28.4. The following two conditions are equivalent: (I) The linear functional $q \circ p$ on $\mathcal{L}(\underset{\sim}{B})$ is positive; (II) S is $\underset{\sim}{B}$-positive. If (I) and (II) hold, the *-representation T of $\underset{\sim}{B}$ generated by $q \circ p$ is unitarily equivalent to $\mathrm{Ind}_{\underset{\sim}{B}_H \uparrow \underset{\sim}{B}}(S)$.

Remark. Since p is continuous in the inductive limit topologies, so is $q \circ p$. Hence (I) and Theorem 16.1 imply that T is well defined.

Proof. This theorem is a topological application of Prop. 5.3.

We are going to apply Prop. 5.3 taking A and B to be $\underset{\sim}{L}(\underset{\sim}{B})$ and $\underset{\sim}{L}(\underset{\sim}{B}_H)$; $(L,[\ ,\ ])$ to be $\underset{\sim}{L}(\underset{\sim}{B})$ together with the mapping $(f,g) \mapsto p(f^* * g)$ and the module structures defined by multiplication in $\underset{\sim}{L}(\underset{\sim}{B})$ and §21(7); and K and V to be $\underset{\sim}{L}(\underset{\sim}{B}_H)$ and $(\varphi,\psi) \mapsto q(\psi^* * \varphi)$. As in the context of Prop. 5.3 let $P = \underset{\sim}{L}(\underset{\sim}{B}) \otimes \underset{\sim}{L}(\underset{\sim}{B}_H)$; and let Γ be the conjugate-bilinear form on P given by

$$\begin{aligned}\Gamma(f \otimes \varphi,\ g \otimes \psi) &= q(\psi^* * p(g^* * f) * \varphi) \\ &= (q \circ p)[(g\psi)^* * (f\varphi)] . \qquad \ldots(6)\end{aligned}$$

(The last equality follows from (6), (9), (10) of §21.) Prop. 5.3 then asserts that Γ is positive if and only if S is $\underset{\sim}{B}$-positive; and that, if this is the case, the integrated form of $\mathrm{Ind}_{\underset{\sim}{B}_H \uparrow \underset{\sim}{B}}(S)$ is unitarily equivalent to the *-representation of $\underset{\sim}{L}(\underset{\sim}{B})$ generated by Γ.

Let N be the linear span in $\underset{\sim}{L}(\underset{\sim}{B})$ of $\{f\varphi\colon f \in \underset{\sim}{L}(\underset{\sim}{B}),\ \varphi \in \underset{\sim}{L}(\underset{\sim}{B}_H)\}$. In view of (6), an easy argument shows that the conclusion of the last paragraph can be rephrased as follows: S is $\underset{\sim}{B}$-positive if and only if $q \circ p$ is positive on N; and, if this is the case, the integrated form of $\mathrm{Ind}_{\underset{\sim}{B}_H \uparrow \underset{\sim}{B}}(S)$ is unitarily equivalent to the *-representation of $\underset{\sim}{L}(\underset{\sim}{B})$ generated by the restriction of $q \circ p$ to N.

To prove the theorem, then, it remains only to justify the replacement of N by $\underset{\sim}{L}(\underset{\sim}{B})$ in the last statement. Since $\underset{\sim}{B}$ has an approximate unit, an easy generalization of the Remark following Theorem 13.1 shows that there is a net $\{\varphi_i\}$ of elements of $\underset{\sim}{L}(\underset{\sim}{B}_H)$ such that (i) $f\varphi_i \to f$ in the inductive limit topology for every f in $\underset{\sim}{L}(\underset{\sim}{B})$, and (ii) the compact supports of the φ_i are all contained in the same compact subset of H.

Now fix an element f of $\underset{\sim}{L}(\underset{\sim}{B})$, and put $f_i = f\varphi_i$ for each index i. By (i) and (ii), $f_i \to f$ in the inductive limit topology and all the f_i vanish outside the same compact set. So $f_i^* * f_i \to f^* * f$ in the inductive limit topology, whence

$$(q\cdot p)(f_i^* * f_i) \to (q\cdot p)(f^* * f) \; . \qquad \ldots(7)$$

Now the f_i belong to N. Thus, if $q\cdot p$ is positive on N, it follows from (7) that $(q\cdot p)(f^* * f) \geq 0$, whence $q\cdot p$ is positive on $\mathcal{L}(\mathcal{B})$. We have shown that $q\cdot p$ is positive (on $\mathcal{L}(\mathcal{B})$) if and only if its restriction to N is positive.

Assume that $q\cdot p$ is positive; and let f and f_i be as in the last paragraph. Then $f_i^* * f_i \to f^* * f$, $f_i^* * f \to f^* * f$, and $f^* * f_i \to f^* * f$ in the inductive limit topology. Denoting by $\tilde{f}$ the image of f in the Hilbert space generated by $q\cdot p$, it follows that $\|\tilde{f}_i - \tilde{f}\|^2 = (q\cdot p)(f_i^* * f_i - f_i^* * f - f^* * f_i + f^* * f) \xrightarrow[i]{} 0$. Thus the Hilbert space generated by $(q\cdot p)|N$ is dense in, and so coincides with, that generated by $q\cdot p$. Now $q\cdot p$ generates a *-representation T of $\mathcal{L}(\mathcal{B})$ in virtue of Theorem 16.1. By the preceding sentence $(q\cdot p)|N$ generates the same *-representation T.

Assembling the facts proved above, we find that S is $\mathcal{B}$-positive if and only if $q\cdot p$ is positive, and that, if this is the case, $\mathrm{Ind}_{\mathcal{B}_H \uparrow \mathcal{B}}(S)$ is unitarily equivalent to T. □

If $\mathcal{B}$ has enough unitary multipliers, then by Cor. 22.4 S is automatically $\mathcal{B}$-positive. So the above theorem implies the following corollary:

Corollary. Assume that $\mathcal{B}$ has an approximate unit and enough unitary multipliers. Let S be the (non-degenerate) *-representation of $\mathcal{B}_H$

generated by a positive linear functional q *on* $\underset{\sim}{L}(\underset{\sim}{B}_H)$ *which is continuous in the inductive limit topology. Then* $q \cdot p$ *is positive on* $\underset{\sim}{L}(\underset{\sim}{B})$ *and generates a* **-representation of* $\underset{\sim}{B}$ *unitarily equivalent to* $\mathrm{Ind}_{\underset{\sim}{B}_H \uparrow \underset{\sim}{B}}(S)$.

If $\underset{\sim}{B}$ is the group bundle of G , the q of the preceding corollary is just a regular Borel measure μ of positive type on H , and $q \cdot p$ is just the measure $\Delta(h)^{\frac{1}{2}} \delta(h)^{-\frac{1}{2}} d\mu h$ on H regarded as a measure on G . So the preceding corollary becomes the following:

Corollary. *Let* μ *be a regular Borel measure of positive type (i.e.,* $\mu(\varphi^* * \varphi) \geq 0$ *for all* φ *in* $\underset{\sim}{L}(H)$*) on* H *; and let* S *be the unitary representation of* H *generated by* μ *. Then* $\Delta(h)^{\frac{1}{2}} \delta(h)^{-\frac{1}{2}} d\mu h$ *, regarded as a regular Borel measure on* G *, is of positive type, and generates a unitary representation of* G *which is unitarily equivalent to* $\mathrm{Ind}_{H \uparrow G}(S)$.

This corollary is due to Blattner [3].

Remark. Applying the first of the preceding corollaries to the case that H is the one-element subgroup $\{e\}$, we see that the generalized regular representations of $\underset{\sim}{B}$ defined in §17 are just those of the form $\mathrm{Ind}_{B_e \uparrow \underset{\sim}{B}}(S)$, where S is a cyclic *-representation of the unit fiber *-algebra B_e .

Inducing in stages.

Theorem 28.5. *Let* H *and* L *be two closed subgroups of* G *with* $H \subset L$ *; and let* S *be a non-degenerate* **-representation of* $\underset{\sim}{B}_H$ *. Then the following two conditions are equivalent:* (I) S *is* $\underset{\sim}{B}$*-positive;*

(II) *S is $\underset{\sim}{B}_L$-positive and $\mathrm{Ind}_{\underset{\sim}{B}_H \uparrow \underset{\sim}{B}_L}(S)$ is $\underset{\sim}{B}$-positive. If these two conditions hold, then*

$$\mathrm{Ind}_{\underset{\sim}{B}_H \uparrow \underset{\sim}{B}}(S) \cong \mathrm{Ind}_{\underset{\sim}{B}_L \uparrow \underset{\sim}{B}}\left(\mathrm{Ind}_{\underset{\sim}{B}_H \uparrow \underset{\sim}{B}_L}(S)\right) . \qquad \ldots(8)$$

Proof. To begin with we shall suppose that $\underset{\sim}{B}$ has an approximate unit. Afterwards this assumption will be dropped.

Let us write $S = \Sigma_i^{\oplus} S^i$, where each S^i is a cyclic *-representation of $\underset{\sim}{B}_H$. By Prop. 28.1 the theorem will hold for S provided it holds for each S^i. Hence we may, and shall, assume from the beginning that S is cyclic, with cyclic vector ξ.

Let δ_H and δ_L be the modular functions of H and L respectively; and define the three basic conditional expectations $p: \underset{\sim}{L}(\underset{\sim}{B}) \to \underset{\sim}{L}(\underset{\sim}{B}_H)$, $p_1: \underset{\sim}{L}(\underset{\sim}{B}) \to \underset{\sim}{L}(\underset{\sim}{B}_L)$, $p_2: \underset{\sim}{L}(\underset{\sim}{B}_L) \to \underset{\sim}{L}(\underset{\sim}{B}_H)$ as in §21(5):

$$p(f)(h) = \Delta(h)^{\frac{1}{2}} \delta_H(h)^{-\frac{1}{2}} f(h) ,$$

$$p_1(f)(m) = \Delta(m)^{\frac{1}{2}} \delta_L(m)^{-\frac{1}{2}} f(m) ,$$

$$p_2(\varphi)(h) = \delta_L(h)^{\frac{1}{2}} \delta_H(h)^{-\frac{1}{2}} \varphi(h)$$

($f \in \underset{\sim}{L}(\underset{\sim}{B})$; $\varphi \in \underset{\sim}{L}(\underset{\sim}{B}_L)$; $m \in L$; $h \in H$). We notice that $p_2 \circ p_1 = p$.

Now by Cor. 22.3 conditions (I) and (II) both imply that S is $\underset{\sim}{B}_L$-positive. So we shall assume this. Put $q(\varphi) = \langle S_\varphi \xi, \xi \rangle$ ($\varphi \in \underset{\sim}{L}(\underset{\sim}{B}_H)$). Then q is a positive linear functional on $\underset{\sim}{L}(\underset{\sim}{B}_H)$ which generates S. Since S is assumed to be $\underset{\sim}{B}_L$-positive, Theorem 28.4 (applied to $\underset{\sim}{B}_L$) tells us that $q \circ p_2$ is positive on $\underset{\sim}{L}(\underset{\sim}{B}_L)$ and generates $V = \mathrm{Ind}_{\underset{\sim}{B}_H \uparrow \underset{\sim}{B}_L}(S)$. Thus Theorem 28.4 (applied this time to $\underset{\sim}{B}$) tells us that $q \circ p$ ($= (q \circ p_2) \circ p_1$) is positive on $\underset{\sim}{L}(\underset{\sim}{B})$ if and only if V is $\underset{\sim}{B}$-positive, and that in that case $q \circ p$ generates $\mathrm{Ind}_{\underset{\sim}{B}_L \uparrow \underset{\sim}{B}}(V)$. On the other hand, a third application of Theorem 28.4 shows that $q \circ p$ is positive on $\underset{\sim}{L}(\underset{\sim}{B})$ if and only if S is $\underset{\sim}{B}$-positive, in which case $q \circ p$ generates $\mathrm{Ind}_{\underset{\sim}{B}_H \uparrow \underset{\sim}{B}}(S)$. Putting these facts together we find that V is $\underset{\sim}{B}$-positive if and only if S is $\underset{\sim}{B}$-positive, and that in that case $\mathrm{Ind}_{\underset{\sim}{B}_H \uparrow \underset{\sim}{B}}(S) \cong \mathrm{Ind}_{\underset{\sim}{B}_L \uparrow \underset{\sim}{B}}(V)$. This is the required result.

We now discard the assumption that $\underset{\sim}{B}$ has an approximate unit (which was needed in the above argument in order to be able to apply Theorem 28.4). Let $\underset{\sim}{C}$ be the bundle C^*-completion of $\underset{\sim}{B}$ and ρ: B —> C the canonical quotient map (see §18). Since $\underset{\sim}{C}$ automatically has an approximate unit (Prop. 11.6), we shall obtain the required result for $\underset{\sim}{B}$ by applying the preceding part of the proof to $\underset{\sim}{C}$.

First I claim that the conditions (I) and (II) both imply that $S|B_e$ is $\underset{\sim}{B}$-positive. Indeed: Condition (I) implies this by Cor. 22.2. Now assume Condition (II); and set $V = \mathrm{Ind}_{\underset{\sim}{B}_H \uparrow \underset{\sim}{B}_L}(S)$, $T = \mathrm{Ind}_{\underset{\sim}{B}_L \uparrow \underset{\sim}{B}}(V)$. By Lemma 27.1 we have $\|S_a\| \leq \|V_a\|$ and also $\|V_a\| \leq \|T_a\|$ for all a in B_e; therefore

$$\|S_a\| \leq \|T_a\| \qquad (a \in B_e) . \qquad \ldots(9)$$

Since $T|B_e$ is $\underset{\sim}{B}$-positive by Cor. 22.2, it follows from (9) and Prop. 22.7 that $S|B_e$ is $\underset{\sim}{B}$-positive. This establishes the claim.

In view of this claim we may as well assume from the beginning that $S|B_e$ is $\underset{\sim}{B}$-positive. Thus by Prop. 27.5 there is a *-representation S' of $\underset{\sim}{C}_H$ such that

$$S_b = S'_{\rho(b)} \quad \text{for} \quad b \in B_H . \qquad \ldots(10)$$

Applying the preceding part of the proof to $\underset{\sim}{C}$ (which has an approximate unit), we conclude that the following two conditions are equivalent: (I′) S' is $\underset{\sim}{C}$-positive; (II′) S' is $\underset{\sim}{C}_L$-positive and $V' = \mathrm{Ind}_{\underset{\sim}{C}_H \uparrow \underset{\sim}{C}_L}(S')$ is $\underset{\sim}{C}$-positive. If these conditions hold, then

$$\mathrm{Ind}_{\underset{\sim}{C}_H \uparrow \underset{\sim}{C}}(S') \cong \mathrm{Ind}_{\underset{\sim}{C}_L \uparrow \underset{\sim}{C}}(V') . \qquad \ldots(11)$$

On the other hand, by (10) and Prop. 28.3, (I′) holds if and only if S is $\underset{\sim}{B}$-positive; and (II′) holds if and only if S is $\underset{\sim}{B}_L$-positive and $V = \mathrm{Ind}_{\underset{\sim}{B}_H \uparrow \underset{\sim}{B}_L}(S)$ is $\underset{\sim}{B}$-positive. This proves the equivalence of (I) and (II) of the theorem. If (I) and (II) hold, that is, if (I′) and (II′) hold, then (11) and Prop. 28.3 imply (8). This completes the proof. □

If $\underset{\sim}{B}$ has enough unitary multipliers, Cor. 22.4 implies the following simpler form of the preceding theorem.

<u>Corollary 28.6</u>. <u>Suppose that $\underset{\sim}{B}$ has enough unitary multipliers. Let H and L be two closed subgroups of G with $H \subset L$. If S is a non-degenerate *-representation of $\underset{\sim}{B}_H$, we have</u>

$$\mathrm{Ind}_{\underset{\sim}{B}_H \uparrow \underset{\sim}{B}}(S) \cong \mathrm{Ind}_{\underset{\sim}{B}_L \uparrow \underset{\sim}{B}}\left(\mathrm{Ind}_{\underset{\sim}{B}_H \uparrow \underset{\sim}{B}_L}(S)\right) .$$

When $\underset{\sim}{B}$ is the group bundle of G, this becomes Mackey's classical theorem on inducing in stages (see Mackey [4]).

<u>Lifting and induced representations</u>.

Let G' be another locally compact group, and $p: G' \to G$ a continuous open homomorphism of G' onto G. Thus we can form the retraction $\underset{\sim}{B}' = (B', \pi', \cdot, ^*)$ of $\underset{\sim}{B}$ by p (see §11); this is a Banach *-algebraic bundle over G'. Let $P: B' \to B$ be the map given by

$$P(x',b) = b \qquad ((x',b) \in B').$$

If T is a *-representation of $\underset{\sim}{B}$, then $T \circ P$ is a *-representation of $\underset{\sim}{B}'$, called the lift of T to $\underset{\sim}{B}'$.

Proposition 28.7. Let H be a closed subgroup of G, and set $H' = p^{-1}(H)$. Let S be a non-degenerate *-representation of $\underset{\sim}{B}_H$, and $S' = S \circ (P|B'_{H'})$ the lift of S to $\underset{\sim}{B}'_{H'}$. Then: (I) S is $\underset{\sim}{B}$-positive if and only if S' is $\underset{\sim}{B}'$-positive; and (II) if this is the case, $\mathrm{Ind}_{\underset{\sim}{B}'_{H'} \uparrow \underset{\sim}{B}'}(S')$ is unitarily equivalent to the lift of $\mathrm{Ind}_{\underset{\sim}{B}_H \uparrow \underset{\sim}{B}}(S)$ to $\underset{\sim}{B}'$.

Proof. (I) follows immediately from Prop. 22.1(III).

Assume that S and S' are positive; and let $\underset{\sim}{Y}$ and $\underset{\sim}{Y}'$ be the Hilbert bundles which they induce. We shall show that $\underset{\sim}{Y}$ and $\underset{\sim}{Y}'$ are essentially the same.

Let ρ be a continuous everywhere positive H-rho function on G. Then $\rho' = \rho \circ p$ is an H'-rho function on G'. (This follows, for example, from Bourbaki [4], Chap. VII, §2, n° 8, Cor. of Prop. 11.) Now let α be a coset in G/H, and $\alpha' = p^{-1}(\alpha)$ the corresponding coset in G'/H'. Let κ_α and $\kappa'_{\alpha'}$ be the quotient maps into Y_α and $Y'_{\alpha'}$ respectively defined just after §23(7). If $\xi, \eta \in X(S)$, $b', c' \in B'_{\alpha'}$, and $\pi'(b') = x'$, $\pi'(c') = y'$, we have by §23(7)

$$\begin{aligned}(\kappa'_{\alpha'}(b' \otimes \xi), \kappa'_{\alpha'}(c' \otimes \eta)) &= (\rho'(x')\rho'(y'))^{-\frac{1}{2}} \langle S'_{c'^*b'}\xi, \eta\rangle \\ &= (\rho(x)\rho(y))^{-\frac{1}{2}} \langle S_{c^*b}\xi, \eta\rangle \\ &= \langle \kappa_\alpha(b \otimes \xi), \kappa_\alpha(c \otimes \eta)\rangle,\end{aligned}$$

where $b = P(b')$, $c = P(c')$, $x = \pi(b) = p(x')$, and $y = \pi(c) =$

$p(y')$. It follows that the equation

$$F_\alpha(\kappa'_{\alpha'}(b' \otimes \xi)) = \kappa_\alpha(P(b') \otimes \xi) \qquad (b' \in B'_{\alpha'}\,;\ \xi \in X(S))$$

defines a linear isometry F_α of $Y'_{\alpha'}$ onto Y_α. The map $F: Y' \to Y$ which coincides with F_α on each $Y'_{\alpha'}$ is thus a bijection; and it follows from Prop. 23.3 that F is a homeomorphism. Thus F identifies $\underset{\sim}{Y}'$ with $\underset{\sim}{Y}$, and so $\underset{\sim}{L}_2(\underset{\sim}{Y}'; \rho'^{\#})$ with $\underset{\sim}{L}_2(\underset{\sim}{Y}; \rho^{\#})$. Under this identification the action of b' on $\underset{\sim}{Y}'$ $(b' \in B')$ coincides with the action of $P(b')$ on $\underset{\sim}{Y}$. Hence the same is true of $\underset{\sim}{L}_2(\underset{\sim}{Y}'; \rho'^{\#})$ and $\underset{\sim}{L}_2(\underset{\sim}{Y}; \rho^{\#})$; and the proof is complete. □

In the group case, of course, Prop. 28.7 becomes:

<u>Corollary 28.8</u>. <u>Let</u> G, G', p, H, H' <u>be as in Prop</u>. 28.7. <u>Let</u> V <u>be a unitary representation of</u> H, <u>and</u> $V' = V \circ (p|H')$ <u>the lifted unitary representation of</u> H'. <u>Then</u>

$$\mathrm{Ind}_{H' \uparrow G'}(V') \cong \left(\mathrm{Ind}_{H \uparrow G}(V)\right) \circ p .$$

In particular this holds when $G' = G$ and p is an automorphism of G. Suppose in fact that p is the inner automorphism $x \mapsto yxy^{-1}$ of G. Then $T \circ p \cong T$ for any unitary representation T of G; and Cor. 28.8 gives:

<u>Corollary 28.9</u>. <u>Let</u> H <u>be a closed subgroup of</u> G, V <u>a unitary representation of</u> H, <u>and</u> y <u>an element of</u> G. <u>Let</u> $H' = y^{-1}Hy$; <u>and let</u> V' <u>be the unitary representation</u> $x \mapsto V_{yxy^{-1}}$ <u>of</u> H'. <u>Then</u>

$$\mathrm{Ind}_{H' \uparrow G}(V') \cong \mathrm{Ind}_{H \uparrow G}(V) .$$

§29. Transformation bundles.

Fix a Banach *-algebraic bundle $\underset{\sim}{B} = (B, \pi, \cdot, ^*)$ over a locally compact group G (with unit e, left Haar measure λ, and modular function Δ), and also a locally compact Hausdorff space M on which G acts continuously to the left as a topological transformation group (the action of G on M being denoted as usual by $(x,m) \mapsto xm$). From these ingredients we are going to construct a new Banach *-algebraic bundle $\underset{\sim}{D}$ over G, called the transformation bundle. The *-representations of $\underset{\sim}{D}$ will be intimately related to the so-called systems of imprimitivity.

For each x in G let D_x be the Banach space $\underset{\sim}{C}_0(M;B_x)$ (with the supremum norm $\| \|_\infty$) of all continuous functions on M to B_x which vanish at infinity. We shall denote by E the important linear space of all functions $f: G \times M \to B$ such that (i) $f(x,m) \in B_x$ for all x in G and m in M, and (ii) f has compact support, that is, $f(x,m) = 0_x$ for all (x,m) outside some compact subset of $G \times M$. If $f \in E$, let $\tilde{f}$ be the function on G assigning to each x in G the element $\tilde{f}(x): m \mapsto f(x,m)$ of D_x. Then it is easy to see that the family $\{\tilde{f}: f \in E\}$ satisfies the hypotheses of Prop. 10.4, and so determines a unique Banach bundle $\underset{\sim}{D} = (D, \rho)$ over G whose fiber over x is D_x, relative to which the cross-sections $\tilde{f}$ $(f \in E)$ are all continuous.

We shall now introduce a multiplication and involution into $\underset{\sim}{D}$ by means of the definitions:

$$(\varphi\psi)(m) = \varphi(m)\psi(x^{-1}m) , \qquad \ldots(1)$$

$$\varphi^*(m) = (\varphi(xm))^* \qquad \ldots(2)$$

$(x,y \in G;\ \varphi \in D_x;\ \psi \in D_y;\ m \in M)$. With these operations, I claim that $\underset{\sim}{D}$ is a Banach *-algebraic bundle over G . Indeed: Since $B_xB_y \subset B_{xy}$ and $(B_x)^* = B_{x^{-1}}$, and since $m \to \infty$ implies $x^{-1}m \to \infty$ and $xm \to \infty$ in M , it is easy to see that the $\varphi\psi$ and φ^* defined in (1) and (2) lie in D_{xy} and $D_{x^{-1}}$ respectively. The postulates of a Banach *-algebraic bundle other than the continuity of multiplication and involution are easily checked. To prove postulate §11(v), the continuity of multiplication in $\underset{\sim}{D}$, we shall make use of the alternative postulate §11(v′), taking Γ to be $\{\tilde{f}: f \in E\}$. Given $f,g,h \in E$, we first show by an easy calculation that $(x,y) \mapsto \|\tilde{h}(xy) - \tilde{f}(x)\tilde{g}(y)\|_\infty$ is continuous on $G \times G$. Since $(x,y) \mapsto \tilde{h}(xy)$ is continuous on $G \times G$ to D for arbitrary h in E , it follows by an easy argument that $(x,y) \mapsto \tilde{f}(x)\tilde{g}(y)$ is also continuous on $G \times G$ to D . So §11(v′) holds, and the product in $\underset{\sim}{D}$ is continuous. Similarly we verify the continuity of the involution by means of §11(xi′).

Definition. This Banach *-algebraic bundle $\underset{\sim}{D}$ over G is called the G,M *transformation bundle derived from* $\underset{\sim}{B}$.

The unit fiber *-algebra of $\underset{\sim}{D}$ is just $\underset{\sim}{C}_0(M; B_e)$, considered as a Banach *-algebra under the pointwise operations.

As an example, assume for the moment that $\underset{\sim}{B}$ is the group bundle of G . Then $\underset{\sim}{D}$ is called simply the G,M *transformation bundle*. The reader will verify that in this case $\underset{\sim}{D}$ is the ι-semidirect product of $\underset{\sim}{C}_0(M)$ and G , where ι_x is the natural action of the element x of G on $\underset{\sim}{C}_0(M)$:

$$(\iota_x\varphi)(m) = \varphi(x^{-1}m) .$$

This special case was discussed in §12(I).

Returning to the general case, we have:

Proposition 29.1. If $\underline{B}$ is saturated, then $\underline{D}$ is also saturated.

Proof. Fix x,y in G; and let L be the linear span in D_{xy} of $\{\varphi\psi: \varphi \in D_x, \psi \in D_y, \varphi$ and ψ have compact support$\}$. Clearly L is closed under multiplication by elements of $\underline{C}(M)$. By the saturation of $\underline{B}$, $\{\chi(m): \chi \in L\}$ is dense in B_{xy} for each m in M. Therefore by Prop. 10.9 L is dense in D_{xy}. □

Proposition 29.2. If $\underline{B}$ has an approximate unit, then so does $\underline{D}$.

Proof. Let M_0 be the one-point compactification of M (M_0 being M if M is compact). We first verify that $(\varphi,m) \mapsto \varphi(m)$ is continuous on $D \times M_0$ to B. (This can be done by approximating the variable φ by values of one of the "standard" cross-sections $\tilde{f}$ $(f \in E)$.)

Now let W be a compact subset of D. By the preceding paragraph, $V = \{\varphi(m): \varphi \in W, m \in M_0\}$ is a compact subset of B. Since $\underline{B}$ has an approximate unit, there is a constant k (independent of W) and an element u of B_e such that (i) $\|u\| \leq k$, and (ii) $\|ub - b\|$ and $\|bu - b\|$ are as small as we want for all b in V. If we now take ψ to be an element of D_e with $\|\psi\|_\infty \leq k$ and $\psi(m) = u$ for all m in a very large compact subset of M, it will follow that $\|\psi\varphi - \varphi\|_\infty$ and $\|\varphi\psi - \varphi\|_\infty$ are as small as we want for all φ in W. □

Now, recalling the definition of E and of the map $f \mapsto \tilde{f}$, we observe that the latter map is an injection of E into $\underline{L}(\underline{D}) \subset \underline{L}_1(\underline{D};\lambda)$. In fact, by Prop. 10.12 the range of $f \mapsto \tilde{f}$ is dense in $\underline{L}_1(\underline{D};\lambda)$. It is natural to ask what the algebraic structure of $\underline{L}_1(\underline{D};\lambda)$ looks like when transferred to E via this map.

Proposition 29.3. E is a normed *-algebra when multiplication $*$, involution *, and norm $\| \ \|$ are defined in E as follows:

$$(f*g)(x,m) = \int_G f(y,m)\, g(y^{-1}x, y^{-1}m)\, d\lambda y \ , \qquad \ldots(3)$$

$$f^*(x,m) = \Delta(x^{-1})[f(x^{-1}, x^{-1}m)]^* \ , \qquad \ldots(4)$$

$$\|f\| = \int_G \sup\{\|f(y,m)\| : m \in M\} d\lambda y \qquad \ldots(5)$$

$(f,g \in E;\ x \in G;\ m \in M)$. With this structure, the map $f \mapsto \tilde{f}$ is an isometric *-isomorphism of E onto a dense *-subalgebra of $L_1(D;\lambda)$.

We leave the reader to verify this.

Definition. E, with the operations (3) and (4), is called the compacted transformation algebra (of G,M derived from B). The completion of E with respect to the norm (5) is called the L_1 transformation algebra.

Sometimes it is convenient to speak of E as if it were actually embedded in $L(D)$ by means of the map $f \mapsto \tilde{f}$.

If M is a one-element space, E and its completion become just the compacted cross-sectional algebra and the L_1 cross-sectional algebra of B defined in §13.

Multipliers in D.

We next observe that multipliers of B give rise to multipliers of D. Indeed, let b be a multiplier of B of order x. It is easy to check that the equations

$$\left.\begin{aligned}(u_b\varphi)(m) &= b\varphi(x^{-1}m) ,\\ (\varphi u_b)(m) &= \varphi(m)b\end{aligned}\right\} \qquad \ldots(6)$$

($\varphi \in D$; $m \in M$) define the left and right actions respectively of a multiplier u_b of $\underset{\sim}{D}$ of the same order x. (The continuity of the left and right actions (6) on D follows from a consideration of the actions of u_b on the "standard" cross-sections $\tilde{f}$, $f \in E$.)

The correspondence $b \mapsto u_b$ preserves multiplication, involution, the linear operations (on each fiber), and norm. In particular, if b is a unitary multiplier of $\underset{\sim}{B}$, u_b is a unitary multiplier of $\underset{\sim}{D}$. Thus we have:

<u>Proposition 29.4</u>. <u>If $\underset{\sim}{B}$ has enough unitary multipliers, so does $\underset{\sim}{D}$</u>.

Since each element b of B can be regarded as a multiplier of $\underset{\sim}{B}$, equations (6) define a map $b \mapsto u_b$ of B into $W(\underset{\sim}{D})$ which is linear on each fiber, preserves multiplication and *, and is norm-decreasing.

<u>Proposition 29.5</u>. <u>The map $b \mapsto u_b$ is continuous on B to $W(\underset{\sim}{D})$ with respect to the strong topology of multipliers (see §11)</u>.

<u>Proof</u>. It is enough to show that $b \mapsto b\tilde{f}(x_0)$ and $b \mapsto \tilde{f}(x_0)b$ are continuous on B to D for each fixed x_0 in G and f in E. For each b in B put $g_b(x,m) = bf(\pi(b)^{-1}x, \pi(b)^{-1}m)$, so that $g_b \in E$. Let $b_i \to b$ in B. An easy compactness argument shows that $g_{b_i} \to g_b$ uniformly on $G \times M$. Hence $\|b_i\tilde{f}(x_0) - (g_b)^{\sim}(\pi(b_i)x_0)\|_\infty = \|(g_{b_i})^{\sim}(\pi(b_i)x_0) - (g_b)^{\sim}(\pi(b_i)x_0)\|_\infty \to 0$. Since $(g_b)^{\sim}(\pi(b_i)x_0) \to (g_b)^{\sim}(\pi(b)x_0) = b\tilde{f}(x_0)$ in D, it therefore follows that $b_i\tilde{f}(x_0) \to b\tilde{f}(x_0)$ in D. So $b \mapsto b\tilde{f}(x_0)$ is continuous. Similarly $b \mapsto \tilde{f}(x_0)b$ is

continuous. □

Another source of natural multipliers of $\underset{\sim}{D}$ is the space of bounded continuous complex functions on M.

Let h be a bounded continuous complex function on M. Putting

$$\left.\begin{aligned} (h\varphi)(m) &= h(m)\ \varphi(m) \\ (\varphi h)(m) &= h(x^{-1}m)\ \varphi(m) \end{aligned}\right\} \qquad \ldots(7)$$

($\varphi \in D_x$; $m \in M$), one checks that the maps $\varphi \mapsto h\varphi$ and $\varphi \mapsto \varphi h$ together form a multiplier v_h of $\underset{\sim}{D}$ of order e. The correspondence $h \mapsto v_h$ is a *-homomorphism, and is norm-decreasing with respect to the supremum norm of the h.

By §13 the integrated form of v_h is a multiplier v'_h of $\underset{\sim}{L}(\underset{\sim}{D})$. The action of v'_h on E is easily seen to be given by

$$(hf)(x,m) = h(m)f(x,m),\ (fh)(x,m) = h(x^{-1}m)f(x,m) \qquad \ldots(8)$$

($f \in E$; $(x,m) \in G \times M$; we write hf and fh for $v'_h f$ and fv'_h).

An important relation holds between the multipliers u_b and v_h of $\underset{\sim}{D}$, namely

$$u_b v_h = v_{yh} u_b \qquad \ldots(9)$$

($b \in B_y$; h is a bounded function in $\underset{\sim}{C}(M)$). Here yh is the y-translate $m \mapsto h(y^{-1}m)$ of h. The routine verification of this is left to the reader.

A similar calculation shows that if $b \in B_y$ and $h \in \underset{\sim}{C}_0(M)$, then the product $v_h u_b$ coincides with the multiplier of $\underset{\sim}{D}$ consisting of left and right multiplication by the following element ψ of D_y:

$$\psi(m) = h(m)b\ . \qquad \ldots(10)$$

§30. Systems of imprimitivity.

There is an intimate relation between the transformation bundles or transformation algebras of §29 and the so-called systems of imprimitivity, in terms of which the Imprimitivity Theorem for groups is usually formulated. In this section we shall present this relation in the more general context of Banach *-algebraic bundles.

We keep the notation of the first paragraph of §29.

Definition. A system of imprimitivity for $\underset{\sim}{B}$ over M is a pair (T,P), where (i) T and P are non-degenerate *-representations of $\underset{\sim}{B}$ and of $\underset{\sim}{C}_0(M)$ respectively acting on the same Hilbert space, and (ii) we have

$$T_b P_\varphi = P_{x\varphi} T_b \qquad \ldots (1)$$

for all φ in $\underset{\sim}{C}_0(M)$, x in G, and b in B_x. (Here as usual $(x\varphi)(m) = \varphi(x^{-1}m)$.)

Remark. It has been more usual in the past to formulate the definition of a system of imprimitivity in terms of projection-valued measures. Let X be a Hilbert space. A regular X-projection-valued measure on M is a map Q assigning to each Borel subset W of M a projection $Q(W)$ on M such that: (i) $Q(M) = 1_X$; (ii) if $W_1, W_2, \cdots$ is any sequence of pairwise disjoint Borel subsets of M, the $Q(W_i)$ are pairwise orthogonal projections whose sum is $Q(\cup_{i=1}^{\infty} W_i)$; (iii) for any Borel subset W of M, $Q(W) = \text{lub}\{Q(C): C$ is a compact subset of $W\}$. It is well known (see for example Naimark [1], Chap. IV, §17, n° 4) that the regular X-projection-valued measures Q on M are in natural one-to-one correspondence with the non-degenerate *-representations P of $\underset{\sim}{C}_0(M)$ on X, the correspondence being given by

$$P_{\varphi} = \int_M \varphi(m)\,dQm \qquad \ldots(2)$$

(the right side being the so-called spectral integral). In view of this correspondence, a system of imprimitivity for $\underset{\sim}{B}$ over M can equally well be regarded as a pair consisting of a non-degenerate *-representation T of $\underset{\sim}{B}$ and a regular $X(T)$-projection-valued measure Q on M such that the following identity (equivalent to (1)) holds:

$$T_b\, Q(W) = Q(xW)\,T_b \qquad \ldots(3)$$

for all x in G and b in B_x, and all Borel subsets W of M. This was the approach followed in Fell [6], §13.

<u>Remark</u>. If $\underset{\sim}{B}$ is the group bundle, the above definition reduces to the more familiar notion of a <u>system of imprimitivity for</u> the group G <u>over</u> M. By this we mean a pair (T,P), where T is a unitary representation of G, P is a non-degenerate *-representation of $\underset{\sim}{C}_0(M)$ acting on the same space as T, and

$$T_x\, P_{\varphi} = P_{x\varphi}\, T_x$$

for all φ in $\underset{\sim}{C}_0(M)$ and x in G.

Returning to the general bundle $\underset{\sim}{B}$, we define concepts such as stable subspaces, Hilbert direct sums, irreducibility, intertwining operators, unitary equivalence, and the commuting algebra for systems of imprimitivity (T,P), just as we do for *-representations. For example, the commuting algebra of T,P is the von Neumann algebra of all bounded linear operators on $X(T)$ which commute with all T_b $(b \in B)$ and all $P_{\varphi}(\varphi \in \underset{\sim}{C}_0(M))$.

It will now be proved that there is a natural one-to-one correspondence between systems of imprimitivity for $\underset{\sim}{B}$ over M and non-degenerate *-representations of the transformation bundle $\underset{\sim}{D}$ constructed in §29.

Theorem 30.1. Let (T,P) be a system of imprimitivity for $\underset{\sim}{B}$ over M, acting in a Hilbert space X. Then there is a unique non-degenerate *-representation S of $\underset{\sim}{D}$ acting in X which satisfies

$$S_{\psi} = P_h T_b \qquad \ldots(4)$$

whenever $b \in B$, $h \in \underset{\sim}{L}(M)$, and ψ is the element of $D_{\pi(b)}$ given by

$$\psi(m) = h(m)b \qquad (m \in M) . \qquad \ldots(5)$$

Conversely, given any non-degenerate *-representation S of $\underset{\sim}{D}$, there is a unique system of imprimitivity (T,P) for $\underset{\sim}{B}$ over M, acting in the same space as S, and related to S by the equation (4).

Proof. Let (T,P) be a system of imprimitivity for $\underset{\sim}{B}$ over M, and let Q be the $X(T)$-projection-valued measure on M corresponding to P (see the preceding Remark), so that (T,Q) is a system of imprimitivity in the sense of Fell [6], §13. In §14 of Fell [6] we constructed from (T,Q) a certain non-degenerate *-representation S of $\underset{\sim}{D}$ satisfying §14(10) of Fell [6]. (In §14 of Fell [6] Q and $\underset{\sim}{D}$ were called P and $\underset{\sim}{C}$ respectively. Notice that the construction of S did not depend at all on the homogeneity of $\underset{\sim}{B}$ which was assumed in Part II of Fell [6].) It follows from §14(8) of Fell [6] that S satisfies (4). Since by Prop. 10.9 (of the present work) the linear span of the set of elements ψ of (5) is dense in each fiber D_x, S is uniquely determined by the condition (4).

Conversely, suppose that S is a non-degenerate *-representation of $\underset{\sim}{D}$. Let S' be the extension of S to the multiplier bundle of $\underset{\sim}{D}$, given by Prop. 14.3. Now in §29(6) we constructed a norm-decreasing map $b \mapsto u_b$ from B to the multiplier bundle of $\underset{\sim}{D}$ which preserved all the algebraic operations and was in addition continuous with respect to the strong multiplier topology (Prop. 29.5). It follows therefore (see Prop. 14.4) that

$$T : b \longrightarrow S'_{u_b} \qquad \ldots(6)$$

is a *-representation of $\underset{\sim}{B}$ acting on $X(S)$. We also constructed in §29(7) a *-homomorphism $h \mapsto v_h$ of $\underset{\sim}{C}_0(M)$ into the *-algebra of multipliers of $\underset{\sim}{D}$ of order e. Thus

$$P : h \mapsto S'_{v_h} \qquad \ldots(7)$$

is a *-representation of $\underset{\sim}{C}_0(M)$ acting on $X(S)$.

By §29(9) the relation (1) holds; and in view of the last paragraph of §29 the relation (4) is valid whenever h, b, ψ are connected by (5).

Suppose now that P were degenerate, with essential space $Y \neq X(S)$. It would then follow from (4) that range$(S_\psi) \subset Y$ for all ψ of the form (5); and since (as we have observed) the latter are dense in each fiber D_x, this would imply that S is degenerate, contradicting our hypothesis. So P is non-degenerate. In view of (1), the same argument shows that T is non-degenerate.

Thus (T,P) is a system of imprimitivity for $\underset{\sim}{B}$ over M, related to S by (4).

It remains only to show that (T,P) is uniquely determined by (4). Let us assume now that (T,P) is <u>any</u> system of imprimitivity for $\underset{\sim}{B}$ over M acting in $X(S)$ and related to S by (4). Let $h,k \in \underset{\sim}{L}(M)$,

$b \in B$, and put $\psi(m) = h(m)b$ $(m \in M)$, $\chi(m) = k(m)\, h(m)b$. Then $\chi = v_k\psi$ by §29(7). Denoting by S' as before the extension of S to the multiplier bundle, we have

$$\begin{aligned} P_k S_\psi &= P_{kh} T_b && \text{(by (4))} \\ &= S_\chi && \text{(by (4))} \\ &= S_{v_k\psi} = S'_{v_k} S_\psi . \end{aligned}$$

By the denseness of the ψ and the non-degeneracy of S this implies that $P_k = S'_{v_k}$; that is, P is uniquely determined by S . A similar argument shows that $T_c = S'_{u_c}$ $(c \in B)$, whence T is uniquely determined by S .

The proof is now complete. □

Definition. The S of the above theorem is called the transformation bundle form of the system of imprimitivity (T,P) .

Let (T,P) and S be as in the theorem above; and let R be the integrated form of S . Thus R is a non-degenerate *-representation of $\mathcal{L}_1(\mathcal{D};\lambda)$. If we identify the compacted transformation algebra E with a dense *-subalgebra of $\mathcal{L}_1(\mathcal{D};\lambda)$ (by means of the map $f \mapsto \tilde{f}$ of §29), $R|E$ becomes a non-degenerate *-representation of E .

Definition. We shall refer to $R|E$ as the (compacted) integrated form of (T,P).

The next proposition gives a simple and useful description of $R|E$.

Let us denote by E_0 the linear span in E of the set of all functions f on $G \times M$ of the form

$$(x,m) \longmapsto \psi(m)\ \varphi(x)\ , \qquad \ldots(8)$$

where $\psi \in \underset{\sim}{L}(M)$ and $\varphi \in \underset{\sim}{L}(\underset{\sim}{B})$.

Proposition 30.2. (I) E_0 *is dense in* E *in the inductive limit topology*. (II) *If* f *is the element of* E_0 *given by* (8), *then*

$$R_f = P_\psi\ T_\varphi \qquad \ldots(9)$$

(T_φ *being of course the operator of the integrated form of* T).

Proof. Let $\bar{E}_0$ be the inductive limit closure of E_0 in E , and denote by $\underset{\sim}{G}$ the set of those functions g in $\underset{\sim}{C}(G \times M)$ such that $\bar{E}_0$ is stable under multiplication by g . Evidently $\underset{\sim}{G}$ is closed in the topology of uniform convergence on compact sets, and contains all products $(x,m) \longmapsto \alpha(x)\ \beta(m)$, where $\alpha \in \underset{\sim}{C}(G)$, $\beta \in \underset{\sim}{C}(M)$. Hence by the Stone-Weierstrass Theorem $\underset{\sim}{G} = \underset{\sim}{C}(G \times M)$. Thus statement (I) follows from the application of Prop. 10.9 to $\bar{E}_0$.

The verification of (9), by means of (4), is left to the reader. □

The results of this section can be summarized as follows:

Theorem 30.3. *The passage from a system of imprimitivity* (T,P) *for* $\underset{\sim}{B}$ *over* M *to its transformation bundle form* S , *and then from* S *to the integrated form* R *of* S *sets up a one-to-one correspondence between the following three sets of objects:* (i) *The set of all systems of imprimitivity for* $\underset{\sim}{B}$ *over* M ; (ii) *the set of all non-degenerate* *-*representations of the transformation bundle* $\underset{\sim}{D}$; (iii) *the set of all non-degenerate* *-*representations of the compacted transformation algebra* E *which are continuous in the norm* §29(5).

This correspondence preserves closed stable subspaces, direct sums, intertwining maps, unitary equivalence, commuting algebras, and irreducibility.

Proof. The bijectivity of the correspondence $(T,P) \mapsto S$ between (i) and (ii) was proved in Theorem 30.1. If S' is the extension of S to the multiplier bundle of $\underset{\sim}{D}$, it follows from the remark preceding Prop. 14.5 that S and S' have identical commuting algebras. Hence by (6) and (7) the commuting algebra of (T,P) contains that of S. Conversely, by (4) (and the denseness of the ψ of (5) in each fiber of $\underset{\sim}{D}$) the commuting algebra of S contains that of (T,P). So the correspondence $(T,P) \mapsto S$ preserves commuting algebras, and hence stable subspaces and irreducibility. A standard extension of this argument shows that it also preserves intertwining maps, and hence also unitary equivalence. It evidently preserves direct sums.

The required properties of the correspondence $S \mapsto R$ follow from §15 (and the denseness of E in $\underset{\sim}{L}_1(\underset{\sim}{D};\lambda)$). □

§31. The canonical imprimitivity bimodule.

In this section, as usual, $\underset{\sim}{B}$ is a Banach *-algebraic bundle over the locally compact group G with unit e, left Haar measure λ, and modular function Δ. We shall also fix a closed subgroup H of G, with left Haar measure ν and modular function δ. We choose once for all an everywhere positive continuous H-rho function ρ on G, and denote by $\rho^{\#}$ the corresponding measure on G/H (see Prop. 21.2).

Our goal in this and the next section is as follows: Let E be the compacted transformation algebra derived from $\underset{\sim}{B}$ and the transitive G-space G/H, as defined in §29. In Theorem 30.3 we established a one-to-one correspondence between the (continuous) non-degenerate *-representations of E and the systems of imprimitivity for $\underset{\sim}{B}$ over G/H. Furthermore, in §21 we made $\underset{\sim}{L}(\underset{\sim}{B})$ into a $\underset{\sim}{L}(\underset{\sim}{B}_H)$-rigged space (with a rigging $[\ ,\ ]$ derived from the basic $\underset{\sim}{L}(\underset{\sim}{B})$, $\underset{\sim}{L}(\underset{\sim}{B}_H)$ conditional expectation p of §21(5)). In this section we are going to define a left E-module structure for $\underset{\sim}{L}(\underset{\sim}{B})$, and also an E-valued rigging $[\ ,\ ]_E$ on $\underset{\sim}{L}(\underset{\sim}{B})$, such that $(\underset{\sim}{L}(\underset{\sim}{B}), [\ ,\ ]_E, [\ ,\ ])$ <u>will become an</u> E, $\underset{\sim}{L}(\underset{\sim}{B}_H)$ <u>imprimitivity bimodule</u>. In fact, if $\underset{\sim}{B}$ is saturated, this imprimitivity bimodule will turn out to be topologically strict in a certain sense, and so, by the abstract Imprimitivity Theorem (§8), will generate an isomorphism between the *-representation theory of $\underset{\sim}{L}(\underset{\sim}{B}_H)$ -- that is, of $\underset{\sim}{B}_H$ -- and the *-representation theory of E -- that is, the systems of imprimitivity for $\underset{\sim}{B}$ over G/H. In other words, if $\underset{\sim}{B}$ is saturated, we shall have a one-to-one correspondence, set up in both directions by the inducing process, between *-representations of $\underset{\sim}{B}_H$ and systems of imprimitivity for $\underset{\sim}{B}$ over G/H. In particular, in the group case we will get a one-to-one correspondence between unitary representations of H and systems of imprimitivity for G over G/H. This latter correspondence is of course expressed by the classical Imprimitivity Theorem of Mackey: <u>Every system of imprimitivity for</u> G <u>over the</u> G-<u>space</u> G/H <u>is obtained by the inducing process from a unique unitary representation of</u> H.

<u>Note</u>. We have already seen in §7(31),(32) how this E, $\underset{\sim}{L}(\underset{\sim}{B}_H)$ imprimitivity bimodule should be constructed in the group case when G is finite. Equations §7(31),(32) serve as an illuminating guide for the much more general construction of this section.

<u>Induced systems of imprimitivity</u>.

We shall begin by observing that the inducing process directly generates not merely a *-representation but a system of imprimitivity for $\underset{\sim}{B}$.

Let S be a non-degenerate $\underset{\sim}{B}$-positive *-representation of $\underset{\sim}{B}_H$, and $\underset{\sim}{Y} = \{Y_\alpha\}$ the Hilbert bundle over G/H induced by S. Thus $\underset{\sim}{L}_2(\underset{\sim}{Y};\rho^\#)$ is the space $X(T)$ of the induced *-representation $T = \mathrm{Ind}_{\underset{\sim}{B}_H \uparrow \underset{\sim}{B}}(S)$ of $\underset{\sim}{B}$. Evidently the equation

$$(P_\varphi f)(\alpha) = \varphi(\alpha)\ f(\alpha) \qquad \ldots(1)$$

($\varphi \in \underset{\sim}{C}_0(G/H)$; $f \in X(T)$; $\alpha \in G/H$) defines a non-degenerate *-representation P of $\underset{\sim}{C}_0(G/H)$ on $X(T)$. If $f \in X(T)$, $b \in B_x$, and $\varphi \in \underset{\sim}{C}_0(G/H)$, then for all α in G/H

$$\begin{aligned}(T_b\ P_\varphi f)(\alpha) &= \tau_b[(P_\varphi f)(x^{-1}\alpha)] \qquad \text{(see §25(1))}\\ &= \varphi(x^{-1}\alpha)\ \tau_b(f(x^{-1}\alpha))\\ &= (x\varphi)(\alpha)\ (T_b f)(\alpha)\\ &= (P_{x\varphi}\ T_b f)(\alpha)\ ,\end{aligned}$$

whence

$$T_b\ P_\varphi = P_{x\varphi}\ T_b\ .$$

Thus (see §30(1)) (T,P) is a system of imprimitivity for $\underset{\sim}{B}$ over G/H.

__Definition__. P and (T,P) are called respectively the __representation of $\underset{\sim}{C}_0(G/H)$ induced by__ S and the __system of imprimitivity induced by__ S.

__Remark 1__. Like T, the preceding P admits another description, in terms of the abstract inducing process. Notice that $\underset{\sim}{L}(\underset{\sim}{B})$ is a left $\underset{\sim}{C}_0(G/H)$-module under the action:

$$(\varphi f)(x) = \varphi(xH)\ f(x) \qquad \ldots(2)$$

($\varphi \in \underset{\sim}{C}_0(G/H)$; $f \in \underset{\sim}{L}(\underset{\sim}{B})$; $x \in G$) . One easily checks that with this left action, and with the right $\underset{\sim}{L}(\underset{\sim}{B}_H)$-module structure of §21(7) and the rigging $[\ ,\ ]:(f,g) \mapsto p(f^* * g)$ derived from the conditional expectation p of §21(5), $\underset{\sim}{L}(\underset{\sim}{B})$ becomes a $\underset{\sim}{L}(\underset{\sim}{B}_H)$-rigged $\underset{\sim}{C}_0(G/H)$-module. Further, if $\varphi \in \underset{\sim}{C}_0(G/H)$, $f \in \underset{\sim}{L}(\underset{\sim}{B})$, and $\xi \in X(S)$, the cross-section $\kappa(\varphi f \otimes \xi)$ of $\underset{\sim}{Y}$ (see §23) is the same as $P_\varphi(\kappa(f \otimes \xi))$. From this and Prop. 23.4 we see that S induces a *-representation of $\underset{\sim}{C}_0(G/H)$ via $(\underset{\sim}{L}(\underset{\sim}{B}), [\ ,\ ])$ (in the abstract sense of §4); and that this induced *-representation is unitarily equivalent to P (under the map E of Prop. 23.4). Thus we had every right to describe P in the preceding definition as <u>induced</u> by S .

Our next step is to make $\underset{\sim}{L}(\underset{\sim}{B})$ into a left E-module. This we do by means of the following action of E on $\underset{\sim}{L}(\underset{\sim}{B})$:

$$(uf)(x) = \int_G u(y,xH)\ f(y^{-1}x)\,d\lambda y \qquad \ldots(3)$$

($u \in E$; $f \in \underset{\sim}{L}(\underset{\sim}{B})$; $x \in G$) .

Notice that (3) is the obvious generalization of §7(31).

Since the integrand in (3), as a function of y , is B_x-valued and continuous with compact support, the right side of (3) exists as a B_x-valued integral. Thus uf as defined by (3) is a cross-section of $\underset{\sim}{B}$ with compact support; and it is continuous by Lemma 10.14. So it belongs to $\underset{\sim}{L}(\underset{\sim}{B})$. Obviously uf is linear in u and in f . Using Prop. 10.13 and the Fubini Theorem 10.15, we have:

$$((u*v)f)(x) = \int (u*v)(y,xH)\, f(y^{-1}x)d\lambda y$$
$$= \iint u(z,xH)\, v(z^{-1}y,z^{-1}xH)\, f(y^{-1}x)d\lambda z\, d\lambda y$$
$$= \iint u(z,xH)\, v(y,z^{-1}xH)\, f(y^{-1}z^{-1}x)d\lambda y\, d\lambda z$$
$$= \int u(z,xH)(vf)(z^{-1}x)d\lambda z$$
$$= (u(vf))(x)$$

($u,v \in E$; $f \in \mathcal{L}(\mathcal{B})$; $x \in G$) . So

$$(u*v)f = u(vf) ,$$

and (3) does define $\mathcal{L}(\mathcal{B})$ as a left E-module.

We next observe that, with the $\mathcal{L}(\mathcal{B}_H)$-valued rigging [,] mentioned in Remark 1 and the module structures (3) and §21(7), $\mathcal{L}(\mathcal{B})$ becomes a $\mathcal{L}(\mathcal{B}_H)$-rigged E-module. Indeed, it is only necessary to verify that

$$[uf,g] = [f,u^*g] \qquad (f,g \in \mathcal{L}(\mathcal{B});\ u \in E) .$$

This results from a routine calculation which we leave to the reader.

Remark 2. Suppose that the element u of E is of the following special form:

$$u(x,\alpha) = \chi(\alpha)\, g(x) \quad (x \in G;\ \alpha \in G/H), \ \ldots(4)$$

where $g \in \mathcal{L}(\mathcal{B})$ and $\chi \in \mathcal{L}(G/H)$. Then (3) becomes

$$uf = \chi(g*f) \qquad \ldots(5)$$

(the left action of χ on $\mathcal{L}(\mathcal{B})$ being given by (2)).

Now let (T,P) be the system of imprimitivity (for $\mathcal{B}$ over G/H) induced by the $\mathcal{B}$-positive non-degenerate *-representation S of $\mathcal{B}_H$. Let R be the compacted integrated form of (T,P) (defined in §30).

Thus R is a non-degenerate *-representation of E. The following proposition is the analogue for (T,P) of the fact observed for P in Remark 1.

Proposition 31.1. S induces a *-representation R' of E via the $\underset{\sim}{L}(\underset{\sim}{B}_H)$-rigged E-module $\underset{\sim}{L}(\underset{\sim}{B})$ (defined just before Remark 2); and $R' \cong R$.

Proof. Let E_0 be the linear span in E of the set of those u which are of the form (4). By Prop. 30.2 E_0 is dense in E in the inductive limit topology of E.

Recall the isometry

$$f \tilde{\otimes} \xi \mapsto \varkappa(f \otimes \xi) \qquad \ldots(6)$$

defined in §23(14). We shall adopt the notation of §23 (except that (6) cannot now be denoted by E). Let R' stand for the *-representation of E on $X(V)$ which corresponds to R via (6). It is enough to show that R' is the *-representation of E abstractly induced via $\underset{\sim}{L}(\underset{\sim}{B})$. This means, after transforming by (6), that it is enough to show that

$$R_u(\varkappa(f \otimes \xi)) = \varkappa(uf \otimes \xi) \qquad \ldots(7)$$

for $u \in E$, $f \in \underset{\sim}{L}(\underset{\sim}{B})$, and $\xi \in X(S)$.

Now, applying (6) to Remark 3 of §21, we see that, for fixed ξ, $g \mapsto \varkappa(g \otimes \xi)$ is continuous in g with respect to the inductive limit topology. Also it is easy to verify that, for fixed f, $u \mapsto uf$ is continuous in u with respect to the inductive limit topologies. Combining the last two facts with the first paragraph of this proof, one concludes that it is enough to verify (7) in the special case that u is of the form (4).

Thus, assume that $\chi \in \underset{\sim}{L}(G/H)$, $g,f \in \underset{\sim}{L}(\underset{\sim}{B})$, and $\xi \in X(S)$; and let u be given by (4). In that case we have by §30(9)

$$R_u = P_\chi T_g ;$$

and so by (5) together with Remark 1 and Prop. 25.3

$$\begin{aligned} \kappa(uf \otimes \xi) &= \kappa(\chi(g*f) \otimes \xi) \\ &= P_\chi(\kappa((g*f) \otimes \xi)) \\ &= P_\chi T_g(\kappa(f \otimes \xi)) \\ &= R_u(\kappa(f \otimes \xi)) . \end{aligned}$$

Thus we have verified (7) for the required special case. □

Remark 3. This proposition shows that it is very appropriate to describe (T,P) , or its integrated form, as induced by S . Compare Remark 1.

Construction of the imprimitivity bimodule.

We have seen that $\underset{\sim}{L}(\underset{\sim}{B})$ is a left module for three different *-algebras, namely $\underset{\sim}{L}(\underset{\sim}{B})$ itself (under left convolution), $\underset{\sim}{C}_0(G/H)$, and E . Of these three we shall next show that E is distinguished by the following important property: By defining a suitable E-valued rigging $[\ ,\ ]_E$ on $\underset{\sim}{L}(\underset{\sim}{B})$, we can make $(\underset{\sim}{L}(\underset{\sim}{B}), [\ ,\ ]_E\ [\ ,\ ])$ into an E , $\underset{\sim}{L}(\underset{\sim}{B}_H)$ imprimitivity bimodule.

Motivated by the formula §7(32), we make the following definition:

Definition. If $f,g \in \underset{\sim}{L}(\underset{\sim}{B})$, let $[f,g]_E$ be the function on $G \times G/H$ to B given by:

$$[f,g]_E(x,yH) = \int_H f(yh)\ g^*(h^{-1}y^{-1}x)\,d\nu h \qquad (x,y \in G) . \quad \ldots(8)$$

To begin with, we note that the right side of (8) exists as a B_x-valued integral, and depends only on x and yH ; by Lemma 10.14 it is continuous as a function of (x,y) , and hence as a function of (x,yH) ; and it vanishes outside a compact subset of $G \times G/H$. So $[f,g]_E \in E$.

Evidently $[f,g]_E$ is linear in f and conjugate-linear in g .

We observe for later use that $[f,g]_E$ is separately continuous in f and g in the inductive limit topologies of $\underset{\sim}{L}(\underset{\sim}{B})$ and E .

As before we denote by $[\ ,\ ]$ the $\underset{\sim}{L}(\underset{\sim}{B}_H)$-valued rigging $(f,g) \mapsto p(f^* * g)$ on $\underset{\sim}{L}(\underset{\sim}{B})$ derived from the basic conditional expectation p . We consider $\underset{\sim}{L}(\underset{\sim}{B})$ as a right $\underset{\sim}{L}(\underset{\sim}{B}_H)$-module by means of §21(7), and as a left E-module by means of (3).

<u>Proposition 31.2</u>. $(\underset{\sim}{L}(\underset{\sim}{B}), [\ ,\ ]_E, [\ ,\ ])$ <u>is an</u> $E, \underset{\sim}{L}(\underset{\sim}{B}_H)$ <u>imprimitivity bimodule</u>.

<u>Proof</u>. The identities that remain to be verified are:

$$[f,g]_E^* = [g,f]_E ,$$

$$u * [f,g]_E = [uf,g]_E ,$$

$$[f\varphi,g]_E = [f,g\varphi^*]_E ,$$

$$[f,g]_E q = f[g,q] \qquad \ldots(9)$$

$(f,g,q \in \underset{\sim}{L}(\underset{\sim}{B});\ u \in E;\ \varphi \in \underset{\sim}{L}(\underset{\sim}{B}_H))$. These verifications are of quite a routine nature. We shall content ourselves with checking (9). By Fubini's Theorem and Prop. 10.13,

$$(f[g,q])(x) = \int_H f(xh^{-1})[g,q](h)(\delta(h)\Delta(h))^{-\frac{1}{2}}\, d\nu h$$

$$= \int_H \int_G \delta(h)^{-1} f(xh^{-1}) g^*(y) q(y^{-1}h) d\lambda y \, d\nu h$$

$$= \int_H \int_G \delta(h^{-1}) f(xh^{-1}) g^*(hx^{-1}y) q(y^{-1}x) d\lambda y \, d\nu h$$

$$= \int_H \int_G f(xh) g^*(h^{-1}x^{-1}y) q(y^{-1}x) d\lambda y \, d\nu h$$

$$= \int_G [f,g]_E(y,xH) q(y^{-1}x) d\lambda y$$

$$= ([f,g]_E q)(x) .$$

So (9) holds. □

Definition. This E, $\underset{\sim}{L}(\underset{\sim}{B}_H)$ imprimitivity bimodule $(\underset{\sim}{L}(\underset{\sim}{B}), [\ ,\ ]_E [\ ,\ ])$ will be called the canonical imprimitivity bimodule derived from $\underset{\sim}{B}$ and H. We shall denote it by $\underset{\sim}{I}(\underset{\sim}{B};H)$, or simply by $\underset{\sim}{I}$.

§32. The Imprimitivity Theorem for Banach *-algebraic bundles.

In this section we reach the climax of the present work, The Imprimitivity Theorem for arbitrary Banach *-algebraic bundles over locally compact groups (Theorem 32.7). The reader will observe that, while Theorem 32.7 holds without any restriction (other than of course the local compactness of the group), it takes on an especially elegant and clear-cut form when the Banach *-algebraic bundle is saturated (Theorem 32.8). In that case it becomes an almost verbatim generalization of the Mackey-Loomis-Blattner Imprimitivity Theorem for groups.

All the notation of §31 will continue in force in this section.

Let E' and B' denote the linear spans in E and $\underset{\sim}{L}(\underset{\sim}{B}_H)$ of range($[\ ,\ ]_E$) and range($[\ ,\ ]$) respectively. Considered as an E',B' imprimitivity bimodule, $\underset{\sim}{I}$ is of course strict, and so by the abstract Imprimitivity Theorem (§8) gives rise to a one-to-one correspondence between the $\underset{\sim}{I}$-positive *-representations of E' and of B'. However, it is not primarily the *-representations of E' and B' that interest us, but the norm-continuous *-representations of the normed *-algebras E (with norm §29(5)) and $L(B_H)$ (with the L_1 norm). In order to pass from the conclusions about E' and B' to the conclusions that interest us, we shall have to ask the following four questions:

(I) Is B' dense in $L(B_H)$?

(II) Is E' dense in E ?

(III) Which *-representations of E are positive with respect to $\underset{\sim}{I}$?

(IV) In the correspondence $S \longleftrightarrow R$ between *-representations S of B' and *-representations R of E', set up by the abstract Imprimitivity Theorem, how is the norm-continuity of S related to the norm-continuity of R ?

Remark 1. In connection with question (IV), notice that the normed *-algebras E and $L(B_H)$ are not complete. We know of no way of extending $\underset{\sim}{I}$ to an imprimitivity bimodule for the completions of E and $L(B_H)$. So Theorem 9.1 is not available for use here. Nor does Prop. 9.3 seem to be immediately applicable.

As regards question (I) we have:

Proposition 32.1. *If B has an approximate unit, then B' is dense in $L(B_H)$ in the inductive limit topology.*

Proof. By the definition of $[\ ,\ ]$, B' is the linear span of $\{p(f*g): f,g \in L(B)\}$. Since p is continuous in the inductive limit topologies and $p(L(B)) = L(B_H)$ (by Theorem 10.7), we have only to show that the linear span of $\{f*g: f,g \in L(B)\}$ is dense in $L(B)$ in the inductive limit topology. But this follows from the hypothesis of an approximate unit and Remark 5 of §13. □

The answer to question (II) is 'no' in general (see Remark 3 of this section); but it is 'yes' if $\underset{\sim}{B}$ is saturated.

Proposition 32.2. If $\underset{\sim}{B}$ is saturated, E' is dense in E in the inductive limit topology.

Proof. Let $\underset{\sim}{C}' = (C', \{C'_{x,y}\})$ be the Banach bundle over $G \times G$ which is the retraction of $\underset{\sim}{B}$ by the continuous map $(x,y) \mapsto xy$ of $G \times G$ onto G (see §10). For each pair f,g of elements of $\underset{\sim}{L}(\underset{\sim}{B})$ we form the element $f \times g: (x,y) \mapsto (x,y,f(x)g(y))$ of $\underset{\sim}{L}(\underset{\sim}{C}')$. I claim that the linear span M' of $\{f \times g: f,g \in \underset{\sim}{L}(\underset{\sim}{B})\}$ is dense in $\underset{\sim}{L}(\underset{\sim}{C}')$ in the inductive limit topology. Indeed: By the saturation of $\underset{\sim}{B}$, $\{F(x,y): F \in M'\}$ is dense in $C'_{x,y}$ for each (x,y) in $G \times G$. Furthermore, an easy Stone-Weierstrass argument with complex-valued functions shows that the inductive limit closure of M' is closed under multiplication by arbitrary complex functions on $G \times G$. Therefore the claim follows from Prop. 10.9.

Let us now transform $\underset{\sim}{C}'$ and M' by the "shear transformation" $(x,y) \mapsto (x,xy)$ of the base space $G \times G$. Under this transformation $\underset{\sim}{C}'$ goes into the Banach bundle $\underset{\sim}{C} = G \times \underset{\sim}{B}$ over $G \times G$ which is the retraction of $\underset{\sim}{B}$ by the map $(x,y) \mapsto y$; and M' goes into the linear span M of the set of all cross-sections of $\underset{\sim}{C}$ of the form $(x,y) \mapsto f(x)g(x^{-1}y)$ $(f,g \in \underset{\sim}{L}(\underset{\sim}{B}))$. By the above claim, M is dense in $\underset{\sim}{L}(\underset{\sim}{C})$ in the inductive limit topology.

We now point out another general fact: For each u in $\underset{\sim}{L}(\underset{\sim}{C})$ let u^0 be the element of E given by:

$$u^0(y,xH) = \int_H u(xh,y)\,d\nu h \qquad (x,y \in G) .$$

If N is any linear subspace of $\underset{\sim}{L}(\underset{\sim}{C})$ which is dense in $\underset{\sim}{L}(\underset{\sim}{C})$ in the inductive limit topology, I claim that $N^0 = \{u^0 : u \in N\}$ is dense in E in the inductive limit topology. This follows from the two easily verified facts that the linear map $u \mapsto u^0$ is onto E and that it is continuous in the inductive limit topologies.

Applying the last paragraph to M, we conclude that M^0 is inductively dense in E. But M^0 is the linear span in E of the set of all functions v of the form

$$v(y,xH) = \int f(xh)\, g(h^{-1}x^{-1}y)\,d\nu h \qquad (x,y \in G) ,$$

that is (by §31(8)),

$$v = [f,g^*]_E ,$$

where $f,g \in \underset{\sim}{L}(\underset{\sim}{B})$. So $M^0 = E'$; and E' is dense in E in the inductive limit topology. □

Suppose that $H = G$. Then G/H is the one-element space, $E \cong \underset{\sim}{L}(\underset{\sim}{B})$, and §31(8) becomes:

$$[f,g]_E = f * g^* .$$

Thus the above proposition has the following useful special case.

Corollary 32.3. *If $\underset{\sim}{B}$ is saturated, the linear span of $\{f*g : f,g \in \underset{\sim}{L}(\underset{\sim}{B})\}$ is dense in $\underset{\sim}{L}(\underset{\sim}{B})$ in the inductive limit topology.*

To answer question (III), we need a lemma.

Lemma 32.4. *Assume that $\underset{\sim}{B}$ has an approximate unit. Then there exists* *a net* $\{g_i\}$ *of elements of* $\underset{\sim}{L}(\underset{\sim}{B})$ *such that*

$$f[g_i,g_i] \to f \quad \textit{for all} \quad f \quad \textit{in} \quad \underset{\sim}{L}(\underset{\sim}{B}) \qquad \ldots(1)$$

in the inductive limit topology.

Proof. Since the involution operation is continuous on $\underset{\sim}{L}(\underset{\sim}{B})$ in the inductive limit topology, it is enough to obtain

$$(f[g_i,g_i])^* \to f^* . \qquad \ldots(2)$$

A simple calculation shows that

$$(f[g_i,g_i])^*(x) = \int_H (g_i^* * g_i)(h)\ f^*(h^{-1}x)d\nu h \qquad \ldots(3)$$

$(f,g_i \in \underset{\sim}{L}(\underset{\sim}{B})$; $x \in G)$.

Now let $\{w_r\}$ be an approximate unit of $\underset{\sim}{B}$. Then evidently

$$\{w_r^* w_r\} \text{ is an approximate unit of } \underset{\sim}{B} . \qquad \ldots(4)$$

For each r choose χ_r in $\underset{\sim}{C}(\underset{\sim}{B})$ such that $\chi_r(e) = w_r$. Next, let $\{\sigma'_\alpha\}$ be an approximate unit on G (in the sense that $\sigma'_\alpha \in (\underset{\sim}{L}(G))_+$, $\int \sigma'_\alpha\, d\lambda = 1$, and the support of σ'_α shrinks down to e); and for each α let

$$\gamma_\alpha = k_\alpha(\sigma'^*_\alpha * \sigma'_\alpha)|H ,$$

where k_α is such a positive constant that $\int_H \gamma_\alpha d\nu = 1$. Thus, setting $\sigma_\alpha = k_\alpha^{\frac{1}{2}}\sigma'_\alpha$, we have

$$\gamma_\alpha = (\sigma_\alpha^* * \sigma_\alpha)|H . \qquad \ldots(5)$$

Since the supports of the γ_α shrink down to e , $\{\gamma_\alpha\}$ is an approximate unit on H .

Now let $\varphi_{r,\alpha}(h) = \gamma_\alpha(h)\chi_r(e)^*\chi_r(h) \quad (h \in H)$, so that $\varphi_{r,\alpha} \in L(B_H)$. In view of (4), an argument exactly similar to that in the proof of Theorem 13.1 (see Prop. 8.2 of Fell [6]) shows that

$$\lim_r \overline{\lim_\alpha} \|\varphi_{r,\alpha} * f - f\|_\infty = 0 \qquad \ldots(6)$$

for all f in $L(B)$. (Here $\varphi_{r,\alpha} * f$ is the element of $L(B)$ given by the familiar modified convolution formula:

$$(\varphi_{r,\alpha} * f)(x) = \int \varphi_{r,\alpha}(h)\ f(h^{-1}x)d\nu h \quad (x\in G)\ .)$$

We now propose to show that (6) remains true when $\varphi_{r,\alpha}$ is replaced by

$$\psi_{r,\alpha} = ((\sigma_\alpha\chi_r)^* * (\sigma_\alpha\chi_r))|H\ .$$

To see this, we observe from (5) that for $h \in H$

$$\psi_{r,\alpha}(h) - \varphi_{r,\alpha}(h)$$

$$= \int_G \sigma_\alpha(y)\sigma_\alpha(yh)[\chi_r(y)^*\chi_r(yh) - \chi_r(e)^*\chi_r(h)]d\lambda y\ . \qquad \ldots(7)$$

I claim that for each r

$$\lim_\alpha \int_H \|\psi_{r,\alpha}(h) - \varphi_{r,\alpha}(h)\|\, d\nu h = 0\ . \qquad \ldots(8)$$

Indeed: Fix r . The bundle element $\chi_r(y)^*\chi_r(yh) - \chi_r(e)^*\chi_r(h)$ is continuous in (y,h) and vanishes when $y = e$. So, given $\varepsilon > 0$, we can find a neighborhood U of e such that

$$\|\chi_r(y)^*\chi_r(yh) - \chi_r(e)^*\chi_r(h)\| < \varepsilon \quad \text{when} \quad y,h \in U\ . \qquad \ldots(9)$$

Now choose an index α_0 so large that, if $\alpha \succ \alpha_0$, $\sigma_\alpha(y)\sigma_\alpha(yh)$

vanishes unless $y,h \in U$. Then, by (7) and (9), for all h and all $\alpha \succ \alpha_0$ $\|\psi_{r,\alpha}(h) - \varphi_{r,\alpha}(h)\|$ is majorized by $\varepsilon(\sigma_\alpha^* * \sigma_\alpha)(h)$; and therefore by (5)

$$\int_H \|\psi_{r,\alpha}(h) - \varphi_{r,\alpha}(h)\| d\nu h \le \varepsilon$$

for all $\alpha \succ \alpha_0$. This proves (8).

Now for any r and α, any f in $\underset{\sim}{L}(\underset{\sim}{B})$, and any x in G,

$$\|(\psi_{r,\alpha} * f - \varphi_{r,\alpha} * f)(x)\| = \left\|\int_H (\psi_{r,\alpha}(h) - \varphi_{r,\alpha}(h)) f(h^{-1}x) d\nu h\right\|$$

$$\le \|f\|_\infty \int_H \|\psi_{r,\alpha}(h) - \varphi_{r,\alpha}(h)\| d\nu h .$$

So by (8), for each r

$$\lim_\alpha \|\psi_{r,\alpha} * f - \varphi_{r,\alpha} * f\|_\infty = 0 . \qquad \ldots(10)$$

Combining (10) with (6), we deduce that

$$\lim_r \overline{\lim_\alpha} \|\psi_{r,\alpha} * f - f\|_\infty = 0 . \qquad \ldots(11)$$

Since the $\{\psi_{r,\alpha}\}$ have their compact supports contained in a single compact set, (11) implies that we can find a net $\{\psi^i\}$ such that each ψ^i is one of the $\psi_{r,\alpha}$, and such that for every f in $\underset{\sim}{L}(\underset{\sim}{B})$

$$\psi^i * f \xrightarrow[i]{} f \qquad \ldots(12)$$

in the inductive limit topology. On the other hand, by (3) and the definition of $\psi_{r,\alpha}$,

$$(\psi_{r,\alpha} * f)(x) = \int ((\sigma_\alpha \chi_r)^* * (\sigma_\alpha \chi_r))(h) \; f(h^{-1}x) d\nu h$$

$$= (f^*[\sigma_\alpha \chi_r, \sigma_\alpha \chi_r])^*(x) . \qquad \ldots(13)$$

It follows from (12) and (13) that (2) holds provided we take g_i to be $\sigma_\alpha \chi_r$, where α and r are so chosen that $\psi^i = \psi_{r,\alpha}$. □

Here is the answer to question (III).

Proposition 32.5. Assume that $\underset{\sim}{B}$ has an approximate unit. Then every *-representation of E' which is continuous with respect to the E-norm $\| \|$ of §29(5) is positive with respect to $\underset{\sim}{I}$.

Proof. By Remark 4 of §9 (applied with A and B reversed), it is enough to find a net $\{g_i\}$ of elements of $\underset{\sim}{L}(\underset{\sim}{B})$ such that

$$\|[f,f[g_i,g_i]]_E - [f,f]_E\| \longrightarrow 0 \qquad \ldots(14)$$

for all f in $\underset{\sim}{L}(\underset{\sim}{B})$.

Take $\{g_i\}$ to be as in Lemma 32.4. We have observed in §31 that $[f,g]_E$ is separately continuous in f and g in the inductive limit topologies. Evidently $\| \|$ is continuous with respect to the inductive limit topology of E. In view of these facts, (14) is implied by the defining property (1) of $\{g_i\}$. □

We shall now answer question (IV) by showing that, if $\underset{\sim}{B}$ has an approximate unit, the $\underset{\sim}{I}$-positive non-degenerate *-representations of B' which are $\underset{\sim}{L}_1$-continuous correspond exactly to the $\underset{\sim}{I}$-positive non-degenerate *-representations of E' which are continuous with respect to the norm of E.

Assume that $\underset{\sim}{B}$ has an approximate unit.

By Prop. 32.1 B' is dense in $\underset{\sim}{L}(\underset{\sim}{B}_H)$. Hence by §15 the non-degenerate $\underset{\sim}{L}_1$-continuous *-representations S' of B' are just the restrictions to B' of the integrated forms of non-degenerate *-representations S of $\underset{\sim}{B}_H$. Further, positivity of S' with respect to $\underset{\sim}{I}$

is the same as $\underset{\sim}{B}$-positivity of the corresponding *-representation S of $\underset{\sim}{B}_H$.

Let S be a non-degenerate $\underset{\sim}{B}$-positive *-representation of $\underset{\sim}{B}_H$, and S' the ($\underset{\sim}{I}$-positive) restriction to B' of the integrated form of S. We shall suppose that, under the correspondence of Theorem 8.3 applied to $\underset{\sim}{I}$, S' corresponds to the non-degenerate *-representation R' of E'. But, by Prop. 31.1, R' is just the restriction to E' of the integrated form of the system of imprimitivity over G/H induced by S. Therefore R' is continuous with respect to the norm of E (see §30).

Thus we have established one direction of the equivalence asserted in the following proposition:

<u>Proposition 32.6.</u> <u>Let</u> S' <u>and</u> R' <u>be</u> <u>$\underset{\sim}{I}$-positive non-degenerate</u> <u>*-representations of</u> B' <u>and</u> E' <u>respectively which correspond under the correspondence of Theorem</u> 8.3 (<u>applied to</u> $\underset{\sim}{I}$). <u>Then</u> R' <u>is continuous in the norm of</u> E <u>if and only if</u> S' <u>is continuous in the</u> <u>$\underset{\sim}{L}_1$-norm</u>.

<u>Proof</u>. To prove the other direction we assume that R' is continuous in the norm of E.

From given elements ξ of $X(R')$ and f of $\underset{\sim}{L}(\underset{\sim}{B})$ we obtain an element of $X(S')$, namely the quotient image $\zeta = f \tilde{\otimes} \xi$ of $f \otimes \xi$. From the definition of the inducing process $\mathrm{Ind}_{E' \uparrow B'}$, we have $S'_\varphi \zeta = f\varphi^* \tilde{\otimes} \xi$, and so

$$\langle S'_\varphi \zeta, \zeta\rangle = \langle R'_{[f, f\varphi^*]_E} \xi, \xi\rangle \qquad (\varphi \in B'). \quad \ldots (15)$$

Now for fixed f it follows from continuity observations in §21 and §31 that $\varphi \mapsto [f, f\varphi^*]_E$ is continuous with respect to the inductive limit topologies of $\underset{\sim}{L}(\underset{\sim}{B}_H)$ and E. Since R' is continuous by hypothesis, (15) implies that $\varphi \mapsto \langle S'_\varphi \zeta, \zeta\rangle$ is continuous on B' with respect to the inductive limit topology. Further, B' is inductive-limit dense in $\underset{\sim}{L}(\underset{\sim}{B}_H)$; and it is easy to check that, for each b in B_H, B' is closed under the left action of the corresponding multiplier on $\underset{\sim}{L}(\underset{\sim}{B}_H)$ (see end of §13). Finally, notice that the vectors ζ of the above form span a dense subspace of $X(S')$.

We have thus verified all the hypotheses of Theorem 16.4 as applied to B' and S'. So Theorem 16.4 implies that S' is $\underset{\sim}{L}_1$-continuous.□

We are now in a position to derive the chief result of this chapter -- the Imprimitivity Theorem for Banach *-algebraic bundles.

Let us say that a system of imprimitivity $\underset{\sim}{T}$ for $\underset{\sim}{B}$ over G/H is <u>non-degenerate on</u> E' if the restriction to E' of the integrated form of $\underset{\sim}{T}$ is non-degenerate. (Recall that E' is the linear span in E of range($[\ ,\ ]_E$).)

In this theorem we assume merely the standing hypotheses of the first paragraph of §31. In particular $\underset{\sim}{B}$ need not have an approximate unit.

<u>Theorem 32.7</u> (Imprimitivity Theorem for Banach *-Algebraic Bundles).
(I) <u>If S is a non-degenerate $\underset{\sim}{B}$-positive *-representation of $\underset{\sim}{B}_H$, then the system of imprimitivity $\underset{\sim}{T}$ for $\underset{\sim}{B}$ over G/H induced by S is non-degenerate on E'.</u>

(II) <u>Conversely, let $\underset{\sim}{T}$ be any system of imprimitivity for $\underset{\sim}{B}$ over G/H which is non-degenerate on E'. Then there is a non-degenerate $\underset{\sim}{B}$-positive *-representation S of $\underset{\sim}{B}_H$ such that $\underset{\sim}{T}$ is unitarily equivalent to the system of imprimitivity induced by S; and this S is unique to within unitary equivalence.</u>

Proof. To begin with we suppose that $\underset{\sim}{B}$ has an approximate unit.

(I) If S' and R' are the restrictions of the integrated forms of S and $\underset{\sim}{T}$ to B' and E' respectively, we have seen in the discussion preceding Prop. 32.6 that S' is non-degenerate, and that R' is the *-representation of E' corresponding to S' as in Theorem 8.3. So R' is non-degenerate.

(II) Let R be the (necessarily continuous) compacted integrated form of $\underset{\sim}{T}$, and R' its (non-degenerate) restriction to E'. By Prop. 32.5 R' is positive with respect to $\underset{\sim}{I}$, and so corresponds by Theorem 8.3 with some non-degenerate $\underset{\sim}{I}$-positive *-representation S' of B'. By Prop. 32.6 S' is continuous with respect to the $\underset{\sim}{L}_1$-norm, and so is the restriction to B' of the integrated form of a non-degenerate *-representation S of $\underset{\sim}{B}_H$ which is positive with respect to $\underset{\sim}{B}$. Let $\underset{\sim}{T}^0$ be the system of imprimitivity induced by S, and R^0 the compacted integrated form of $\underset{\sim}{T}^0$. By Prop. 31.1

$$R^0|E' \cong R' = R|E' . \qquad \ldots(16)$$

Since E' is a *-ideal of E and the two sides of (16) are non-degenerate, it follows from (16) that $R^0 \cong R$. By Theorem 30.3 this implies that $\underset{\sim}{T}^0 \cong \underset{\sim}{T}$. This shows the existence of an S having the required property.

The uniqueness of S follows from the biuniqueness of the correspondence of Theorem 8.3. To be more specific, S is determined by the restriction S' of its integrated form to B'; and S' is determined to within unitary equivalence by the representation R' of E' to which it corresponds by Theorem 8.3. On the other hand R' is the restriction to E' of the system of imprimitivity $\underset{\sim}{T}$ induced by S. So S is determined to within unitary equivalence by $\underset{\sim}{T}$. This completes the proof if $\underset{\sim}{B}$ has an approximate unit.

We now discard the assumption that $\underset{\sim}{B}$ has an approximate unit.

Let $\underset{\sim}{C} = (C,\{C_x\})$ be the bundle C^*-completion of $\underset{\sim}{B}$, and $\rho:B \to C$ the canonical quotient map (see §18). Let F be the compacted transformation *-algebra for $\underset{\sim}{C}$ and G/H; let $[\ ,\ ]_F$ be the F-valued rigging on $\underset{\sim}{L}(\underset{\sim}{C})$ defined as in §31(8) (with $\underset{\sim}{B}$ replaced by $\underset{\sim}{C}$); and let F' be the linear span of range$([\ ,\ ]_F)$. The map ρ gives rise to *-homomorphisms $\rho':\underset{\sim}{L}(\underset{\sim}{B}) \to \underset{\sim}{L}(\underset{\sim}{C})$ and $\tilde{\rho}: E \to F$:

$$\rho'(f)(x) = \rho(f(x)),\ \tilde{\rho}(u)(x,\alpha) = \rho(u(x,\alpha))\ ;$$

and we have

$$\tilde{\rho}([f,g]_E) = [\rho'(f),\rho'(g)]_F \quad (f,g \in \underset{\sim}{L}(\underset{\sim}{B})). \quad \ldots(17)$$

By Prop. 10.9 $\rho'(\underset{\sim}{L}(\underset{\sim}{B}))$ and $\tilde{\rho}(E)$ are dense in $\underset{\sim}{L}(\underset{\sim}{C})$ and F respectively in the inductive limit topologies. Hence by (17) and the separate continuity of $[\ ,\ ]_F$, $\tilde{\rho}(E')$ is dense in F'.

By Prop. 11.6 $\underset{\sim}{C}$ has an approximate unit; so our theorem holds for $\underset{\sim}{C}$ by the first part of the proof.

To prove (I) for $\underset{\sim}{B}$, let S be a non-degenerate $\underset{\sim}{B}$-positive *-representation of $\underset{\sim}{B}_H$. By Prop. 27.5 there is a non-degenerate $\underset{\sim}{C}$-positive *-representation S^0 of $\underset{\sim}{C}_H$ such that $S = S^0\cdot(\rho|B_H)$. If R and R^0 are the integrated forms of the systems of imprimitivity induced by S and S^0 respectively, one verifies (see Prop. 28.3) that

$$R_u = R^0_{\tilde{\rho}(u)} \qquad (u \in E)\ . \qquad \ldots(18)$$

Now by (I) of the present theorem applied to $\underset{\sim}{C}$, $R^0|F'$ is non-degenerate. So (18) and the denseness of $\tilde{\rho}(E')$ in F' imply that

$R|E'$ is non-degenerate; and (I) is proved for $\underset{\sim}{B}$.

Since S^0 is determined to within unitary equivalence by R^0, the above argument also shows that S is determined to within unitary equivalence by R. Thus the uniqueness statement in (II) holds for $\underset{\sim}{B}$.

Finally, let $\underset{\sim}{T} = (T,P)$ be a system of imprimitivity for $\underset{\sim}{B}$ over G/H non-degenerate on E'. By the definition of $\underset{\sim}{C}$, we have $T = T^0 \circ \rho$ for some $*$-representation T^0 of $\underset{\sim}{C}$; and $\underset{\sim}{T}^0 = (T^0,P)$ is evidently a system of imprimitivity for $\underset{\sim}{C}$ over G/H non-degenerate on F'. By (II) applied to $\underset{\sim}{C}$, $\underset{\sim}{T}^0$ is induced by some non-degenerate $*$-representation S^0 of $\underset{\sim}{C}_H$. Hence, in view of Prop. 28.3, $\underset{\sim}{T}$ is induced by $S = S^0 \circ (\rho|B_H)$. This completes the proof of (II) for $\underset{\sim}{B}$. □

If $\underset{\sim}{B}$ is saturated, then by Prop. 32.2 E' is dense in E in the inductive limit topology. In that case <u>every</u> system of imprimitivity for $\underset{\sim}{B}$ over G/H is non-degenerate on E'; and Theorem 32.7 becomes:

<u>Theorem 32.8</u>. <u>Assume that $\underset{\sim}{B}$ is saturated. Then, given any system of imprimitivity $\underset{\sim}{T}$ for $\underset{\sim}{B}$ over G/H, there is a non-degenerate $\underset{\sim}{B}$-positive $*$-representation S of $\underset{\sim}{B}_H$ such that $\underset{\sim}{T}$ is unitarily equivalent to the system of imprimitivity induced by S. Further, the S having this property is unique to within unitary equivalence.</u>

In the group case $\underset{\sim}{B}$ (being the group bundle) is automatically saturated; and Theorem 32.8 becomes the classical Imprimitivity Theorem of Mackey:

<u>Theorem 32.9</u> (Mackey). <u>Given any system of imprimitivity $\underset{\sim}{T}$ for G over G/H, there is a unitary representation S of H (unique to within unitary equivalence) such that the system of imprimitivity</u>

induced by S *is unitarily equivalent to* T .

It is of interest to write down the explicit form of the Imprimitivity Theorem for *semidirect product bundles*. Let A be a saturated Banach *-algebra; and suppose that $\underset{\sim}{C} = A \times_{\iota} G$ is a ι-semidirect product of A and G as in §12(I). Thus $\underset{\sim}{C}$ is saturated. Combining Theorem 32.8 (applied to $\underset{\sim}{C}$) with the description of *-representations of $\underset{\sim}{C}$ given in Prop. 14.6, we easily obtain:

Theorem 32.10. *Let* (V,P) *be a system of imprimitivity for the group* G *over* G/H ; *and let* Q *be a non-degenerate* **-representation of* A , *acting on the same space as* V *and* P , *and satisfying:*

(i) $\quad Q_a P_\varphi = P_\varphi Q_a \qquad (a \in A;\ \varphi \in \underset{\sim}{C}_0(G/H))$,

(ii) $\quad V_x Q_a V_x^{-1} = Q_{\iota_x(a)} \qquad (a \in A;\ x \in G)$.

Then there exists a unitary representation W *of* H *and a non-degenerate* **-representation* R *of* A *(acting in the same space as* W *) such that* (a)

$$W_h R_a W_h^{-1} = R_{\iota_h(a)} \qquad (a \in A;\ h \in H) ,$$

and (b) *the triple* (Q, V, P) *is unitarily equivalent to* (Q', V', P') , *where* (V', P') *is the Blattner formulation of the system of imprimitivity induced by* W , *and* Q' *is the* Q *of* §26(16). *Furthermore, the pair* (W,R) *is unique to within unitary equivalence.*

The case of discrete G/H .

If G/H is discrete, the recovery of S from the system of imprimitivity $\underset{\sim}{T}$ induced by S is very much simpler than in the general case. Also, the rather mysterious condition of non-degeneracy on E'

which appears in Theorem 32.7 becomes much more transparent if G/H is discrete.

As regards the recovery of S from $\underset{\sim}{T}$ we have:

Proposition 32.11. Assume that G/H is discrete. Let S be a $\underset{\sim}{B}$-positive non-degenerate *-representation of $\underset{\sim}{B}_H$, and (T,P) the system of imprimitivity for $\underset{\sim}{B}$ over G/H induced by S. Then S is unitarily equivalent to the subrepresentation of $T|\underset{\sim}{B}_H$ which acts on range(P_H).

Note. If α is a coset in G/H (such as H), by P_α we mean of course P_φ, where $\varphi(\alpha) = 1$ and $\varphi(\beta) = 0$ for $\beta \neq \alpha$.

Proof. Let $\underset{\sim}{Y} = \{Y_\alpha\}$ $(\alpha \in G/H)$ be the Hilbert bundle over G/H induced by S. Since G/H is discrete, range(P_H) is just the fiber Y_H of $\underset{\sim}{Y}$. Also, since G/H being discrete has a G-invariant measure, it follows that $\delta(h) = \Delta(h)$ for $h \in H$. Hence, identifying Y_H with X(S) by means of the F of §23(6), we have from §24(14)

$$T_b\xi = \iota_b\xi = S_b\xi \quad (b \in B_H;\ \xi \in X(S)).$$

This shows that $T|B_H$ coincides on range(P_H) with S. □

Proposition 32.12. Suppose that G/H is discrete; and let (T,P) be a system of imprimitivity for $\underset{\sim}{B}$ over G/H. Then (T,P) is non-degenerate on E′ (in the sense of Theorem 32.7) if and only if range(P_H) generates X(T) under T, that is, if and only if $\{T_b\xi : b \in B,\ \xi \in \text{range}(P_H)\}$ spans a dense subspace of X(T).

Proof. Since G/H is discrete, H is open in G; and we may as well assume that ν coincides on H with λ. As usual, we denote by R the integrated form of (T,P) .

Suppose that $f,g \in \underset{\sim}{L}(\underset{\sim}{B})$ and that f vanishes outside the coset zH $(z \in G)$. Then, for $x,y \in G$,

$$[f,g]_E(x,yH) = \int_H f(yh)\ g^*(h^{-1}y^{-1}x)d\lambda h$$

$$= \begin{cases} 0 & \text{if } yH \neq zH \\ (f*g^*)(x) & \text{if } yH = zH\ . \end{cases}$$

It follows that, for $\xi \in X(T)$,

$$R_{[f,g]_E}\xi = P_{zH}\int_G T_{(f*g^*)(x)}\xi\, d\lambda x \qquad \text{(by §30(9))}$$

$$P_{zH}\ T_{f*g^*}\xi\ ,$$

or

$$R_{[f,g]_E} = P_{zH}\ T_f(T_g)^*\ . \qquad \ldots(19)$$

Now for $x \in zH$ we have by the definition of a system of imprimitivity $T_{f(x)}P_H = P_{zH}T_{f(x)}$. Integrating this with respect to x (over zH) gives

$$T_f\ P_H = P_{zH}\ T_f\ . \qquad \ldots(20)$$

Together, (19) and (20) imply that

$$R_{[f,g]_E} = T_f\ P_H(T_g)^*\ . \qquad \ldots(21)$$

Now any f in $\underset{\sim}{L}(\underset{\sim}{B})$ is a sum of elements of $\underset{\sim}{L}(\underset{\sim}{B})$ each of which vanishes outside some coset zH . Hence by linearity (21) holds for all f,g in $\underset{\sim}{L}(\underset{\sim}{B})$.

Now (T,P) is non-degenerate on E' if and only if $\{R_{[f,g]_E} : f,g \in \underset{\sim}{L}(\underset{\sim}{B})\}$ acts non-degenerately on $X(T)$. By (21) and the non-degeneracy of the integrated form of T , this happens if and only if the linear span of $\{T_f(\text{range}(P_H)) : f \in \underset{\sim}{L}(\underset{\sim}{B})\}$ is dense in $X(T)$. □

Remark 2. The last two propositions make it very simple to prove the Imprimitivity Theorem in case G/H is discrete. Indeed, given a system of imprimitivity (T,P) for $\underset{\sim}{B}$ over G/H , all we have to do in that case is to define S in accordance with Prop. 32.11, that is, as the subrepresentation of $T|\underset{\sim}{B}_H$ acting on $\text{range}(P_H)$, and then to verify (using Prop. 32.12) that (T,P) is canonically equivalent to the system of imprimitivity induced by S . We suggest that the reader carry out this verification for his own instruction.

If G/H is not discrete, then $P_H = 0$, and the above approach breaks down. In fact the main purpose of this chapter has been to provide machinery strong enough to handle the case that G/H is not discrete.

Remark 3. Prop. 32.12 suggests almost trivial examples of the failure of the property of non-degeneracy on E' . Suppose for instance that G is the two-element group $\{e,u\}$, and that $\underset{\sim}{B}$ is a Banach *-algebraic bundle over G with $B_u = \{0\}$. Let T be a non-zero non-degenerate *-representation of $\underset{\sim}{B}$ (i.e., of B_e), and P the *-representation of $\underset{\sim}{L}(G)$ given by $P_\varphi = \varphi(u)\mathbb{1}$. Then (T,P) is a system of imprimitivity for $\underset{\sim}{B}$ over G which by Prop. 32.12 obviously fails to be non-degenerate on E' .

Thus, in general E' will not be dense in the compacted transformation *-algebra E .

The commuting algebra of a system of imprimitivity.

We now drop the assumption that G/H is discrete.

Let S be a non-degenerate $\underset{\sim}{B}$-positive *-representation of $\underset{\sim}{B}_H$, and $\underset{\sim}{T} = (T,P)$ the system of imprimitivity induced by S. Let $\underset{\sim}{I}(S)$ and $\underset{\sim}{I}(\underset{\sim}{T})$ be the commuting algebras of S and $\underset{\sim}{T}$ respectively. We are going to set up a canonical *-isomorphism between $\underset{\sim}{I}(S)$ and $\underset{\sim}{I}(\underset{\sim}{T})$. This result should be considered as an integral part of the Imprimitivity Theorem.

Let $\underset{\sim}{Y} = (Y,\{Y_\alpha\})$ be the Hilbert bundle over G/H induced by S; and let us readopt the rest of the notation of §23.

Fix an element γ of $\underset{\sim}{I}(S)$. By Prop. 22.1(III) and Theorem 1.2, for each α in G/H the equation

$$\gamma_\alpha(\kappa_\alpha(b \otimes \xi)) = \kappa_\alpha(b \otimes \gamma(\xi)) \qquad \ldots(22)$$

$(\xi \in X(S);\ b \in B_\alpha)$ defines a bounded linear operator γ_α on Y_α satisfying

$$\|\gamma_\alpha\| \leq \|\gamma\| . \qquad \ldots(23)$$

It follows from (22) and §23(9) that

$$\gamma_\alpha(\kappa_\alpha(\varphi \otimes \xi)) = \kappa_\alpha(\varphi \otimes \gamma(\xi)) \qquad \ldots(24)$$

for $\varphi \in \underset{\sim}{L}(\underset{\sim}{B}_\alpha)$, $\xi \in X(S)$. Let $\tilde{\gamma}: Y \rightarrow Y$ be the map coinciding with γ_α on Y_α (for each α). If f is a cross-section of $\underset{\sim}{Y}$ of the form $\alpha \mapsto \kappa_\alpha((\varphi|\alpha) \otimes \xi)$ $(\varphi \in \underset{\sim}{L}(\underset{\sim}{B});\ \xi \in X(S))$, then by (24) $(\tilde{\gamma}\cdot f)(\alpha) = \kappa_\alpha((\varphi|\alpha) \otimes \gamma(\xi))$. From this, (23), and the definition of the topology of $\underset{\sim}{Y}$, we deduce that $\tilde{\gamma}$ is continuous. This implies (by (23) and the definition of local measurability in §10) that, if f is a locally $\rho^\#$-measurable cross-section of $\underset{\sim}{Y}$, then so is $\tilde{\gamma}\cdot f$. Thus, again using (23), we see that $\gamma^0: f \rightarrow \tilde{\gamma}\cdot f$ $(f \in \underset{\sim}{L}_2(\underset{\sim}{Y};\rho^\#))$ is a bounded linear operator on $\underset{\sim}{L}_2(\underset{\sim}{Y};\rho^\#)$ satisfying $\|\gamma^0\| \leq \|\gamma\|$. It is evident that γ^0 commutes with all T_b $(b \in B)$ and all P_φ $(\varphi \in \underset{\sim}{L}(G/H))$, and so belongs to $\underset{\sim}{I}(\underset{\sim}{T})$.

Theorem 32.13. *The map* $\gamma \mapsto \gamma^0$ *just defined is a* *-*isomorphism of* $\underset{\sim}{I}(S)$ *onto* $\underset{\sim}{I}(\underset{\sim}{T})$.

Proof. Let B', E' be as at the beginning of this section. I first claim that the restriction S' to B' of the integrated form of S is non-degenerate. Indeed: Let $\underset{\sim}{C}$ be the bundle C^*-completion of $\underset{\sim}{B}$. As in the latter part of the proof of Theorem 32.7, S is lifted from a non-degenerate *-representation S^0 of $\underset{\sim}{C}_H$. Since $\underset{\sim}{C}$ has an approximate unit, the analogue C' of B' in $\underset{\sim}{C}$ is dense in $\underset{\sim}{L}(\underset{\sim}{C}_H)$ in the inductive limit topology by Prop. 32.1; and so $S^0|C'$ is non-degenerate. On the other hand, by an argument similar to one found in the proof of Theorem 32.7, the image of B' in $\underset{\sim}{L}(\underset{\sim}{C}_H)$ is dense in C'. Therefore $S' = S|B'$ is non-degenerate; and the claim is proved.

As in Theorem 32.7 let R' be the restriction to E' of the integrated form R of the system of imprimitivity $\underset{\sim}{T} = (T,P)$ induced by S. By part (I) of Theorem 32.7 R', like S', is non-degenerate; and by Prop. 31.1 R' corresponds to S' under the correspondence of Theorem 8.3 applied to the canonical strict E',B' imprimitivity bimodule $\underset{\sim}{I}$ of §31. Therefore by Theorem 8.4 the commuting algebras of S' and R' are *-isomorphic under the *-isomorphism $\Phi: F \mapsto \tilde{F}$ of Theorem 8.4. Passing from the abstract to the concrete description of the induced representation via the isometry of Prop. 23.4 we check that the Φ of Theorem 8.4 becomes just the mapping $\gamma \mapsto \gamma^0$.

To complete the proof we need only to show that the commuting algebras of S and S' are the same, and that those of R and R' are the same. But this follows from Prop. 14.5. □

Corollary 32.14. Let S and T be as in the preceding theorem. Then T is irreducible if and only if S is irreducible.

Compact induced representations.

Definition. A *-representation T of B is said to be compact if the integrated form of T is compact, that is, if T_f is a compact operator for every f in $L(B)$.

Theorem 32.15. Assume that B is saturated; and let S be a compact non-degenerate B-positive *-representation of B_H. Then the integrated form R of the system of imprimitivity (T,P) induced by S is a compact *-representation of E. In particular, the product $P_\varphi T_f$ is a compact operator whenever $\varphi \in L(G/H)$ and $f \in L(B)$.

Proof. Let B', E' be as at the beginning of this section. Applying Prop. 8.6 to I considered as a strict E', B' imprimitivity bimodule, we conclude that R_u is compact whenever $u \in E'$. But by Prop. 32.2 this implies that R_u is compact for all u in E. Taking u to be of the special form $(x,\alpha) \mapsto \varphi(\alpha)f(x)$, where $\varphi \in L(G/H)$ and $f \in L(B)$, we have $R_u = P_\varphi T_f$ by §30(9); and from this the last statement of the theorem follows. □

If G/H is compact, we can take $\varphi \equiv 1$ in the last statement of the preceding theorem, in which case $P_\varphi T_f = T_f$. Thus the following interesting corollary emerges:

Corollary 32.16. Assume that B is saturated and G/H is compact. Then, if S is a compact non-degenerate B-positive *-representation of B_H, $\mathrm{Ind}_{B_H \uparrow B}(S)$ is a compact *-representation of B.

This corollary in the group case is due to Schochetman [1].

§33. Conjugation of *-representations.

Let G be a group, H a subgroup of G, and S a unitary representation of H. For each element x of G the formula

$$S'_h = S_{x^{-1}hx} \qquad (h \in xHx^{-1}) \qquad \ldots(1)$$

defines a unitary representation S' of the conjugate subgroup xHx^{-1}. We call S' the representation conjugate to S under x.

Is there a bundle generalization of this construction? Given a Banach *-algebraic bundle $\underset{\sim}{B}$ over G, a closed subgroup H of G, an element x of G, and a *-representation S of $\underset{\sim}{B}_H$, is there a natural way to construct a "conjugate" *-representation S' of $\underset{\sim}{B}_{xHx^{-1}}$? Formula (1) is of no use as it stands, since xhx^{-1} has no meaning for elements h of the bundle space. In spite of this, we shall be able in the present section to answer this question affirmatively. It will turn out that the operation of conjugation so obtained is another special case of Rieffel's abstract inducing process (see §4).

The most satisfactory results are obtained when $\underset{\sim}{B}$ is saturated; and we shall restrict ourselves to this case.

The conjugation process described in this section will be of the utmost importance when we come to generalize the Mackey normal subgroup analysis to saturated Banach *-algebraic bundles (in a subsequent publication).

Throughout this section $\underset{\sim}{B} = (B, \pi, \cdot, ^*)$ is a saturated Banach *-algebraic bundle over the locally compact group G with unit e, left Haar measure λ, and modular function Δ. We fix a closed subgroup H of G, with left Haar measure ν and modular function δ, and choose once for all an everywhere positive continuous H-rho-function

ρ on G.

Take a $\underset{\sim}{B}$-positive non-degenerate *-representation S of $\underset{\sim}{B}_H$. Let $\underset{\sim}{Y} = (Y, \{Y_\alpha\})$ be the Hilbert bundle over G/H induced by S as in §23; and let us adopt without further ado the rest of the notation of §§23-25. In particular ι_b is the action of b on Y defined in §24.

Fix an element x of G. If $c \in B_{xHx^{-1}}$, so that the group element $h = x^{-1}\pi(c)x$ belongs to H, then by §24(9)

$$\iota_c(Y_{xH}) \subset Y_{\pi(c)xH} = Y_{xhH} = Y_{xH} ;$$

and the equation

$$S'_c(\zeta) = \delta(h)^{\frac{1}{2}}\Delta(h)^{-\frac{1}{2}}\iota_c(\zeta) \qquad (\zeta \in Y_{xH}) \qquad \ldots(2)$$

defines a bounded linear operator S'_c on Y_{xH}.

<u>Proposition 33.1</u>. <u>$S' : c \mapsto S'_c$ is a non-degenerate *-representation of $\underset{\sim}{B}_{xHx^{-1}}$ on the Hilbert space Y_{xH}.</u>

<u>Proof</u>. S' is obviously linear on fibers and is multiplicative by §24(10). It is continuous in the strong operator topology by Prop. 24.6. To show that it preserves adjoints, let $\xi, \eta \in Y_{xH}$, $c \in B_{xHx^{-1}}$, and put $h = x^{-1}\pi(c)x \in H$. By (2) and §24(11)

$$\begin{aligned}
\langle S'_c\xi, \eta\rangle_{Y_{xH}} &= \delta(h)^{\frac{1}{2}}\Delta(h)^{-\frac{1}{2}} \langle \iota_c\xi, \eta\rangle_{Y_{xH}} \\
&= \delta(h)^{\frac{1}{2}}\Delta(h)^{-\frac{1}{2}}\sigma(\pi(c), xH) \langle \xi, \iota_{c^*}\eta\rangle_{Y_{xH}} \\
&= \delta(h)^{\frac{1}{2}}\Delta(h)^{-\frac{1}{2}}\rho(x)\,\rho(xh)^{-1} \langle \xi, \iota_{c^*}\eta\rangle_{Y_{xH}} \\
&= \delta(h^{-1})^{\frac{1}{2}}\Delta(h^{-1})^{-\frac{1}{2}} \langle \xi, \iota_{c^*}\eta\rangle_{Y_{xH}}
\end{aligned}$$

$$= \langle \xi, S'_{c^*} \eta \rangle_{Y_{xH}} \; .$$

Thus $(S'_c)^* = S'_{c^*}$; and we have shown that S' is a *-representation. Its non-degeneracy follows from Prop. 24.5. □

Definition. The above S' is called the *-representation conjugate to S under x, or simply the x-conjugate of S, and is denoted by ${}^xS$.

Remark 1. ${}^xS$ evidently depends only on the H-coset xH to which x belongs.

Remark 2. It follows from (2) and §24(14) that

$${}^eS \cong S \; .$$

Remark 3. ${}^xS$ can be constructed as above, and will be non-degenerate, even if $\underset{\sim}{B}$ is not saturated.

From Remarks 1 and 2 applied to the case that $H = G$ we obtain:

Proposition 33.2. If T is a non-degenerate *-representation of $\underset{\sim}{B}$, then ${}^xT \cong T$ for all x in G.

Returning to the case of a general subgroup H and fixed element x of G, let α denote the coset xH ; and recall the description of Y_α given in connection with §23(7). When the Z_α of that description is subjected to the linear automorphism $b \otimes \xi \mapsto \rho(y)^{-\frac{1}{2}} b \otimes \xi$ ($\xi \in X$; $b \in B_y$; $y \in \alpha$), the inner product §23(7) is transformed into

$$\langle b \otimes \xi, c \otimes \eta \rangle_\alpha'' = \langle S_{c^*b} \xi, \eta \rangle \; ; \qquad \ldots(3)$$

and one verifies that (2) takes the simpler form

$$S''_c(\kappa''_\alpha(b \otimes \xi)) = \kappa''_\alpha(cb \otimes \xi) \qquad \ldots(4)$$

(where κ''_α is the quotient map with respect to the transformed inner product). Thus the latter formula defines an equivalent version S'' of ${}^xS$.

Notice that in this new formulation of the operation of conjugation there is no reference to modular functions and rho-functions. We conclude that, as in the group case (1), <u>the definition of the conjugation operation is unaltered when the topology of G is replaced by the discrete topology</u>.

<u>Proposition 33.3</u>. <u>Let K be another closed subgroup of G with</u> $K \subset H$. <u>Let S be a $\underset{\sim}{B}$-positive non-degenerate *-representation of</u> $\underset{\sim}{B}_H$; <u>and let $x \in G$. Then ${}^x(S|\underset{\sim}{B}_K)$ and $({}^xS)|\underset{\sim}{B}_{xKx^{-1}}$ are unitarily equivalent *-representations of $\underset{\sim}{B}_{xKx^{-1}}$.</u>

<u>Proof</u>. Let $\underset{\sim}{Z}$ be the Hilbert bundle over G/K induced by $S|\underset{\sim}{B}_K$; and put $\alpha = xH$, $\beta = xK$. On using the description (3) of Y_α and Z_β, we see that $Z_\beta \subset Y_\alpha$, and that (by (4)) ${}^x(S|\underset{\sim}{B}_K)$ is the same as the subrepresentation of $({}^xS)|\underset{\sim}{B}_{xKx^{-1}}$ acting on Z_β. Hence the proposition will be established if we show that $Z_\beta = Y_\alpha$.

Now Z_β certainly contains $\kappa''_\alpha(B_x \otimes X(S))$. Furthermore, the definition (3) of $\langle\ ,\ \rangle''_\alpha$ implies that

$$\kappa''_\alpha(b \otimes S_d\xi) = \kappa''_\alpha(bd \otimes \xi)$$

for all $b \in B_x$, $d \in B_H$, and $\xi \in X(S)$. It follows that Z_β contains $\kappa''_\alpha(B_xB_H \otimes X(S))$. But by the saturation of $\underset{\sim}{B}$ the linear span of B_xB_h is dense in B_{xh} for every h in H. Therefore Z_β contains, and so is equal to, Y_α . □

Applying Prop. 33.3 with H and K replaced by G and H, and using Prop. 33.2, we obtain:

Corollary 33.4. *If* $x \in G$ *and* T *is a non-degenerate* **-representation of* $\underset{\sim}{B}$, *then* ${}^{x}(T|\underset{\sim}{B}_H)$ *is unitarily equivalent to* $T|\underset{\sim}{B}_{xHx^{-1}}$.

Remark 4. It is useful to observe the explicit form of the unitary equivalence claimed in this corollary. Let $\alpha = xH$; put $S = T|\underset{\sim}{B}_H$; and adopt the notation of §23. Then the equation

$$F(\varkappa_\alpha(b \otimes \xi)) = \rho(\pi(b))^{-\frac{1}{2}} T_b \xi \qquad \ldots(5)$$

$(b \in B_\alpha ; \xi \in X(T))$ defines F as a linear isometry of $X({}^{x}S)$ onto $X(T)$ which intertwines ${}^{x}S$ and $T|\underset{\sim}{B}_{xHx^{-1}}$.

Conjugation and imprimitivity bimodules.

Now the operation of conjugation of representations can also be described as a special case of the abstract inducing process of §4, with respect to an $\underset{\sim}{L}(\underset{\sim}{B}_{xHx^{-1}})$, $\underset{\sim}{L}(\underset{\sim}{B}_H)$ imprimitivity bimodule.

Fix an element x of G . We shall make $\underset{\sim}{L}(\underset{\sim}{B}_{xH})$ into a left $\underset{\sim}{L}(\underset{\sim}{B}_{xHx^{-1}})$-module and a right $\underset{\sim}{L}(\underset{\sim}{B}_H)$-module by means of the actions : described as follows:

$$(\varphi : f)(xh) = \int_H \delta(k)^{\frac{1}{2}} \Delta(k)^{-\frac{1}{2}} \varphi(xkx^{-1}) f(xk^{-1}h) d\nu k , \qquad \ldots(6)$$

$$(f : \psi)(xh) = \int_H \Delta(k)^{\frac{1}{2}} \delta(k)^{-\frac{1}{2}} f(xhk) \psi(k^{-1}) d\nu k \qquad \ldots(7)$$

$(f \in \underset{\sim}{L}(\underset{\sim}{B}_{xH}) ; \varphi \in \underset{\sim}{L}(\underset{\sim}{B}_{xHx^{-1}}) ; \psi \in \underset{\sim}{L}(\underset{\sim}{B}_H) ; h \in H)$. We also equip $\underset{\sim}{L}(\underset{\sim}{B}_{xH})$

with an $\underset{\sim}{L}(\underset{\sim}{B}_{xHx^{-1}})$-valued rigging $[\ ,\ ]'_x$ and an $\underset{\sim}{L}(\underset{\sim}{B}_H)$-valued rigging $[\ ,\ ]_x$ defined by:

$$[f,g]'_x(xhx^{-1}) = \Delta(h)^{\frac{1}{2}}\delta(h)^{-\frac{1}{2}} \int_H \Delta(k)f(xhk)g(xk)^* d\nu k\ , \quad \ldots(8)$$

$$[f,g]_x(h) = \Delta(h)^{\frac{1}{2}}\delta(h)^{-\frac{1}{2}} \int_H \Delta(k)\,\delta(k)^{-1}f(xk)^* g(xkh)d\nu k \quad \ldots(9)$$

$(f,g \in \underset{\sim}{L}(\underset{\sim}{B}_{xH});\ h \in H)$. We leave to the reader the tedious but routine verification of the fact that, when $\underset{\sim}{L}(\underset{\sim}{B}_{xH})$ is equipped with the module structures (6) and (7) and the riggings (8) and (9) it becomes an $\underset{\sim}{L}(\underset{\sim}{B}_{xHx^{-1}})$, $\underset{\sim}{L}(\underset{\sim}{B}_H)$ imprimitivity bimodule. This imprimitivity bimodule will be denoted throughout this section by $\underset{\sim}{J}_x$.

Remark 5. It is assumed above that in defining the convolution in $\underset{\sim}{L}(\underset{\sim}{B}_{xHx^{-1}})$ one uses the left Haar measure ν' on xHx^{-1} obtained by x-conjugation of ν: $\nu'(W) = \nu(x^{-1}Wx)$.

Notice that the $\underset{\sim}{L}(\underset{\sim}{B}_H)$-valued rigging (9) on $\underset{\sim}{L}(\underset{\sim}{B}_{xH})$ differs from the rigging §22(2) only by the linear automorphism $f \mapsto f'$ of $\underset{\sim}{L}(\underset{\sim}{B}_{xH})$, where $f'(xh) = \Delta(h)^{\frac{1}{2}}\delta(h)^{-\frac{1}{2}}f(xh)$. It follows therefore from §23(2) that the integrated form of a $\underset{\sim}{B}$-positive *-representation of $\underset{\sim}{B}_H$ is positive with respect to $\underset{\sim}{J}_x$.

Proposition 33.5. *The linear spans of the ranges of* $[\ ,\ ]_x$ *and* $[\ ,\ ]'_x$ *are dense in* $\underset{\sim}{L}(\underset{\sim}{B}_H)$ *and* $\underset{\sim}{L}(\underset{\sim}{B}_{xHx^{-1}})$ *respectively in the inductive limit topologies.*

Proof. We shall first show that range$([\ ,\ ]'_x)$ has dense linear span in $\underset{\sim}{L}(\underset{\sim}{B}_{xHx^{-1}})$.

Let us denote by M the inductive limit closure of the linear span of range($[\ ,\ ]_x'$) , and by L the linear span in $\underset{\sim}{L}(\underset{\sim}{B}_{xHx^{-1}})$ of the set of all cross-sections of $\underset{\sim}{B}_{xHx^{-1}}$ of the form

$$xhx^{-1} \mapsto b\varphi(h)c^* \qquad (h \in H) , \qquad \ldots(10)$$

where $\varphi \in \underset{\sim}{L}(\underset{\sim}{B}_H)$ and $b,c \in B_x$. I claim that

$$L \subset M . \qquad \ldots(11)$$

To prove (11), take $b,c \in B_x$ and $\psi,\chi \in \underset{\sim}{L}(\underset{\sim}{B}_H)$. Then the equations

$$f(xh) = \delta(h)^{\frac{1}{2}}\Delta(h)^{-\frac{1}{2}}b\,\psi(h) ,$$

$$g(xh) = \delta(h)^{\frac{1}{2}}\Delta(h)^{-\frac{1}{2}}c\,\chi(h)$$

($h \in H$) define elements f and g of $\underset{\sim}{L}(\underset{\sim}{B}_{xH})$; and, for $h \in H$, we have by (8)

$$\begin{aligned} [f,g]_x'(xhx^{-1}) &= \int b\,\psi(hk)\chi(k)^*c^*\,\delta(k)d\nu k \\ &= b\left[\int \psi(hk)\chi(k)^*\delta(k)d\nu k\right]c^* \\ &= b(\psi*\chi^*)(h)c^* . \end{aligned}$$

It follows from this that (10) belongs to M whenever φ is of the form $\psi * \chi^*$ $(\psi,\chi \in \underset{\sim}{L}(\underset{\sim}{B}_H))$. By Cor. 32.3 such φ span a dense subset of $\underset{\sim}{L}(\underset{\sim}{B}_H)$. Since the map carrying φ into (10) is clearly continuous in the inductive limit topologies of $\underset{\sim}{L}(\underset{\sim}{B}_H)$ and $\underset{\sim}{L}(\underset{\sim}{B}_{xHx^{-1}})$, the last two sentences imply that (10) belongs to M for all φ in $\underset{\sim}{L}(\underset{\sim}{B}_H)$ and all b,c in B_x . Thus (11) is established.

Now L is clearly closed under multiplication by continuous complex functions on xHx^{-1} . Also the saturation of $\underset{\sim}{B}$ implies that $\{f(k): f \in L\}$ is dense in B_k for every k in xHx^{-1} . So by Prop.

10.9 $\underset{\sim}{L}$ is dense in $\underset{\sim}{L}(\underset{\sim}{B}_{xHx^{-1}})$ in the inductive limit topology. This and (11) show that $M = \underset{\sim}{L}(\underset{\sim}{B}_{xHx^{-1}})$. Hence, by the definition of M, the linear span of range($[\ ,\ ]'_x$) is dense in $\underset{\sim}{L}(\underset{\sim}{B}_{xHx^{-1}})$ in the inductive limit topology.

The proof that the linear span of range($[\ ,\ ]_x$) is dense in $\underset{\sim}{L}(\underset{\sim}{B}_H)$ is very similar, and is left to the reader. □

The next proposition shows that x-conjugation of a *-representation of $\underset{\sim}{B}_H$ is the same as inducing with respect to $\underset{\sim}{J}_x$.

Proposition 33.6. *Let* x *be an element of* G, *and* S *a non-degenerate* $\underset{\sim}{B}$*-positive* **-representation of* $\underset{\sim}{B}_H$. *Then the integrated form of* ${}^xS$ *is unitarily equivalent to the* **-representation of* $\underset{\sim}{L}(\underset{\sim}{B}_{xHx^{-1}})$ *abstractly induced from* S *via* $\underset{\sim}{J}_x$.

Proof. Denote ${}^xS$ and its integrated form by S'. Writing α for xH, and adopting the notation of §23, we have only to show that

$$S'_\varphi(\kappa_\alpha(f \otimes \xi)) = \kappa_\alpha(\varphi : f \otimes \xi) \qquad \ldots(12)$$

whenever $\varphi \in \underset{\sim}{L}(\underset{\sim}{B}_{xHx^{-1}})$, $f \in \underset{\sim}{L}(\underset{\sim}{B}_\alpha)$, and $\xi \in X(S)$. But for any ζ in Y_α we have

$$\langle S'_\varphi \kappa_\alpha(f \otimes \xi), \zeta\rangle_{Y_\alpha}$$

$$= \int \langle S'_{\varphi(xkx^{-1})}\kappa_\alpha(f \otimes \xi), \zeta\rangle_{Y_\alpha}\, d\nu k$$

$$= \int \delta(k)^{\frac{1}{2}}\Delta(k)^{-\frac{1}{2}} \langle \iota_{\varphi(xkx^{-1})}\kappa_\alpha(f \otimes \xi), \zeta\rangle_{Y_\alpha}\, d\nu k \text{ (by (2))}$$

$$= \int \delta(k)^{\frac{1}{2}}\Delta(k)^{-\frac{1}{2}} \langle \kappa_\alpha(\varphi(xkx^{-1})f \otimes \xi), \zeta\rangle_{Y_\alpha}\, d\nu k \quad \text{(by §24(7))} \qquad \ldots(13)$$

(where $(\varphi(xkx^{-1})f)(xh) = \varphi(xkx^{-1})f(xk^{-1}h)$). Now, just as in Prop. 13.2, the definition (6) of $\varphi{:}f$ can be written in the form

$$\varphi{:}f = \int \delta(k)^{\frac{1}{2}}\Delta(k)^{-\frac{1}{2}}(\varphi(xkx^{-1})f)d\nu k\ , \qquad \ldots(14)$$

where the right side of (14) is an $L(B_\alpha)$-valued Bochner integral with respect to the inductive limit topology.

Applying to both sides of (14) the linear functional $g \mapsto \langle \kappa_\alpha(g \otimes \xi), \zeta\rangle_{Y_\alpha}$ (which is continuous on $L(B_\alpha)$ in the inductive limit topology by Prop. 23.1), we find

$$\langle \kappa_\alpha(\varphi{:}f \otimes \xi), \zeta\rangle_{Y_\alpha}$$

$$= \int \delta(k)^{\frac{1}{2}}\Delta(k)^{-\frac{1}{2}} \langle \kappa_\alpha((\varphi(xkx^{-1})f) \otimes \xi), \zeta\rangle_{Y_\alpha} d\nu k\ . \qquad \ldots(15)$$

Combining (13) and (15), by the arbitrariness of ζ one obtains (12).□

Properties of conjugation.

The process of conjugation preserves Hilbert direct sums.

Proposition 33.7. *Let x be an element of G, and $\{S^i\}$ an indexed collection of non-degenerate B-positive *-representations of B_H. Then*

$$^x\Big(\sum_i^{\oplus} S^i\Big) \cong \sum_i^{\oplus} {}^x(S^i)\ .$$

Proof. Combine Props. 33.6 and 5.1. □

Proposition 33.8. *Let x be an element of G, and S a non-degenerate B-positive *-representation of B_H. The commuting algebras of S and ^xS are *-isomorphic under the *-isomorphism $F \mapsto \tilde{F}$ given by*

$$\tilde{F}(\kappa_{xH}(f \otimes \xi)) = \kappa_{xH}(f \otimes F(\xi)) \qquad \ldots(16)$$

$(f \in \underset{\sim}{L}(\underset{\sim}{B}_{xH});\ \xi \in X(S))$. In particular ${}^{x}S$ is irreducible if and only if S is irreducible.

Proof. Let B and E be the linear spans of the ranges of $[\ ,\ ]_x$ and $[\ ,\ ]'_x$ respectively (see (8), (9)); and let T and T' be the restrictions of S and ${}^{x}S$ to B and E respectively. Thus T and T' are non-degenerate by Prop. 33.5, and by Prop. 33.6 T' is induced from T via $\underset{\sim}{J}_x$ when the latter is considered as a strict E,B imprimitivity bimodule. Hence by Theorem 8.4 the commuting algebras of T and T' are *-isomorphic under (16). But, again by Prop. 33.5, the commuting algebras of S and T are the same, and likewise those of ${}^{x}S$ and T' . This completes the proof. □

Corollary 33.9. Suppose that B_e has the property that all its irreducible *-representations are finite-dimensional. If S is a finite-dimensional $\underset{\sim}{B}$-positive *-representation of $\underset{\sim}{B}_H$, then every fiber of the Hilbert bundle $\underset{\sim}{Y}$ induced by S is finite-dimensional.

Proof. Let $x \in G$. The fiber Y_{xH} of $\underset{\sim}{Y}$ is the space of ${}^{x}S$. Now $S|B_e = \Sigma^{\oplus}_{i\in I} R^i$, where I is a finite index set and each R^i is an irreducible (finite-dimensional) *-representation of B_e . So by Props. 33.3 and 33.7 $({}^{x}S)|B_e \cong {}^{x}(S|B_e) \cong \Sigma^{\oplus}_{i\in I}\ {}^{x}(R^i)$. By Prop. 33.8 each ${}^{x}(R^i)$ is irreducible, and so by hypothesis is finite-dimensional. It follows that ${}^{x}S$ is finite-dimensional, whence by the opening remark of the proof Y_{xH} is finite-dimensional. □

Corollary 33.10. In addition to the hypotheses of Cor. 33.9 suppose that G/H is finite. Then $\mathrm{Ind}_{\underset{\sim}{B}_H \uparrow \underset{\sim}{B}}(S)$ is finite-dimensional.

Remark 6. Without the hypothesis that all the elements of $(B_e)^{\wedge}$ are finite-dimensional, the conclusions of Cor. 33.9 and Cor. 33.10 would be false. Indeed, we shall see in the class of examples at the end of this section that in general the process of conjugation will carry a finite-dimensional *-representation of B_e into an infinite-dimensional one.

In the group case one would like to know that the conjugation defined above coincides (at least up to unitary equivalence) with the more obvious conjugation given by (1). More generally we will prove:

Proposition 33.11. *Let S be a non-degenerate $\mathcal{B}$-positive *-representation of B_H, and x an element of G; and suppose that $\mathcal{B}$ has an approximate unit and that there exists a unitary multiplier u of $\mathcal{B}$ of order x. Then ^xS is unitarily equivalent to the *-representation T of $B_{xHx^{-1}}$ defined by*

$$T_b = S_{u^*bu} \qquad (b \in B_{xHx^{-1}}) . \qquad \ldots(17)$$

Remark 7. We have assumed an approximate unit in $\mathcal{B}$ in order that $\mathcal{B}$ should have no annihilators. All associative laws between elements and multipliers of $\mathcal{B}$ then conveniently hold. However the proposition remains true, in slightly modified form, even if there is no approximate unit.

Proof of Proposition. In this argument we will use the description §23(7) of Y_{xH}.

Let ξ, η be in $X(S)$ and b, c in B_{xH}; and put $\pi(b) = y$, $\pi(c) = z$, $\alpha = xH$. Then

$$<\kappa_\alpha(b \otimes \xi), \kappa_\alpha(c \otimes \eta)>_{Y_\alpha}$$

$$= (\rho(y)\rho(z))^{-\frac{1}{2}} \langle S_{c*b}\xi, \eta\rangle \quad \text{(by §23(10))}$$

$$= (\rho(y)\rho(z))^{-\frac{1}{2}} \langle S_{(u*c)*u*b}\xi, \eta\rangle$$

$$= (\rho(y)\rho(z))^{-\frac{1}{2}} \langle S_{u*b}\xi, S_{u*c}\eta\rangle$$

(since u^*b, u^*c are in B_H). It follows that the equation

$$F(\kappa_\alpha(b \otimes \xi)) = (\rho(\pi(b)))^{-\frac{1}{2}} S_{u*b}\xi$$

($b \in B_{xH}$; $\xi \in X(S)$) defines a linear isometry of $X({}^xS)$ into $X(S)$. Since $u^*(B_{xh}) = B_h$ for each h in H and S is non-degenerate, F is onto $X(S)$. If $\xi \in X(S)$, $h,k \in H$, $b \in B_{xh}$, and $c \in B_{xkx^{-1}}$, then

$$F[({}^xS)_c\, \kappa_\alpha(b \otimes \xi)]$$

$$= \delta(k)^{\frac{1}{2}}\Delta(k)^{-\frac{1}{2}} F(\kappa_\alpha(cb \otimes \xi)) \quad \text{(by (2) and §24(3))}$$

$$= \delta(k)^{\frac{1}{2}}\Delta(k)^{-\frac{1}{2}} \rho(xkh)^{-\frac{1}{2}} S_{u*cb}\xi$$

$$= \rho(xh)^{-\frac{1}{2}} S_{u*cb}\xi$$

$$= \rho(\pi(b))^{-\frac{1}{2}} S_{u*cu} S_{u*b}\xi$$

$$= T_c\, F(\kappa_\alpha(b \otimes \xi)) \quad \text{(by (17))} .$$

This shows that F intertwines ${}^xS$ and T , completing the proof. □

We come now to a very important result (which is certainly to be expected by analogy with (1)).

Proposition 33.12. *Let* S *be a non-degenerate* $\mathcal{B}$*-positive* **-representation of* $\mathcal{B}_H$ *, and let* x,y *be two elements of* G *. Then* (I) ${}^xS$ *is* $\mathcal{B}$*-positive (so that we can form* ${}^y({}^xS)$) ; *and* (II) ${}^y({}^xS) \cong {}^{(yx)}S$.

Proof. Put $H' = xHx^{-1}$; and notice that the formula

$$\rho'(z) = \rho(zx) \qquad (z \in G) \qquad \ldots(18)$$

defines ρ' as a continuous everywhere positive H'-rho function on G.

Take vectors $\xi_1, \cdots, \xi_n$ in $X = X(S)$, elements $b_1, \cdots, b_n$ of B_{xH}, and elements $c_1, \cdots, c_n$ of $B_{yH'}$. Thus $\pi(b_i) = xh_i$ and $\pi(c_i) = yxk_i x^{-1}$, where h_i and k_i are in H. Putting $\alpha = xH$, we have by (2), §24(3), and §23(10),

$$\sum_{i,j=1}^{n} (\rho'(\pi(c_i))\rho'(\pi(c_j)))^{-\frac{1}{2}} \langle ({}^{x}S)_{c_j^* c_i} \varkappa_\alpha(b_i \otimes \xi_i), \varkappa_\alpha(b_j \otimes \xi_j)\rangle$$

$$= \sum_{i,j} (\rho(yxk_i)\rho(yxk_j))^{-\frac{1}{2}} \delta(k_j^{-1}k_i)^{\frac{1}{2}} \Delta(k_j^{-1}k_i)^{-\frac{1}{2}} \langle \varkappa_\alpha(c_j^* c_i b_i \otimes \xi_i), \varkappa_\alpha(b_j \otimes \xi_j)\rangle$$

$$= \sum_{i,j=1}^{n} N_{ij} \langle S_{b_j^* c_j^* c_i b_i} \xi_i, \xi_j\rangle , \qquad \ldots(19)$$

where $N_{ij} = (\rho(yxk_i)\rho(yxk_j))^{-\frac{1}{2}} \delta(k_j^{-1}k_i)^{\frac{1}{2}} \Delta(k_j^{-1}k_i)(\rho(xk_j^{-1}k_i h_i)\rho(xh_j))^{-\frac{1}{2}} = \rho(x)^{-1}(\rho(yxk_i h_i)\rho(yxk_j h_j))^{-\frac{1}{2}} = \rho(x)^{-1}(\rho(\pi(c_i b_i))\rho(\pi(c_j b_j)))^{-\frac{1}{2}}$. Therefore the right side of (19) becomes

$$\rho(x)^{-1} \sum_{i,j=1}^{n} \langle \varkappa_{y\alpha}(c_i b_i \otimes \xi_i), \varkappa_{y\alpha}(c_j b_j \otimes \xi_j)\rangle. \qquad \ldots(20)$$

Now (20) is non-negative by the $\underset{\sim}{B}$-positivity of S. So the left side of (19) is non-negative. Since the $\varkappa_\alpha(b \otimes \xi)$ span a dense subspace of Y_α, this implies (by the arbitrariness of y) that ${}^{x}S$ is $\underset{\sim}{B}$-positive. So statement (I) of the proposition is proved.

Thus we can form the Hilbert bundle $\underset{\sim}{Y}' = (Y', \{Y'_\beta\})$ over G/H' generated by $S' = {}^{x}S$ (using the H'-rho function ρ'). Let $\varkappa'_\beta$ $(\beta \in G/H')$ denote the canonical quotient map of $\underset{\sim}{L}(\underset{\sim}{B}_\beta) \otimes X(S')$ into Y'_β. The equality of (19) and (20) asserts that there is a linear isometry $F_{yH'}: Y'_{yH'} \to Y_{yxH}$ satisfying

$$F_\beta(\kappa'_\beta(c \otimes \kappa_\alpha(b \otimes \xi))) = \rho(x)^{-\frac{1}{2}}\kappa_{y\alpha}(cb \otimes \xi) \qquad \ldots(21)$$

($\xi \in X$; $b \in B_\alpha$; $c \in B_\beta$; $\beta = yH'$) . Since $\underset{\sim}{B}$ is saturated, the set of all such products cb spans a dense subspace of each fiber within $B_{y\alpha}$. Therefore by (21) the range of F_β is dense in $Y_{y\alpha}$ and so equal to $Y_{y\alpha}$. Using (2) and (21) one verifies easily that F_β intertwines $^y({}^xS)$ (whose space is Y'_β) and ^{yx}S (whose space is $Y_{y\alpha}$). So statement (II) is proved. □

Prop. 33.12 (together with Remark 2) asserts that G <u>acts as a left transformation group on the set of all pairs</u> (K,S) , <u>where</u> K <u>is a closed subgroup of</u> G <u>and</u> S <u>is a unitary equivalence class of non-degenerate</u> $\underset{\sim}{B}$-<u>positive</u> *-<u>representations of</u> $\underset{\sim}{B}_K$; <u>the element</u> x <u>of</u> G <u>sends</u> (K,S) <u>into</u> $(xKx^{-1}, {}^xS)$. In particular, if H is <u>normal</u>, $(x,S) \mapsto {}^xS$ is a left action of G on the space of all unitary equivalence classes of non-degenerate $\underset{\sim}{B}$-positive *-representations of $\underset{\sim}{B}_H$. We shall see that this action is continuous (with respect to the regional topology of the space of *-representations of $\underset{\sim}{B}_H$). In view of Prop. 33.8 <u>this action leaves stable the space</u> $(\underset{\sim}{B}_H)^\wedge$ <u>of all equivalence classes of irreducible</u> $\underset{\sim}{B}$-<u>positive</u> *-<u>representations of</u> $\underset{\sim}{B}_H$.

The proof of Prop. 33.12 gives further interesting information. Thus we can prove:

<u>Proposition 33.13</u>. <u>Let</u> x <u>be an element of</u> G , <u>and</u> S <u>a non-degenerate</u> $\underset{\sim}{B}$-<u>positive</u> *-<u>representation of</u> $\underset{\sim}{B}_H$. <u>Then</u>

$$\mathrm{Ind}_{\underset{\sim}{B}_{xHx^{-1}}\uparrow \underset{\sim}{B}}({}^{x}S) \cong \mathrm{Ind}_{\underset{\sim}{B}_H \uparrow \underset{\sim}{B}}(S) \quad . \qquad \ldots(22)$$

Proof. We adopt without further ado all the notation of the proof of Prop. 33.12. Notice that the correspondence $\beta \mapsto \beta x$ $(\beta \in G/H')$ is a homeomorphism of G/H' onto G/H.

Let $F: Y' \to Y$ be the bijection coinciding on each Y'_β with the bijection $F_\beta: Y'_\beta \to Y_{\beta x}$ constructed in the proof of Prop. 33.12. Observe that (20) amounts to

$$F_\beta(\kappa'_\beta(c \otimes \zeta)) = \rho(x)^{-\frac{1}{2}} \iota_c(\zeta) \qquad \ldots(23)$$

$(\beta \in G/H'; \ c \in B_\beta; \ \zeta \in Y_{xH})$, where ι of course is the action of B on Y set up in §24.

I claim that $F: Y' \to Y$ is a homeomorphism. Indeed: Suppose $\underset{\sim}{T}$ stands for that topology of Y' which makes F a homeomorphism. Then, equipped with $\underset{\sim}{T}$ instead of its original topology, $\underset{\sim}{Y}'$ is still a Hilbert bundle over G/H'; and by (23) and the continuity of ι (Prop. 24.6), for each fixed ζ in $X({}^{x}S)$, the map $c \mapsto \kappa'_{\pi(c)H'}(c\otimes\zeta)$ is continuous on B with respect to $\underset{\sim}{T}$. By Prop. 23.3 this implies that $\underset{\sim}{T}$ is the same as the original topology of Y'. So F is a homeomorphism.

Thus, to prove the proposition, it remains only to show that F carries the action ι of B on Y into the corresponding action ι' of B on Y'. If β, c, ζ are as in (23) and $d \in B_z$, we have by (23)

$$\begin{aligned} F[\iota'_d \kappa'_\beta(c \otimes \zeta)] &= F[\kappa'_{z\beta}(dc \otimes \zeta)] \\ &= \rho(x)^{-\frac{1}{2}} \iota_{dc}(\zeta) = \rho(x)^{-\frac{1}{2}} \iota_d \iota_c(\zeta) \\ &= \iota_d F[\kappa'_\beta(c \otimes \zeta)] \ . \end{aligned}$$

So $F \circ \iota'_d = \iota_d \circ F$; and the proof is complete. □

An equivalent definition of conjugation.

A closer look at the proof of Prop. 33.13, combined with the Imprimitivity Theorem 32.8, suggests a useful equivalent definition of the process of conjugation.

Fix an element x of G and write $H' = xHx^{-1}$; and let us denote by Φ the homeomorphism $\beta \mapsto \beta x$ of G/H' onto G/H. Of course Φ is an isomorphism of G/H' and G/H as left G-spaces. Thus if (T,P) is any system of imprimitivity for $\underset{\sim}{B}$ over G/H, and if we write P' for the *-representation $\varphi \mapsto P_{\varphi \circ \Phi^{-1}}$ of $\underset{\sim}{C}_0(G/H')$, then clearly (T,P') is a system of imprimitivity for $\underset{\sim}{B}$ over G/H'. Let us refer to (T,P') as the x-conjugate of (T,P).

Proposition 33.14. Let S be a non-degenerate $\underset{\sim}{B}$-positive *-representation of $\underset{\sim}{B}_H$, and x an element of G. Let (T,P) and (T',P') be the systems of imprimitivity for $\underset{\sim}{B}$ over G/H and $G/(xHx^{-1})$ induced by S and ${}^xS$ respectively. Then (T',P') is equivalent to the x-conjugate of (T,P).

Proof. The map F of the proof of Prop. 33.13 is an isometric isomorphism of $\underset{\sim}{Y}$ and $\underset{\sim}{Y}'$ covariant with Φ, and carries the action ι of B on Y into the action ι' of B on Y'. From this the result follows immediately. □

In view of the uniqueness statement in Theorem 32.8, the preceding proposition leads to the following equivalent description of ${}^xS$.

Let S be a non-degenerate $\underset{\sim}{B}$-positive *-representation of $\underset{\sim}{B}_H$, and (T,P) the system of imprimitivity for $\underset{\sim}{B}$ over G/H induced by

S. Let x be an element of G, and (T,P') the x-conjugate of (T,P). Then ${}^{x}S$ is that non-degenerate $\underset{\sim}{B}$-positive *-representation of $\underset{\sim}{B}_{xHx^{-1}}$ (unique to within unitary equivalence) such that the system of imprimitivity induced by ${}^{x}S$ is equivalent to (T,P').

The continuity of the conjugation operation.

If H is normal, it turns out that the conjugation operation $(x,S) \mapsto {}^{x}S$ is continuous in the regional topology. For simplicity we shall prove this only for the case that $H = \{e\}$. The general case can be reduced to the case $H = \{e\}$ by means of so-called partial cross-sectional bundles (see our subsequent publication).

Proposition 33.15. Let $\underset{\sim}{S}$ be the set of all non-degenerate $\underset{\sim}{B}$-positive *-representations of B_e. Then the conjugation map $(x,S) \mapsto {}^{x}S$ is continuous on $G \times \underset{\sim}{S}$ to $\underset{\sim}{S}$ with respect to the regional topology of $\underset{\sim}{S}$.

Proof. We have seen in Prop. 33.12(I) that the range of the conjugation map lies in $\underset{\sim}{S}$.

We shall write A for B_e. Suppose that $S^i \to S$ in $\underset{\sim}{S}$ regionally and $x^i \to x$ in G. We must show that ${}^{x^i}(S^i) \to {}^{x}S$ regionally. For this it is enough to show that, for each U in a basis of regional neighborhoods of ${}^{x}S$, some subnet of $\{{}^{x^i}(S^i)\}$ eventually lies in U.

As usual let κ_x be the canonical quotient map of $B_x \otimes X(S)$ into $X({}^{x}S)$ (the x-fiber of the Hilbert bundle induced by S).

Choose finitely many vectors $\zeta^1,\dots,\zeta^n$ in $X({}^{x}S)$ of the form

$$\zeta^r = \kappa_x(b^r \otimes \xi^r) \qquad (b^r \in B_x;\ \xi^r \in X(S)) . \qquad \dots(24)$$

Passing to a subnet (and recalling §6(6), (7)), we can find vectors

ξ_i^r in $X(S^i)$ such that for $r,s = 1,\cdots,n$ and $a \in A$

$$\langle \xi_i^r, \xi_i^s \rangle \xrightarrow[i]{} \langle \xi^r, \xi^s \rangle , \qquad \ldots(25)$$

$$\langle S_a^i \xi_i^r, \xi_i^s \rangle \xrightarrow[i]{} \langle S_a \xi^r, \xi^s \rangle . \qquad \ldots(26)$$

Now pick continuous cross-sections γ^r of $\underset{\sim}{B}$ such that $\gamma^r(x) = b^r$ $(r = 1,\cdots,n)$, and set

$$\zeta_i^r = \kappa_{x^i}(\gamma^r(x^i) \otimes \xi_i^r) \in X({}^{x^i}(S^i)) .$$

From (26) and the fact that

$$\|\gamma^s(x^i)^* \gamma^r(x^i) - (b^s)^* b^r\| \xrightarrow[i]{} 0$$

we obtain

$$|\langle S^i_{\gamma^s(x^i)^*\gamma^r(x^i)} \xi_i^r, \xi_i^s \rangle - \langle S_{(b^s)^* b^r} \xi^r, \xi^s \rangle|$$

$$\leq |\langle S^i_{(b^s)^* b^r} \xi_i^r, \xi_i^s \rangle - \langle S_{(b^s)^* b^r} \xi^r, \xi^s \rangle|$$

$$+ |\langle S^i_{(\gamma^s(x^i)^*\gamma^r(x^i) - (b^s)^* b^r)} \xi_i^r, \xi_i^s \rangle|$$

$$\xrightarrow[i]{} 0 ,$$

whence

$$\langle \zeta_i^r, \zeta_i^s \rangle = \langle S^i_{\gamma^s(x^i)^*\gamma^r(x^i)} \xi_i^r, \xi_i^s \rangle$$

$$\xrightarrow[i]{} \langle S_{(b^s)^* b^r} \xi^r, \xi^s \rangle = \langle \zeta^r, \zeta^s \rangle . \qquad \ldots(27)$$

Further, if $a \in A$, a similar argument yields

$$\langle ({}^{x^i}(S^i))_a \zeta_i^r, \zeta_i^s \rangle = \langle S^i_{\gamma^s(x^i)^* a \gamma^r(x^i)} \xi_i^r, \xi_i^s \rangle$$

$$\xrightarrow[i]{} \langle S_{(b^s)^* a b^r} \xi^r, \xi^s \rangle$$

$$= \langle ({}^{x}S)_a \zeta^r, \zeta^s \rangle . \qquad \ldots (28)$$

Now suppose that U is a regional neighborhood of ${}^{x}S$ of the form §6(1), using vectors in $X({}^{x}S)$ of the form (24). It follows from (27) and (28) that ${}^{x^i}(S^i)$ is i-eventually in U. By Prop. 6.1 such U form a basis of regional neighborhoods of ${}^{x}S$. So, as we observed earlier, it has been established that the original net $\{{}^{x^i}(S^i)\}$ converges to ${}^{x}S$. □

It follows from Props. 33.8 and 33.15 that G acts (continuously), by the operation of conjugation, as a left topological transformation group on the intersection of $\underset{\sim}{S}$ with the structure space of B_e (that is, on the space of all unitary equivalence classes of $\underset{\sim}{B}$-positive irreducible *-representations of B_e).

A class of examples.

To conclude this work we will sketch the construction of an interesting class of saturated C^*-algebraic bundles, and determine explicitly the action of their base group by conjugation on the structure space of their unit fiber *-algebra. Many routine verifications will be left to the reader.

This construction generalizes the finite-dimensional Example VII of §12.

Fix a locally compact Hausdorff space M and a non-zero Hilbert space X; and for each m in M let a non-zero closed linear subspace

Y_m of X be given. We shall assume that the map $m \mapsto Y_m$ is "upper-semicontinuous" in the following sense: For each m in M and ξ in Y_m, there is a continuous function $\varphi: M \to X$ such that (i) $\varphi(r) \in Y_r$ for all r in M, and (ii) $\varphi(m) = \xi$. Let $Y = \{(\xi,m) \in X \times M: \xi \in Y_m\}$. The above assumption implies (in fact by Theorem 10.5 is equivalent to) the openness of $\pi|Y: Y \to M$, where $\pi: X \times M \to M$ is the projection map $(\xi,m) \mapsto m$. Thus $\underset{\sim}{Y} = (Y,\{Y_m\})$, equipped with the relativized topology, norm, and operations of the trivial bundle $X \times M$ over M, becomes a Hilbert bundle over M.

For each (m,r) in $M \times M$, put

$$D_{m,r} = \{\alpha \in \underset{\sim}{O}_c(X): \alpha(Y_r^{\perp}) = \{0\},\ \text{range}(\alpha) \subset Y_m\}.$$

We have

$$D_{m,r} D_{p,q} \subset D_{m,q},$$

$$(D_{m,r})^* = D_{r,m}.$$

It will often (but not always) be convenient to identify elements of $D_{m,r}$ with maps $\alpha: Y_r \to Y_m$ (neglecting the part $Y_r^{\perp}$ of X which is annihilated by α).

Let $D = \{(\alpha,m,r) \in \underset{\sim}{O}_c(X) \times M \times M: \alpha \in D_{m,r}\}$. I claim that $(m,r) \mapsto D_{m,r}$ is "upper-semicontinuous" in the same sense as $m \mapsto Y_m$. That is, I claim that $\underset{\sim}{D} = (D,\{D_{m,r}\})$ (with the relativized topology, norm, and operations of the trivial bundle $\underset{\sim}{O}_c(X) \times M \times M$ over $M \times M$) is a Banach bundle over $M \times M$.

To see this, it is enough to show that, given any α in $D_{m,r}$, there is a continuous map $F: M \times M \to \underset{\sim}{O}_c(X)$ such that (i) $F(p,q) \in D_{p,q}$ for all (p,q) in $M \times M$, and (ii) $F(m,r) = \alpha$. By linearity and the denseness of $\underset{\sim}{O}_F(X)$ in $\underset{\sim}{O}_c(X)$, it is enough to assume that α is of rank 1 and hence of the form

$$\alpha(\zeta) = \langle\zeta,\eta\rangle\xi \qquad (\zeta \in X),$$

where $\xi \in Y_m$ and $\eta \in Y_r$. Choose elements φ and ψ of $\underset{\sim}{C}(\underset{\sim}{Y})$ such that $\varphi(m) = \xi$, $\psi(r) = \eta$. Then the map $F\colon M \times M \mapsto \underset{\sim}{O}_c(X)$ given by

$$F(p,q)(\zeta) = \langle\zeta,\psi(q)\rangle\varphi(p)$$

$(p,q \in M;\ \zeta \in X)$ has the required properties. So $\underset{\sim}{D}$ is a Banach bundle over $M \times M$.

Now let G be a fixed locally compact group acting as a topological transformation group on M (the action being denoted as usual by $(x,m) \mapsto xm$). Let $\underset{\sim}{E}$ be the Banach bundle retraction of $\underset{\sim}{D}$ by the continuous map $\Phi\colon (x,m) \mapsto (m,x^{-1}m)$ of $G \times M$ into $M \times M$. So $\underset{\sim}{E}$ is a Banach bundle over $G \times M$ whose fiber over (x,m) is $D_{m,x^{-1}m}$. For each x in G let $\underset{\sim}{E}^x$ be the Banach bundle retraction of $\underset{\sim}{E}$ by the homeomorphism $m \mapsto (x,m)$ of M into $G \times M$; and let B_x be the Banach space $\underset{\sim}{C}_0(\underset{\sim}{E}^x)$ (with the supremum norm). Thus B_x consists of all f in $\underset{\sim}{C}_0(M;\underset{\sim}{O}_c(X))$ such that $f(m) \in D_{m,x^{-1}m}$ for each m. Next we define B as the disjoint union $\cup_{x\in G}(B_x \times \{x\})$ of the B_x $(x \in G)$, and give to B the unique topology (see Prop. 10.4) such that (a) $\underset{\sim}{B} = (B,\{B_x\})$ is a Banach bundle, and (b) for each F in $\underset{\sim}{L}(\underset{\sim}{E})$ the corresponding cross-section $x \mapsto (m \mapsto F(x,m))$ of $\underset{\sim}{B}$ is continuous. The topology of B has in fact a simpler description: It is the relativized topology which B inherits as a subset of $P = \underset{\sim}{C}_0(M;\underset{\sim}{O}_c(X)) \times G$. (This fact is non-trivial. It results from the openness of the restriction to B of the projection $P \to G$; and this in turn can be proved by applying Tietze's Extension Theorem in $\underset{\sim}{E}$.)

Now $\underset{\sim}{C}_0(M;\underset{\sim}{O}_c(X))$ is a C^*-algebra (under the pointwise operations and supremum norm); and the action of G on M generates an action ι of G by *-automorphisms on $\underset{\sim}{C}_0(M;\underset{\sim}{O}_c(X))$:

$$\iota_x(f)(m) = f(x^{-1}m)$$

$(x \in G;\ f \in \underset{\sim}{C}_0(M;\underset{\sim}{O}_c(X));\ m \in M)$. Let $\underset{\sim}{P}$ be the ι-semidirect product C^*-algebraic bundle $\underset{\sim}{C}_0(M;\underset{\sim}{O}_c(X)) \underset{\iota}{\times} G$, with bundle space P . The multiplication and *-operation in $\underset{\sim}{P}$ are thus given by

$$(f,x)(g,y) = (f\iota_x(g), xy) ,$$
$$(f,x)^* = (\iota_{x^{-1}}(f^*), x^{-1})$$

$(f,g \in \underset{\sim}{C}_0(M;\underset{\sim}{O}_c(X));\ x,y \in G)$. One verifies without difficulty that B , as a subset of P , is stable under these operations. Therefore, equipped with the restriction of these operations to B , $\underset{\sim}{B}$ becomes a C^*-algebraic bundle. (The continuity of these operations on B results from their continuity on P and the last statement of the preceding paragraph.)

Let us denote the unit fiber C^*-algebra B_e of $\underset{\sim}{B}$ by A . Thus A is the C^*-subalgebra of $\underset{\sim}{C}_0(M;\underset{\sim}{O}_c(X))$ consisting of those f such that $f(m) \in \underset{\sim}{O}_c(Y_m)$ $(\subset \underset{\sim}{O}_c(X))$ for all m . For each m in M , $S^{(m)}: f \mapsto f(m)$ $(f \in A)$ is an irreducible *-representation of A . In fact it is well known (see for example Fell [2], Theorem 1.1 and Corollary of Theorem 1.2) that under these circumstances $m \mapsto S^{(m)}$ is a homeomorphism of M onto $\hat{A}$.

If we regard the homeomorphism $m \mapsto S^{(m)}$ as an identification of M with $\hat{A}$, the original action of G on M becomes an action of G on $\hat{A}$.

Since each Y_m is non-zero, it is easy to see that $\underset{\sim}{B}$ is saturated. We may refer to $\underset{\sim}{B}$ as <u>the saturated</u> C^*-<u>algebraic bundle canonically constructed from the ingredients</u> G, M, <u>and</u> $\{Y_m\}$.

Finally, we assert the following proposition:

<u>Proposition 33.16</u>. <u>If</u> $x \in G$ <u>and</u> $m \in M$, <u>then</u> $S^{(xm)}$ <u>is unitarily equivalent to the</u> x-<u>conjugate of</u> $S^{(m)}$ (<u>with respect to</u> $\underset{\sim}{B}$). <u>That is to say, if</u> $m \mapsto S^{(m)}$ <u>is regarded as an identification, the action of</u> G <u>on</u> $\hat{A}$ <u>by conjugation</u> (<u>in</u> $\underset{\sim}{B}$) <u>coincides with the original action of</u> G <u>on</u> M.

<u>Sketch of Proof</u>. Let S' be the x-conjugate of $S^{(m)}$. The unitary equivalence between S' and $S^{(xm)}$ is realized by the isometry $f \otimes \xi \mapsto f(xm)\xi$ ($f \in B_x$; $\xi \in Y_m$). □

<u>Remark 8</u>. The class of examples worked out above shows that the conjugation operation in a Banach *-algebraic bundle $\underset{\sim}{B}$ has a great deal more latitude when $\underset{\sim}{B}$ is merely saturated than in the more special case when $\underset{\sim}{B}$ has an approximate unit and enough unitary multipliers (see Prop. 11.4). Indeed, in the latter case, Prop. 33.11 shows that the action of a group element x by conjugation on *-representations of B_e is derived from the *-automorphism $b \mapsto ubu^*$ of B_e (u being a unitary multiplier of order x); in particular, for any $\underset{\sim}{B}$-positive *-representation S of B_e, S and ${}^xS$ have the same Hilbert dimension. However, in the preceding class of examples, the spaces Y_m and Y_{xm} of $S^{(m)}$ and ${}^x(S^{(m)})$ will in general be of different dimension; one might well be of finite and the other of infinite dimension. This shows in particular that, if $\underset{\sim}{B}$ is merely saturated, the action of G by conjugation on *-representations of B_e need not be derived from

any underlying action of G by *-automorphisms on B_e.

Remark 9. Let A be any C^*-algebra of the sort which could arise as the unit fiber *-algebra of one of the saturated C^*-algebraic bundles $\underset{\sim}{B}$ occurring in the above class of examples. Thus A must be CCR (liminaire) and must have a Hausdorff structure space. A could be any C^*-subalgebra of the C^*-algebra of compact operators on a Hilbert space, in particular any finite-dimensional C^*-algebra. Now according to the above construction, given any continuous action of G on $M \cong \hat{A}$, there is a natural way to build a saturated C^*-algebraic bundle $\underset{\sim}{B}$ over G such that (i) A is the unit fiber *-algebra of $\underset{\sim}{B}$, and (ii) the action of G on $\hat{A}$ by conjugation in $\underset{\sim}{B}$ is precisely the given action. This leads us to the following important conjecture: Let A be any C^*-algebra whatsoever, G a locally compact group, and α any continuous left action of G on the structure space $\hat{A}$ of A. Then there exists a saturated C^*-algebraic bundle $\underset{\sim}{B}$ over G, with unit fiber *-algebra A, such that the action of G on $\hat{A}$ by conjugation in $\underset{\sim}{B}$ is exactly α.

At present we do not know whether this is true or not.

Remark 10. If the above conjecture is true, it suggests future applications of saturated C^*-algebraic bundles to physics. Indeed, suppose that A is a C^*-algebra whose irreducible *-representations correspond to possible elementary particles; and suppose that $\hat{A}$ is acted upon by a group G in such a way that the orbits in $\hat{A}$ under G are exactly the "clusters" of elementary particles which we wish to regard as "states" of some "super-particle". Then the saturated C^*-algebraic bundle constructed from A and G as in the above conjecture (or rather its cross-sectional algebra) might well serve as the "extension" of A appropriate for describing transitions from one elementary particle state to another.

This approach might possibly be flexible enough to obviate the difficulties that arise from such theorems as that of O'Raifeartaigh [1] (see also Segal [1]) on the non-existence of group extensions adequate to cope with the combination of space-time symmetries and internal symmetries.

APPENDIX

THE EXISTENCE OF CONTINUOUS CROSS-SECTIONS OF BANACH BUNDLES

In this Appendix we present, with their permission, the following remarkable unpublished result of A. Douady and L. dal Soglio-Hérault [1]: <u>A Banach bundle over a base space</u> X <u>has enough continuous cross-sections whenever</u> X <u>is either paracompact or locally compact</u>. The proof presented below is theirs (with some modifications of style).

Let $\underset{\sim}{B} = (B,\pi)$ be a fixed Banach bundle over the Hausdorff space X. In addition to keeping the notation of the definition of a Banach bundle in §10, we will write D for $\{(b,c) \in B \times B : \pi(b) = \pi(c)\}$. If $U \subset X$ and $\varepsilon > 0$, we define $B(U,\varepsilon) = \{b \in B: \pi(b) \in U, \|b\| < \varepsilon\}$. Postulate (iv) of the definition of a Banach bundle says that, if ε runs over all positive numbers and U runs over a basis of neighborhoods of a point x of X, then the $B(U,\varepsilon)$ run over a basis of neighborhoods of 0_x in B.

As a first step toward constructing continuous cross-sections of $\underset{\sim}{B}$, we shall construct cross-sections whose discontinuities are "small". To see what this means, we make a definition.

<u>Definition</u>. Let ε be a positive number. A subset U of B is ε-<u>thin</u> if $\|b-b'\| < \varepsilon$ whenever $b,b' \in U$ and $\pi(b) = \pi(b')$.

<u>Proposition 1</u>. <u>If</u> $\varepsilon > 0$ <u>and</u> $b \in B$, <u>then</u> b <u>has an</u> ε-<u>thin neighborhood</u>.

<u>Proof</u>. Since $\sigma: (b',b'') \mapsto b' - b''$ is continuous on D to B, there is a D-neighborhood W of the pair (b,b) such that

$$\sigma(W) \subset B(X,\varepsilon) \; . \qquad \ldots(1)$$

Since D carries the relativized topology of $B \times B$, we may as well assume that

$$W = (U \times U) \cap D \; , \qquad \ldots(2)$$

where U is a neighborhood of b in B. The combination of (1) and (2) shows that U is ε-thin. □

<u>Proposition 2</u>. <u>Let</u> b <u>be an element of</u> B ; <u>and let</u> $\{U_i\}$ <u>be a decreasing net of neighborhoods of</u> b <u>with the following two properties</u>: (i) $\{\pi(U_i)\}$ <u>shrinks to</u> $\pi(b)$ (<u>that is,</u> $\{\pi(U_i)\}$ <u>is a basis of neighborhoods of</u> $\pi(b)$); (ii) U_i <u>is</u> ε_i<u>-thin, where</u> $\{\varepsilon_i\}$ <u>is a net of positive numbers such that</u> $\lim_i \varepsilon_i = 0$. <u>Then</u> $\{U_i\}$ <u>is a basis of neighborhoods of</u> b .

<u>Proof</u>. Put $x = \pi(b)$, $V_i = \pi(U_i)$. Thus the $B(V_i,\varepsilon_i)$ form a basis of neighborhoods of O_x .

Now let W be any neighborhood of b . Since $\rho\colon (b',b'') \mapsto b' + b''$ is continuous on D , in particular at (b,O_x) , we can find a neighborhood W' of b and an index i such that

$$\rho[(W' \times B(V_i,\varepsilon_i)) \cap D] \subset W \; . \qquad \ldots(3)$$

Noting that $\pi(W' \cap U_i)$ is a neighborhood of x , we next choose an index $j \succ i$ such that

$$V_j \subset \pi(W' \cap U_i) \; . \qquad \ldots(4)$$

I now claim that

$$U_j \subset W \; . \qquad \ldots(5)$$

Indeed: Since $j \succ i$ and the $\{U_k\}$ are decreasing, (4) gives

$$U_j \subset U_i \cap \pi^{-1}\pi(W' \cap U_i) \ . \qquad \ldots(6)$$

Let $c \in U_j$. By (6) $c \in U_i$ and there exists an element c' of $W' \cap U_i$ such that $\pi(c') = \pi(c)$. Since U_i is ε_i-thin, this says that $\|c-c'\| < \varepsilon_i$. So $c-c' \in B(V_i, \varepsilon_i)$, and $c = c' + (c-c') \in \rho[(W' \times B(V_i, \varepsilon_i)) \cap D] \subset W$ by (3). Since c was any element of U_j , we have proved (5). Thus, by the arbitrariness of W , $\{U_i\}$ is a basis of neighborhoods of b . □

<u>Remark</u>. The above proposition fails if we omit the hypothesis that the $\{U_i\}$ are decreasing. Consider for example the simple Banach bundle $(\mathbb{C} \times \mathbb{R}, \pi)$ over $\mathbb{R}$, where $\pi(z,t) = t$. One can easily construct a sequence $\{U_n\}$ of neighborhoods of $(0,0)$ such that U_n is n^{-1}-thin and $\{\pi(U_n)\}$ shrinks down to 0 , while the image of U_n under the projection $(z,t) \mapsto z$ contains $\{z: |z| \leq 1\}$ for all n . Of course the $\{U_n\}$ will not be decreasing in n .

We are now ready to define what it means for a cross-section to have only "small" discontinuities.

<u>Definition</u>. Let $\varepsilon > 0$. A cross-section f of $\underset{\sim}{B}$ is called ε-<u>continuous at</u> a point x of X if there is a neighborhood V of x and an ε-thin neighborhood U of $f(x)$ such that $f(V) \subset U$. If f is ε-continuous at all points of X , it is called simply ε-<u>continuous</u>.

Taking U and V to be open and replacing U by $U \cap \pi^{-1}(V)$, we can rephrase the above definition as follows: f is ε-continuous at x if and only if there is an open ε-thin neighborhood U of $f(x)$ such that $f(\pi(U)) \subset U$.

<u>Proposition 3</u>. <u>If</u> $\varepsilon > 0$ <u>and</u> f <u>is a cross-section of</u> $\underset{\sim}{B}$ <u>which is</u>

ε-continuous at a point x of X , then y $\mapsto$ $\|f(y)\|$ is bounded on some neighborhood of x .

Proof. Choose U and V as in the definition of ε-continuity; and set

$$U' = \{b \in U: \|b\| < \|f(x)\| + 1\} ,$$

$$V' = \pi(U') \cap V .$$

Since U′ is an open neighborhood of f(x) , V′ is an open neighborhood of x . Let $y \in V'$. Then $f(y) \in U$ (since $V' \subset V$); and there is an element b of U′ such that $\pi(b) = y$. Since $U' \subset U$ and U is ε-thin, this implies that $\|f(y) - b\| < \varepsilon$. Also $\|b\| < \|f(x)\| + 1$ (since $b \in U'$). Therefore $\|f(y)\| \leq \|b\| + \|f(y) - b\| < \|f(x)\| + 1 + \varepsilon$; and this is true for all y in the neighborhood V′ of x . □

Proposition 4. Let f be a cross-section of $\underset{\sim}{B}$ which is ε-continuous for all $\varepsilon > 0$. Then f is continuous.

Proof. Fix a point x of X . For each positive integer n let U_n be an n^{-1}-thin open neighborhood of f(x) such that

$$f(\pi(U_n)) \subset U_n . \qquad \qquad \ldots (7)$$

We notice that $U_n' = U_n \cap U_{n-1}$ is also n^{-1}-thin and satisfies (7). Indeed, to see that U_n' satisfies (7), observe that $y \in \pi(U_n') \Rightarrow y \in$ $\pi(U_n)$ and $y \in \pi(U_{n-1}) \Rightarrow f(y) \in U_n \cap U_{n-1} = U_n'$. Thus, replacing U_n by $U_1 \cap \cdots \cap U_n$, we may as well assume from the beginning that the $\{U_n\}$ are decreasing. From this and Proposition 2 it follows that the $U_n \cap \pi^{-1}(V)$, where $n = 1, 2, \cdots$ and V is an X-neighborhood of x , form a basis of neighborhoods of f(x) .

Let W be any neighborhood of f(x) . By the preceding paragraph,

$$U_n \cap \pi^{-1}(V) \subset W \qquad \ldots(8)$$

for some $n = 1,2,\cdots$ and some open neighborhood V of x. Thus, if $y \in V \cap \pi(U_n)$, it follows from (7) and (8) that $f(y) \in W$. Since $V \cap \pi(U_n)$ is a neighborhood of x, this proves the proposition. □

Lemma 5. Let V be an open subset of X; and let $U_1,\cdots,U_n$ be open subsets of B such that $\pi(U_i) = V$ for all $i = 1,\cdots,n$. Furthermore, let $\varphi_1,\cdots\varphi_n$ be continuous complex functions on X; and assume that $\varphi_1(x) \neq 0$ for all x in V. Define W to be the set of all points of B of the form

$$\sum_{i=1}^{n} \varphi_i(x)b_i \ ,$$

where $x \in V$ and $b_i \in U_i \cap B_x$ for each $i = 1,2,\cdots,n$. Then W is open in B.

Proof. Set $E = \{(b_1,\cdots,b_n) \in B^n\colon \pi(b_1) = \pi(b_2) = \cdots = \pi(b_n) \in V\}$; and define $\rho\colon E \to E$ as follows:

$$\rho(b_1,\cdots b_n) = \left(\sum_{i=1}^{n} \varphi_i(x)b_i, b_2, b_3,\cdots,b_n\right)$$

(where $x = \pi(b_1) = \cdots = \pi(b_n)$). Evidently ρ is continuous on E and has a continuous inverse ρ^{-1}:

$$\rho^{-1}(b_1,\cdots b_n) = \left(\varphi_1(x)^{-1}\left[b_1 - \sum_{i=2}^{n} \varphi_i(x)b_i\right], b_2,\cdots,b_n\right)$$

(where $x = \pi(b_1) = \cdots = \pi(b_n)$). So ρ is a homeomorphism of E onto itself.

Furthermore, let us define $\beta\colon E \to \pi^{-1}(V)$ and $\gamma\colon E \to \pi^{-1}(V)$:

$$\beta(b_1,\cdots,b_n) = \sum_{i=1}^{n} \varphi_i(x) b_i \qquad (x = \pi(b_1) = \cdots = \pi(b_n)) ,$$

$$\gamma(b_1,\cdots,b_n) = b_1 .$$

Notice that $\gamma: E \longrightarrow \pi^{-1}(V)$ is an open surjection. Also $\gamma \circ \rho = \beta$. Since ρ is a homeomorphism, it follows that β is also open on E. Now $W = \beta((U_1 \times \cdots \times U_n) \cap E)$. Therefore W is open. □

Proposition 6. *Let $\varepsilon > 0$ and $x \in X$. Let $\varphi_1,\cdots,\varphi_n$ be continuous non-negative real functions on X such that $\Sigma_{i=1}^{n} \varphi_i(y) \equiv 1$. Let $f_1,\cdots,f_n$ be cross-sections of $\mathcal{B}$ which are ε-continuous at x. Then $f = \Sigma_{i=1}^{n} \varphi_i f_i$ is ε-continuous at x.*

Proof. Since $\varphi_i(x) \neq 0$ for some i, we may as well assume that $\varphi_1(x) \neq 0$, and then restrict attention to an open neighborhood of x on which φ_1 is never 0.

For each i we take an open ε-thin neighborhood U_i of $f_i(x)$ such that

$$f_i(\pi(U_i)) \subset U_i . \qquad \ldots(9)$$

Cutting down the U_i, we may assume that the $\pi(U_i)$ are all the same open neighborhood V of x. Using these U_i, we define W as in Lemma 5. By Lemma 5 W is an open neighborhood of $f(x)$. Further, if $b,b' \in W$ and $\pi(b) = \pi(b') = y$, we can write

$$b = \sum_{i=1}^{n} \varphi_i(y) b_i ,$$

$$b' = \sum_{i=1}^{n} \varphi_i(y) b_i' ,$$

where $b_i, b_i' \in U_i \cap B_y$; and hence $\|b-b'\| \leq \Sigma_{i=1}^{n} \varphi_i(y) \|b_i - b_i'\| <$

$\sum_{i=1}^{n} \varphi_i(y)\varepsilon = \varepsilon$ (since $\Sigma_i \varphi_i \equiv 1$ and U_i is ε-thin). Therefore W is ε-thin. From (9) it follows that $f(V) \subset W$. Consequently f is ε-continuous at x . □

Recall that X is <u>completely regular</u> if, given a neighborhood U of a point x of X , there is a continuous real function φ on X such that $\varphi(x) = 1$ and $\varphi \equiv 0$ outside U .

<u>Proposition 7</u>. <u>Assume that</u> X <u>is completely regular</u>. <u>Given</u> $\varepsilon > 0$ <u>and</u> $b \in B$, <u>there exists an</u> ε-<u>continuous cross-section</u> f <u>of</u> $\underset{\sim}{B}$ <u>such that</u> $f(\pi(b)) = b$.

<u>Proof</u>. By Prop. 1 b has an open ε-thin neighborhood U . Put $x = \pi(b)$, $V = \pi(U)$. Now there obviously exists a cross-section g of the reduced bundle $\underset{\sim}{B}_V$ such that $g(y) \in U$ for all y in V and $g(x) = b$. By the complete regularity of X there is an open neighborhood W of x such that $\bar{W} \subset V$, and a continuous function $\varphi: X \to [0,1]$ such that $\varphi(x) = 1$ and $\varphi \equiv 0$ outside $\bar{W}$. Now define

$$f(y) = \begin{cases} \varphi(y)g(y) & \text{for } y \in V , \\ 0_y & \text{for } y \notin V . \end{cases}$$

Since g is ε-continuous on V , it follows from Prop. 6 that f $(= \varphi g + (1-\varphi)0)$ is ε-continuous on V . Being identically 0 on $X-\bar{W}$, f is trivially ε-continuous on $X-\bar{W}$. So f is ε-continuous on X . Evidently $f(x) = b$. □

We recall that the topological space X is <u>paracompact</u> if it is Hausdorff and every open covering of X has a locally finite open refinement (see Bourbaki [1], Chap. I, §9, no. 10). Every paracompact space is normal (see Bourbaki [2], Chap. IX, §4, no. 4), hence completely regular.

We come now to the crucial step in the development of this appendix.

__Proposition 8__. __Assume that__ X __is paracompact__. __Suppose that__ $\varepsilon > 0$ __and__ $x_0 \in X$; __and let__ f __be an__ ε-__continuous cross-section of__ $\underset{\sim}{B}$. __Then there exists an__ $\frac{\varepsilon}{2}$-__continuous cross-section__ f' __of__ $\underset{\sim}{B}$ __satisfying__:

$$\|f'(x)-f(x)\| \leq \frac{3}{2}\varepsilon \quad \text{\underline{for all} } x \text{ \underline{in} } X, \qquad \ldots(10)$$

$$f'(x_0) = f(x_0) . \qquad \ldots(11)$$

__Proof__. For the moment we fix any element x of X . By Prop. 7 there is a $\frac{\varepsilon}{2}$-continuous cross-section $\tilde{f}$ of $\underset{\sim}{B}$ such that $\tilde{f}(x) = f(x)$. I claim that there exists a neighborhood U of x such that

$$\|\tilde{f}(y)-f(y)\| < \frac{3}{2}\varepsilon \quad \text{for all } y \text{ in } U . \qquad \ldots(12)$$

Indeed: Choose an open ε-thin neighborhood V of $f(x)$ such that $f(\pi(V)) \subset V$, and an open $\frac{\varepsilon}{2}$-thin neighborhood W of $\tilde{f}(x) = f(x)$ such that $\tilde{f}(\pi(W)) \subset W$. Let $U = \pi(V \cap W)$. If $y \in U$, there exists an element b of $V \cap W$ such that $\pi(b) = y$; and thus $\|f(y)-b\| < \varepsilon$ and $\|\tilde{f}(y)-b\| < \frac{1}{2}\varepsilon$, giving (12). This proves the claim.

By the above claim we can find an open covering $\{U_i\}$ $(i \in I)$ of X and for each i an $\frac{\varepsilon}{2}$-continuous cross-section g_i of $\underset{\sim}{B}$ satisfying

$$\|g_i(y) - f(y)\| < \frac{3}{2}\varepsilon \quad \text{for all } y \text{ in } U_i . \qquad \ldots(13)$$

Suppose that the point x_0 belongs to some fixed U_k . From the preceding construction we can suppose that

$$g_k(x_0) = f(x_0) . \qquad \ldots(14)$$

Choose an open neighborhood Z of x_0 with $\bar{Z} \subset U_k$; and replace each U_i with $i \neq k$ by $U_i - \bar{Z}$. The result is an open covering with the same properties as before, and with the additional property that

$$x_0 \notin \bar{U}_i \quad \text{for} \quad i \neq k . \qquad \ldots(15)$$

Now by Bourbaki [2], Chap. IX, §4, no. 4, Corollary 1, there is a continuous partition of unity $\{\varphi_i\}$ $(i \in I)$, subordinate to $\{U_i\}$. We can therefore form the cross-section

$$f' = \sum_{i \in I} \varphi_i g_i .$$

Proposition 6, applied to a small neighborhood of each point on which only finitely many of the φ_i do not vanish, shows that f' is $\frac{\varepsilon}{2}$-continuous.

Given x in X , we have

$$\|f'(x) - f(x)\| = \|\sum_{i \in I} \varphi_i(x)(g_i(x) - f(x))\|$$
$$\leq \sum_{i \in I} \varphi_i(x) \|g_i(x) - f(x)\| . \qquad \ldots(16)$$

Now for each index i , either $x \notin U_i$, in which case $\varphi_i(x) = 0$, or $x \in U_i$, in which case $\|g_i(x) - f(x)\| < \frac{3}{2}\varepsilon$ by (13). So for all i $\varphi_i(x)\|g_i(x) - f(x)\| \leq \frac{3}{2}\varepsilon\,\varphi_i(x)$; and (16) gives $\|f'(x) - f(x)\| \leq \sum_i \frac{3}{2}\varepsilon\,\varphi_i(x)$ $= \frac{3}{2}\varepsilon$, which is (10).

Finally, in view of (14) and (15) we have $\varphi_k(x_0) = 1$ and $f'(x_0)$ $= g_k(x_0) = f(x_0)$; and this is (11). □

Two more simple propositions complete the preparation for the main theorem.

<u>Proposition 9</u>. <u>Let</u> $\varepsilon > 0$; <u>and let</u> $\{f_n\}$ <u>be a sequence of</u> ε-<u>continuous cross-sections of</u> $\underset{\sim}{B}$ <u>converging uniformly on</u> X <u>to a cross-section</u> f . <u>Then</u> f <u>is</u> ε'-<u>continuous for every</u> $\varepsilon' > \varepsilon$.

<u>Proof</u>. Let $\varepsilon' > \varepsilon$. By hypothesis there is a positive integer n such that

$$\|f_n(x) - f(x)\| < \tfrac{1}{2}(\varepsilon'-\varepsilon) \quad \text{for all } x . \qquad \dots(17)$$

Fix $x_0 \in X$; and choose an open ε-thin neighborhood U of $f_n(x_0)$ such that

$$f_n(V) \subset U , \text{ where } V = \pi(U) . \qquad \dots(18)$$

Now set $W = U + B(V,\frac{1}{2}(\varepsilon'-\varepsilon)) = \{b+c\colon (b,c) \in D,\ b \in U,\ c \in B(V,\frac{1}{2}(\varepsilon'-\varepsilon))\}$. By Lemma 5 W is an open neighborhood of $f_n(x_0)$. Also W is ε'-thin; for if $b,c \in U$ and $u,v \in B(V,\frac{1}{2}(\varepsilon'-\varepsilon))$, where b,c,u,v are all in B_x $(x \in V)$, then

$$\begin{aligned}\|(b+u)-(c+v)\| &\le \|b-c\| + \|u-v\| \\ &< \varepsilon + 2\cdot\tfrac{1}{2}(\varepsilon'-\varepsilon) = \varepsilon' .\end{aligned}$$

Further, if $x \in V$ we have

$$\begin{aligned} f(x) &= f_n(x) + (f(x)-f_n(x)) \\ &\in U + B(V,\tfrac{1}{2}(\varepsilon'-\varepsilon)) \qquad \text{(by (17),(18)} \\ &= W . \end{aligned}$$

Thus $f(V) \subset W$; and f is ε'-continuous at x_0 for all x_0 in X.□

Proposition 10. Let $\{f_n\}$ *be a sequence of cross-sections of* $\mathbf{B}$ *such that* (i) $\{f_n\}$ *converges uniformly on* X *to a cross-section* f *of* $\mathbf{B}$ *, and* (ii) f_n *is* ε_n*-continuous, where* $\lim_{n\to\infty} \varepsilon_n = 0$ *. Then the cross-section* f *is continuous.*

Proof. By Prop. 9 f is ε-continuous for every $\varepsilon > 0$. So by Prop. 4 f is continuous. □

We are now ready for the main results.

Theorem 11. Assume that X *is paracompact. Then* $\mathbf{B}$ *has enough continuous cross-sections.*

Proof. Given b_0 in B , we must find a continuous cross-section f of $\mathbf{B}$ such that $f(x_0) = b_0$, where $x_0 = \pi(b_0)$.

Let f_0 be the cross-section of $\mathbf{B}$ defined by: $f_0(x_0) = b_0$, $f_0(x) = 0_x$ for $x \neq x_0$. Thus f_0 is ε_0-continuous, where $\varepsilon_0 = 2\|b_0\|+1$. Now set $\varepsilon_n = 2^{-n}\varepsilon_0$. Applying Prop. 8 repeatedly, we obtain a sequence $\{f_n\}$ of cross-sections of $\mathbf{B}$ with the following three properties: (i) f_n is ε_n-continuous for all n ; (ii) $\|f_{n+1}(x) - f_n(x)\| \leq \frac{3}{2}\varepsilon_n$ for all x and all n ; (iii) $f_n(x^0) = b_0$ for all n . It follows from (ii) that $\{f_n\}$ converges uniformly on X to some cross-section f of $\mathbf{B}$. By (i) and Prop. 10 f is continuous. By (iii) $f(x_0) = b_0$. Thus the required continuous cross-section has been constructed. □

Theorem 12. Assume that X *is locally compact. Then* $\mathbf{B}$ *has enough continuous cross-sections.*

Proof. If X is compact it is paracompact, and so Theorem 11 is applicable. Assume then that X is not compact; and let X_0 be the one-point compactification of X (the adjoined point being called ∞). Since X_0 is paracompact it will be enough by Theorem 11 to show that $\underset{\sim}{B}$ is the reduction to X of some Banach bundle $\underset{\sim}{B}^0$ over X_0.

We shall make the fiber B^0_∞ of $\underset{\sim}{B}^0$ over ∞ to be zero-dimensional: $B^0_\infty = \{0_\infty\}$ (where $0_\infty \notin B$). Put $B^0 = B \cup \{0_\infty\}$; and give to B^0 that Hausdorff topology such that (i) the topology of B^0 relativized to B is just that of B, and (ii) a net $\{b_i\}$ of elements of B converges to 0_∞ in B^0 if and only if $\pi(b_i) \to \infty$ in X_0 and $\|b_i\| \to 0$. If we define $\pi^0: B^0 \to X_0$ to coincide with π on B and to send 0_∞ into ∞, we see without difficulty that (B^0, π^0) is a Banach bundle over X_0 whose reduction to X is $\underset{\sim}{B}$. □

BIBLIOGRAPHY

V. Bargmann [1], "Irreducible unitary representations of the Lorentz group," Ann. Math. 48 (1947), 548-640.

J. G. Bennett [1], "Induced representations and positive linear maps of C*-algebras," Thesis, Washington University, 1976.

R. J. Blattner [1], "On induced representations," Amer. J. Math. 83 (1961), 79-98.

______________ [2], "On induced representations II. Infinitesimal induction," Amer. J. Math. 83 (1961), 499-512.

______________ [3], "Positive definite measures," Proc. Amer. Math. Soc. 14 (1963), 423-428.

______________ [4], "Group extension representations and the structure space," Pac. J. Math. 15 (1965), 1101-1113.

N. Bourbaki [1], "General Topology," Part I, Addison-Wesley, 1966.

___________ [2), "General Topology," Part II, Addison-Wesley, 1966.

___________ [3], Livre VI, "Intégration," Chaps. 1-4, 2nd Edn., Hermann, Paris, 1965.

___________ [4], Livre VI, "Intégration," Chaps. 7-8, Hermann, Paris, 1963.

I. D. Brown [1], "Representations of finitely generated nilpotent groups," Pac. J. Math. 45 (1973), 13-26.

R. C. Busby [1], "On the equivalence of twisted group algebras and Banach *-algebraic bundles," Proc. Amer. Math. Soc. 37 (1973), 142-148.

R. C. Busby and H. A. Smith [1], "Representations of twisted group algebras," Trans. Amer. Math. Soc. 149 (1970), 503-537.

A. H. Clifford [1], "Representations induced in an invariant subgroup," Ann. Math. 38 (1937), 533-550.

C. W. Curtis and I. Reiner [1], "Representation theory of finite groups and associative algebras," Interscience Publishers, 1962.

E. C. Dade [1], "Compounding Clifford's Theory," Ann. Math. 91 (1970), 236-290.

J. Dauns and K. H. Hofmann [1], "Representations of rings by sections," Memoir of the Amer. Math. Soc. No. 83 (1968).

J. Dixmier [1], "Les C*-algèbres et leurs représentations," Gauthier-Villars, Paris, 1964.

J. Dixmier [2], "Représentations induites holomorphes des groupes résolubles algébriques," Bull. Soc. Math. France 94 (1966), 181-206.

S. Doplicher, D. Kastler, and D. Robinson [1], "Covariance algebras in field theory and statistical mechanics," Comm. Math. Phys. 3 (1966), 1-28.

A. Douady and L. dal Soglio-Hérault [1], "Existence de sections pour un fibré de Banach au sens de Fell," unpublished manuscript.

C. M. Edwards and J. T. Lewis [1], "Twisted group algebras I, II," Comm. Math. Phys. 13 (1969), 119-141.

E. Effros and F. Hahn [1], "Locally compact transformation groups and C*-algebras," Memoir of the Amer. Math. Soc. No. 75 (1967).

J. M. G. Fell [1], "C*-algebras with smooth dual," Illinois J. Math. 4 (1960), 221-230.

____________ [2], "The structure of algebras of operator fields," Acta Math. 106 (1961), 233-280.

____________ [3], "Weak containment and induced representations of groups," Can. J. Math. 14 (1962), 237-268.

____________ [4], "Weak containment and induced representations of groups II," Trans. Amer. Math. Soc. 110 (1964), 424-447.

____________ [5], "An extension of Mackey's method to algebraic bundles over finite groups," Amer. J. Math. 91 (1969), 203-238.

____________ [6], "An extension of Mackey's method to Banach *-algebraic bundles," Memoir of the Amer. Math. Soc. No. 90 (1969).

____________ [7], "A new look at Mackey's imprimitivity theorem," Conference on Harmonic Analysis, College Park, Maryland, 1971, Lecture Notes in Mathematics, Vol. 266, Springer-Verlag, 1972; pp. 43-58.

____________ [8], "The dual spaces of C*-algebras," Trans. Amer. Math. Soc. 94 (1960), 365-403.

G. Frobenius [1], "Über Relationen zwischen den Charakteren einer Gruppe und denen ihrer Untergruppen," Sitzber. Preuss. Akad. Wiss. (1898), 501-515.

I. M. Gelfand and M. A. Naimark [1], "Unitary representations of the group of linear transformations of the line," Dokl. Akad. Nauk SSSR 55 (1947), 571-574.

________________________ [2], "Unitary representations of the Lorentz group," Izvestiya Akad. Nauk SSSR Ser. Mat. 11 (1947), 411-504.

I. M. Gelfand and M. A. Naimark [3], "Unitary representations of the classical groups," Trudy Mat. Instituta Im. V. A. Steklova 36 (1950).

J. Glimm [1], "Families of induced representations," Pac. J. Math. 12 (1962), 885-911.

________ [2], "On a certain class of operator algebras," Trans. Amer. Math. Soc. 95 (1960), 318-340.

R. Godement [1], "Théorie générale des sommes continues d'espaces de Banach," C. R. Acad. Sci. Paris 228 (1949), 1321-1323.

__________ [2], "Théorie des représentations unitaires," Ann. Math. 53 (1951), 68-124.

__________ [3], "A theory of spherical functions, I" Trans. Amer. Math. Soc. 73 (1952), 496-556.

D. G. Higman [1], "Induced and produced modules," Can. J. Math. 7 (1955), 490-508.

G. Hochschild [1], "Cohomology and representations of associative algebras," Duke Math. J. 14 (1947), 921-948.

I. Kaplansky [1], "Modules over operator algebras," Amer. J. Math. 75 (1953), 839-858.

G. P. Johnson [1], "Spaces of functions with values in a Banach algebra," Trans. Amer. Math. Soc. 92 (1959), 411-429.

J. L. Kelley, I. Namioka and co-authors [1], "Linear topological spaces," van Nostrand, 1963.

M. Landstad [1], "Duality theory for covariant systems," Thesis, University of Pennsylvania, 1974.

S. Lang [1], "Algebra," Addison-Wesley, 1965.

H. Leptin [1], "Verallgemeinerte L^1-Algebren," Math. Annalen 159 (1965), 51-76.

_________ [2], "Verallgemeinerte L^1-Algebren und projektive Darstellungen lokal kompakter Gruppen, I," Inventiones Math. 3 (1967), 257-281.

_________ [3], "Verallgemeinerte L^1-Algebren und projektive Darstellungen lokal kompakter Gruppen, II," Inventiones Math. 4 (1967), 68-86.

_________ [4], "Darstellungen verallgemeinerter L^1-Algebren," Inventiones Math. 5 (1968), 192-215.

_________ [5], "Darstellungen verallgemeinerter L^1-Algebren," Lectures on Operator Algebras (dedicated to the memory of David Topping, Tulane University Ring and Operator Theory year, 1970-1971, Vol. II) pp. 251-307, Lecture Notes in Mathematics, Springer, Vol. 247, 1972.

L. H. Loomis [1], "Positive definite functions and induced representations," Duke Math. J. 27 (1960), 569-580.

G. W. Mackey [1], "Imprimitivity for representations of locally compact groups, I," Proc. Nat. Acad. Sci. U.S.A. 35 (1949), 537-545.

__________ [2], "On a theorem of Stone and von Neumann," Duke Math. J. 16 (1949), 313-326.

__________ [3], "On induced representations of groups," Amer. J. Math. 73 (1951), 576-592.

__________ [4], "Induced representations of locally compact groups, I," Ann. Math. 55 (1952), 101-139.

__________ [5], "Unitary representations of group extensions, I," Acta Math. 99 (1958), 265-311.

__________ [6], "Infinite-dimensional group representations," Bull. Amer. Math. Soc. 69 (1963), 628-686.

O. Maréchal [1], "Champs mesurables d'espaces hilbertiens," Bull. Sci. Math. 93 (1969), 113-143.

K. Morita [1], "Duality for modules and its applications to the theory of rings with minimum condition," Tokyo Kyoiku Daigaku Sec. A 6 (1958), 83-142.

H. Moscovici [1], "Generalized induced representations," Rev. Roumaine Math. Pures et Appl. 14 (1969), 1539-1551.

M. A. Naimark [1], "Normed Rings," Trans. by L. F. Boron, P. Noordhoff, Groningen, 1964.

L. O'Raifeartaigh [1], "Mass differences and Lie algebras of finite order," Phys. Rev. Letters 14 (1965), 575-577.

W. L. Paschke [1], "Inner product modules over B*-algebras," Trans. Amer. Math. Soc. 182 (1973), 443-468.

R. T. Prosser [1], "On the ideal structure of operator algebras," Memoir of the Amer. Math. Soc. No. 45, 1963.

C. E. Rickart [1], "General theory of Banach algebras," van Nostrand, 1960.

M. A. Rieffel [1], "Induced representations of C*-algebras," Advances in Mathematics 13, No. 2 (1974), 176-257.

__________ [2], "Morita equivalence for C*-algebras and W*-algebras," J. Pure and Applied Algebra 5 (1974), 51-96.

H. H. Schaefer [1], "Topological vector spaces," Macmillan, 1966.

I. Schochetman [1], "Topology and the duals of certain locally compact groups," Trans. Amer. Math. Soc. 150 (1970), 477-489.

I. Segal [1], "An extension of a theorem of L. O'Raifeartaigh," J. Functional Analysis 1 (1967), 1-21.

W. F. Stinespring [1], "Positive functions on C^*-algebras," Proc. Amer. Math. Soc. 6 (1955), 211-216.

M. Takesaki [1], "Covariant representations of C^*-algebras and their locally compact automorphism groups," Acta Math. 119 (1967), 273-303.

__________ [2], "Duality and von Neumann algebras," Lecture Notes in Mathematics, Springer, Vol. 247 (1972), 665-786.

__________ [3], "A liminal crossed product of a uniformly hyperfinite C^*-algebra by a compact Abelian group," J. Functional Analysis 7 (1971), 140-146.

T. Turumaru [1], "Crossed product of operator algebra," Tôhoku Math. J. (2) 10 (1958), 355-365.

L. I. Vainerman and G. I. Kac [1], "Non-unimodular ring groups and Hopf-von Neumann algebras," Mat. Sbornik 94 (1974), 194-225.

H. N. Ward [1], "The analysis of representations induced from a normal subgroup," Michigan J. 15 (1968), 417-428.

A. Weil [1], "L'intégration dans les groupes topologiques et ses applications," Hermann, Paris, 1940.

E. Wigner [1], "On unitary representations of the inhomogeneous Lorentz group," Ann. Math. 40 (1939), 149-204.

G. Zeller-Meier [1], "Produits croisés d'une C^*-algèbre par un groupe d'automorphismes," C. R. Acad. Sci. Paris 263 (1966), 20-23.

INDEX OF TERMINOLOGY

Vol. 309: D. H. Sattinger, Topics in Stability and Bifurcation Theory. VI, 190 pages. 1973. DM 20,–

Vol. 310: B. Iversen, Generic Local Structure of the Morphisms in Commutative Algebra. IV, 108 pages. 1973. DM 18,–

Vol. 311: Conference on Commutative Algebra. Edited by J. W. Brewer and E. A. Rutter. VII, 251 pages. 1973. DM 24,–

Vol. 312: Symposium on Ordinary Differential Equations. Edited by W. A. Harris, Jr. and Y. Sibuya. VIII, 204 pages. 1973. DM 22,–

Vol. 313: K. Jörgens and J. Weidmann, Spectral Properties of Hamiltonian Operators. III, 140 pages. 1973. DM 18,–

Vol. 314: M. Deuring, Lectures on the Theory of Algebraic Functions of One Variable. VI, 151 pages. 1973. DM 18,–

Vol. 315: K. Bichteler, Integration Theory (with Special Attention to Vector Measures). VI, 357 pages. 1973. DM 29,–

Vol. 316: Symposium on Non-Well-Posed Problems and Logarithmic Convexity. Edited by R. J. Knops. V, 176 pages. 1973. DM 20,–

Vol. 317: Séminaire Bourbaki – vol. 1971/72. Exposés 400–417. IV, 361 pages. 1973. DM 29,–

Vol. 318: Recent Advances in Topological Dynamics. Edited by A. Beck. VIII, 285 pages. 1973. DM 27,–

Vol. 319: Conference on Group Theory. Edited by R. W. Gatterdam and K. W. Weston. V, 188 pages. 1973. DM 20,–

Vol. 320: Modular Functions of One Variable I. Edited by W. Kuyk. V, 195 pages. 1973. DM 20,–

Vol. 321: Séminaire de Probabilités VII. Edité par P. A. Meyer. VI, 322 pages. 1973. DM 29,–

Vol. 322: Nonlinear Problems in the Physical Sciences and Biology. Edited by I. Stakgold, D. D. Joseph and D. H. Sattinger. VIII, 357 pages. 1973. DM 29,–

Vol. 323: J. L. Lions, Perturbations Singulières dans les Problèmes aux Limites et en Contrôle Optimal. XII, 645 pages. 1973. DM 46,–

Vol. 324: K. Kreith, Oscillation Theory. VI, 109 pages. 1973. DM 18,–

Vol. 325: C.-C. Chou, La Transformation de Fourier Complexe et L'Equation de Convolution. IX, 137 pages. 1973. DM 18,–

Vol. 326: A. Robert, Elliptic Curves. VIII, 264 pages. 1973. DM 24,–

Vol. 327: E. Matlis, One-Dimensional Cohen-Macaulay Rings. XII, 157 pages. 1973. DM 20,–

Vol. 328: J. R. Büchi and D. Siefkes, The Monadic Second Order Theory of All Countable Ordinals. VI, 217 pages. 1973. DM 22,–

Vol. 329: W. Trebels, Multipliers for (C, α)-Bounded Fourier Expansions in Banach Spaces and Approximation Theory. VII, 103 pages. 1973. DM 18,–

Vol. 330: Proceedings of the Second Japan-USSR Symposium on Probability Theory. Edited by G. Maruyama and Yu. V. Prokhorov. VI, 550 pages. 1973. DM 40,–

Vol. 331: Summer School on Topological Vector Spaces. Edited by L. Waelbroeck. VI, 226 pages. 1973. DM 22,–

Vol. 332: Séminaire Pierre Lelong (Analyse) Année 1971-1972. V, 131 pages. 1973. DM 18,–

Vol. 333: Numerische, insbesondere approximationstheoretische Behandlung von Funktionalgleichungen. Herausgegeben von R. Ansorge und W. Törnig. VI, 296 Seiten. 1973. DM 27,–

Vol. 334: F. Schweiger, The Metrical Theory of Jacobi-Perron Algorithm. V, 111 pages. 1973. DM 18,–

Vol. 335: H. Huck, R. Roitzsch, U. Simon, W. Vortisch, R. Walden, B. Wegner und W. Wendland, Beweismethoden der Differentialgeometrie im Großen. IX, 159 Seiten. 1973. DM 20,–

Vol. 336: L'Analyse Harmonique dans le Domaine Complexe. Edité par E. J. Akutowicz. VIII, 169 pages. 1973. DM 20,–

Vol. 337: Cambridge Summer School in Mathematical Logic. Edited by A. R. D. Mathias and H. Rogers. IX, 660 pages. 1973. DM 46,–

Vol. 338: J. Lindenstrauss and L. Tzafriri, Classical Banach Spaces. IX, 243 pages. 1973. DM 24,–

Vol. 339: G. Kempf, F. Knudsen, D. Mumford and B. Saint-Donat, Toroidal Embeddings I. VIII, 209 pages. 1973. DM 22,–

Vol. 340: Groupes de Monodromie en Géométrie Algébrique. (SGA 7 II). Par P. Deligne et N. Katz. X, 438 pages. 1973. DM 44,–

Vol. 341: Algebraic K-Theory I, Higher K-Theories. Edited by H. Bass. XV, 335 pages. 1973. DM 29,–

Vol. 342: Algebraic K-Theory II, "Classical" Algebraic K-Theory, and Connections with Arithmetic. Edited by H. Bass. XV, 527 pages. 1973. DM 40,–

Vol. 343: Algebraic K-Theory III, Hermitian K-Theory and Geometric Applications. Edited by H. Bass. XV, 572 pages. 1973. DM 40,–

Vol. 344: A. S. Troelstra (Editor), Metamathematical Investigation of Intuitionistic Arithmetic and Analysis. XVII, 485 pages. 1973. DM 38,–

Vol. 345: Proceedings of a Conference on Operator Theory. Edited by P. A. Fillmore. VI, 228 pages. 1973. DM 22,–

Vol. 346: Fučík et al., Spectral Analysis of Nonlinear Operators. II, 287 pages. 1973. DM 26,–

Vol. 347: J. M. Boardman and R. M. Vogt, Homotopy Invariant Algebraic Structures on Topological Spaces. X, 257 pages. 1973. DM 24,–

Vol. 348: A. M. Mathai and R. K. Saxena, Generalized Hypergeometric Functions with Applications in Statistics and Physical Sciences. VII, 314 pages. 1973. DM 26,–

Vol. 349: Modular Functions of One Variable II. Edited by W. Kuyk and P. Deligne. V, 598 pages. 1973. DM 38,–

Vol. 350: Modular Functions of One Variable III. Edited by W. Kuyk and J.-P. Serre. V, 350 pages. 1973. DM 26,–

Vol. 351: H. Tachikawa, Quasi-Frobenius Rings and Generalizations. XI, 172 pages. 1973. DM 20,–

Vol. 352: J. D. Fay, Theta Functions on Riemann Surfaces. V, 137 pages. 1973. DM 18,–

Vol. 353: Proceedings of the Conference on Orders, Group Rings and Related Topics. Organized by J. S. Hsia, M. L. Madan and T. G. Ralley. X, 224 pages. 1973. DM 22,–

Vol. 354: K. J. Devlin, Aspects of Constructibility. XII, 240 pages. 1973. DM 24,–

Vol. 355: M. Sion, A Theory of Semigroup Valued Measures. V, 140 pages. 1973. DM 18,–

Vol. 356: W. L. J. van der Kallen, Infinitesimally Central-Extensions of Chevalley Groups. VII, 147 pages. 1973. DM 18,–

Vol. 357: W. Borho, P. Gabriel und R. Rentschler, Primideale in Einhüllenden auflösbarer Lie-Algebren. V, 182 Seiten. 1973. DM 20,–

Vol. 358: F. L. Williams, Tensor Products of Principal Series Representations. VI, 132 pages. 1973. DM 18,–

Vol. 359: U. Stammbach, Homology in Group Theory. VIII, 183 pages. 1973. DM 20,–

Vol. 360: W. J. Padgett and R. L. Taylor, Laws of Large Numbers for Normed Linear Spaces and Certain Fréchet Spaces. VI, 111 pages. 1973. DM 18,–

Vol. 361: J. W. Schutz, Foundations of Special Relativity: Kinematic Axioms for Minkowski Space Time. XX, 314 pages. 1973. DM 26,–

Vol. 362: Proceedings of the Conference on Numerical Solution of Ordinary Differential Equations. Edited by D. Bettis. VIII, 490 pages. 1974. DM 34,–

Vol. 363: Conference on the Numerical Solution of Differential Equations. Edited by G. A. Watson. IX, 221 pages. 1974. DM 20,–

Vol. 364: Proceedings on Infinite Dimensional Holomorphy. Edited by T. L. Hayden and T. J. Suffridge. VII, 212 pages. 1974. DM 20,–

Vol. 365: R. P. Gilbert, Constructive Methods for Elliptic Equations. VII, 397 pages. 1974. DM 26,–

Vol. 366: R. Steinberg, Conjugacy Classes in Algebraic Groups (Notes by V. V. Deodhar). VI, 159 pages. 1974. DM 18,–

Vol. 367: K. Langmann und W. Lütkebohmert, Cousinverteilungen und Fortsetzungssätze. VI, 151 Seiten. 1974. DM 16,–

Vol. 368: R. J. Milgram, Unstable Homotopy from the Stable Point of View. V, 109 pages. 1974. DM 16,–

Vol. 369: Victoria Symposium on Nonstandard Analysis. Edited by A. Hurd and P. Loeb. XVIII, 339 pages. 1974. DM 26,–

Vol. 370: B. Mazur and W. Messing, Universal Extensions and One Dimensional Crystalline Cohomology. VII, 134 pages. 1974. DM 16,–

Vol. 371: V. Poenaru, Analyse Différentielle. V, 228 pages. 1974. DM 20,–

Vol. 372: Proceedings of the Second International Conference on the Theory of Groups 1973. Edited by M. F. Newman. VII, 740 pages. 1974. DM 48,–

Vol. 373: A. E. R. Woodcock and T. Poston, A Geometrical Study of the Elementary Catastrophes. V, 257 pages. 1974. DM 22,–

Vol. 374: S. Yamamuro, Differential Calculus in Topological Linear Spaces. IV, 179 pages. 1974. DM 18,–

Vol. 375: Topology Conference 1973. Edited by R. F. Dickman Jr. and P. Fletcher. X, 283 pages. 1974. DM 24,–

Vol. 376: D. B. Osteyee and I. J. Good, Information, Weight of Evidence, the Singularity between Probability Measures and Signal Detection. XI, 156 pages. 1974. DM 16.-

Vol. 377: A. M. Fink, Almost Periodic Differential Equations. VIII, 336 pages. 1974. DM 26,–

Vol. 378: TOPO 72 – General Topology and its Applications. Proceedings 1972. Edited by R. Alò, R. W. Heath and J. Nagata. XIV, 651 pages. 1974. DM 50,–

Vol. 379: A. Badrikian et S. Chevet, Mesures Cylindriques, Espaces de Wiener et Fonctions Aléatoires Gaussiennes. X, 383 pages. 1974. DM 32,–

Vol. 380: M. Petrich, Rings and Semigroups. VIII, 182 pages. 1974. DM 18,–

Vol. 381: Séminaire de Probabilités VIII. Edité par P. A. Meyer. IX, 354 pages. 1974. DM 32,–

Vol. 382: J. H. van Lint, Combinatorial Theory Seminar Eindhoven University of Technology. VI, 131 pages. 1974. DM 18,–

Vol. 383: Séminaire Bourbaki – vol. 1972/73. Exposés 418-435 IV, 334 pages. 1974. DM 30,–

Vol. 384: Functional Analysis and Applications, Proceedings 1972. Edited by L. Nachbin. V, 270 pages. 1974. DM 22,–

Vol. 385: J. Douglas Jr. and T. Dupont, Collocation Methods for Parabolic Equations in a Single Space Variable (Based on C^1-Piecewise-Polynomial Spaces). V, 147 pages. 1974. DM 16,–

Vol. 386: J. Tits, Buildings of Spherical Type and Finite BN-Pairs. IX, 299 pages. 1974. DM 24,–

Vol. 387: C. P. Bruter, Eléments de la Théorie des Matroïdes. V, 138 pages. 1974. DM 18,–

Vol. 388: R. L. Lipsman, Group Representations. X, 166 pages. 1974. DM 20,–

Vol. 389: M.-A. Knus et M. Ojanguren, Théorie de la Descente et Algèbres d' Azumaya. IV, 163 pages. 1974. DM 20,–

Vol. 390: P. A. Meyer, P. Priouret et F. Spitzer, Ecole d'Eté de Probabilités de Saint-Flour III – 1973. Edité par A. Badrikian et P.-L. Hennequin. VIII, 189 pages. 1974. DM 20,–

Vol. 391: J. Gray, Formal Category Theory: Adjointness for 2-Categories. XII, 282 pages. 1974. DM 24,–

Vol. 392: Géométrie Différentielle, Colloque, Santiago de Compostela, Espagne 1972. Edité par E. Vidal. VI, 225 pages. 1974. DM 20,–

Vol. 393: G. Wassermann, Stability of Unfoldings. IX, 164 pages. 1974. DM 20,–

Vol. 394: W. M. Patterson 3rd, Iterative Methods for the Solution of a Linear Operator Equation in Hilbert Space – A Survey. III, 183 pages. 1974. DM 20,–

Vol. 395: Numerische Behandlung nichtlinearer Integrodifferential- und Differentialgleichungen. Tagung 1973. Herausgegeben von R. Ansorge und W. Törnig. VII, 313 Seiten. 1974. DM 28,–

Vol. 396: K. H. Hofmann, M. Mislove and A. Stralka, The Pontryagin Duality of Compact O-Dimensional Semilattices and its Applications. XVI, 122 pages. 1974. DM 18,–

Vol. 397: T. Yamada, The Schur Subgroup of the Brauer Group. V, 159 pages. 1974. DM 18,–

Vol. 398: Théories de l'Information, Actes des Rencontres de Marseille-Luminy, 1973. Edité par J. Kampé de Fériet et C. Picard. XII, 201 pages. 1974. DM 23,–

Vol. 399: Functional Analysis and its Applications, Proceedings 1973. Edited by H. G. Garnir, K. R. Unni and J. H. Williamson. XVII, 569 pages. 1974. DM 44,–

Vol. 400: A Crash Course on Kleinian Groups – San Francisco 1974. Edited by L. Bers and I. Kra. VII, 130 pages. 1974. DM 18,–

Vol. 401: F. Atiyah, Elliptic Operators and Compact Groups. V, 93 pages. 1974. DM 18,–

Vol. 402: M. Waldschmidt, Nombres Transcendants. VIII, 277 pages. 1974. DM 25,–

Vol. 403: Combinatorial Mathematics – Proceedings 1972. Edited by D. A. Holton. VIII, 148 pages. 1974. DM 18,–

Vol. 404: Théorie du Potentiel et Analyse Harmonique. Edité par J. Faraut. V, 245 pages. 1974. DM 25,–

Vol. 405: K. Devlin and H. Johnsbråten, The Souslin Problem. VIII, 132 pages. 1974. DM 18,–

Vol. 406: Graphs and Combinatorics – Proceedings 1973. Edited by R. A. Bari and F. Harary. VIII, 355 pages. 1974. DM 30,–

Vol. 407: P. Berthelot, Cohomologie Cristalline des Schémas de Caracteristique $p > o$. VIII, 598 pages. 1974. DM 44,–

Vol. 408: J. Wermer, Potential Theory. VIII, 146 pages. 1974. DM 18,–

Vol. 409: Fonctions de Plusieurs Variables Complexes, Séminaire François Norguet 1970–1973. XIII, 612 pages. 1974. DM 47,–

Vol. 410: Séminaire Pierre Lelong (Analyse) Année 1972–1973. VI, 181 pages. 1974. DM 18,–

Vol. 411: Hypergraph Seminar. Ohio State University, 1972. Edited by C. Berge and D. Ray-Chaudhuri. IX, 287 pages. 1974. DM 28,–

Vol. 412: Classification of Algebraic Varieties and Compact Complex Manifolds. Proceedings 1974. Edited by H. Popp. V, 333 pages. 1974. DM 30,–

Vol. 413: M. Bruneau, Variation Totale d'une Fonction. XIV, 332 pages. 1974. DM 30,–

Vol. 414: T. Kambayashi, M. Miyanishi and M. Takeuchi, Unipotent Algebraic Groups. VI, 165 pages. 1974. DM 20,–

Vol. 415: Ordinary and Partial Differential Equations, Proceedings of the Conference held at Dundee, 1974. XVII, 447 pages. 1974. DM 37,–

Vol. 416: M. E. Taylor, Pseudo Differential Operators. IV, 155 pages. 1974. DM 18,–

Vol. 417: H. H. Keller, Differential Calculus in Locally Convex Spaces. XVI, 131 pages. 1974. DM 18,–

Vol. 418: Localization in Group Theory and Homotopy Theory and Related Topics Battelle Seattle 1974 Seminar. Edited by P. J. Hilton. VI, 171 pages. 1974. DM 20,–

Vol. 419: Topics in Analysis – Proceedings 1970. Edited by O. E. Lehto, I. S. Louhivaara, and R. H. Nevanlinna. XIII, 391 pages. 1974. DM 35,–

Vol. 420: Category Seminar. Proceedings, Sydney Category Theory Seminar 1972/73. Edited by G. M. Kelly. VI, 375 pages. 1974. DM 32,–

Vol. 421: V. Poénaru, Groupes Discrets. VI, 216 pages. 1974. DM 23,–

Vol. 422: J.-M. Lemaire, Algèbres Connexes et Homologie des Espaces de Lacets. XIV, 133 pages. 1974. DM 23,–

Vol. 423: S. S. Abhyankar and A. M. Sathaye, Geometric Theory of Algebraic Space Curves. XIV, 302 pages. 1974. DM 28,–

Vol. 424: L. Weiss and J. Wolfowitz, Maximum Probability Estimators and Related Topics. V, 106 pages. 1974. DM 18,–

Vol. 425: P. R. Chernoff and J. E. Marsden, Properties of Infinite Dimensional Hamiltonian Systems. IV, 160 pages. 1974. DM 20,–

Vol. 426: M. L. Silverstein, Symmetric Markov Processes. IX, 287 pages. 1974. DM 28,–

Vol. 427: H. Omori, Infinite Dimensional Lie Transformation Groups. XII, 149 pages. 1974. DM 18,–

Vol. 428: Algebraic and Geometrical Methods in Topology, Proceedings 1973. Edited by L. F. McAuley. XI, 280 pages. 1974. DM 28,–